Hearing the Crimean War

Hearing the Crimean War

Wartime Sound and the Unmaking of Sense

EDITED BY GAVIN WILLIAMS

OXFORD
UNIVERSITY PRESS

OXFORD
UNIVERSITY PRESS

Oxford University Press is a department of the University of Oxford. It furthers the University's objective of excellence in research, scholarship, and education by publishing worldwide. Oxford is a registered trade mark of Oxford University Press in the UK and certain other countries.

Published in the United States of America by Oxford University Press
198 Madison Avenue, New York, NY 10016, United States of America.

Library of Congress Cataloging-in-Publication Data
Names: Williams, Gavin, 1986–
Title: Hearing the Crimean War : wartime sound and the unmaking of sense / edited by Gavin Williams
Description: New York, NY : Oxford University Press, [2019] |
Includes bibliographical references and index.
Identifiers: LCCN 2018017157 | ISBN 9780190916756 (pbk.) |ISBN 9780190916749 (hardcover)
Subjects: LCSH: Crimean War, 1853–1856—Music and the war. |
Music—Political aspects—History—19th century. |
Sound—Political aspects—History—19th century.
Classification: LCC ML3916 .H43 2019 | DDC 781.5/ 99— dc23
LC record available at https:// lccn.loc.gov/2018017157

CONTENTS

LIST OF ILLUSTRATIONS

ACKNOWLEDGMENTS

This book began life as a reading group on music on tape, then took a swerve toward the sounds of nineteenth-century wartime. The effects of that swerve, for which we have yet to find a cause, have played themselves out in meetings, study days, conferences, and conversations, sustained by the efforts of many people, in shifting permutations, over many years. There are accordingly many people who need to be thanked, as well as many people who must do the thanking.

The contributors would like to express thanks to Harvard's music department for supporting its graduate students in an initial phase of research. The next stop was a study day held at the University of North Carolina, Chapel Hill in March 2014, kindly supported by the Joint Projects Scheme between the University of North Carolina, Chapel Hill, and King's College London. Several participants in that study day were to become contributors to this book; we would also like to thank the participants who gave their time and energy freely over a windy weekend in Chapel Hill, particularly Christopher Bowen, Annegret Fauser, Louise McReynolds, and Andrew Ringlee.

Another happy station along the way was the conference, "Theatres of the Crimean War: Sound, Affect, and Media in the Production of Wartime," held the following October at King's College London. This conference—as well as the study day at UNC—received funding from the European Research Council (ERC) under the European Union's Seventh Framework Programme (FP7/2007–2013)/ERC grant agreement no. 323404. The contributors want to express particular gratitude to Roger Parker, whose "Music in London" project generated an important context for us, as well as for many others working on nineteenth-century music history, broadly conceived. Roger has personally, and always selflessly, aided the publication of this book in countless ways. We would also like to thank Angela Waplington for assisting in the organization of the conference, and many others for their participation and continued support: Tamsin Alexander, Laura Protano-Biggs, Harriet Boyd-Bennett, Rory Finnin, Holly

Furneaux, James Grande, Jonathan Hicks, Tariq Jazeel, Oskar Cox Jensen, Axel Körner, Ellen Lockhart, Stephen Lovell, Nazan Maksudyan, Adile Namazova, Arman Schwartz, Leo Shtutin, Martin Stokes, and Wiebke Thormählen.

Finally, special thanks are due to unsung heroes: the anonymous reviewers who gave such generous and constructive feedback, improving the book in many important ways.

CONTRIBUTORS

Andrea F. Bohlman is Assistant Professor of Music at the University of North Carolina at Chapel Hill. Her publications on European popular song, music and displacement, and social movements treat sound media as a music historical archive, cultural practice, and documentary trail, making use of the toolboxes offered by sound and media studies. She works to shape a place for music and sound in the cultural history of East Central Europe. Her research on music and political action in recent Polish history was distinguished with the Alfred Einstein Award from the American Musicological Society and was supported by fellowships from the Institute of Advanced Study in Berlin, the American Council of Learned Societies, and the National Endowment for the Humanities. Most recently, she and Peter McMurray co-edited a special issue of *Twentieth-Century Music* devoted to tape and tape recording, and she is preparing a project on amateur sound recording under state socialism.

Delia Casadei is an Assistant Professor of Musicology at the University of California, Berkeley. She obtained her PhD in Musicology from the University of Pennsylvania in 2015 with a thesis entitled "The Crowded Voice: Speech, Music and Community in Milan, 1955–1974." She was a recipient of the AMS 50 Fellowship for the academic year 2014–15, and between 2015 and 2017, a postdoctoral fellow at the University of Cambridge. She has published in the *Cambridge Opera Journal, The Opera Quarterly, Journal of the Royal Musical Association,* and has an article entitled "Sound Evidence, 1969: Recording a Milanese Riot" forthcoming in *Representations*. Her current book project is about laughter as a vocal and political technique.

Dina Gusejnova is a Lecturer in Modern History at the University of Sheffield. After receiving her PhD from the University of Cambridge, she has taught at the University of Chicago, University College London, and Queen Mary University of London. Her research interests comprise European intellectual history and, more broadly, the social history of ideas in the long twentieth century. She is the author of *European Elites and Ideas of Empire, 1917–57* (Cambridge University Press, 2016), which explores connections between ideas of Europe and imperial memory in elite intellectual milieus, and the editor of *Cosmopolitanism in Conflict: Imperial Encounters from the Seven Years' War to the Cold War* (Palgrave, 2018). Dina's current research concentrates on ideas of citizenship and the experience of statelessness during the Second World War.

Kevin C. Karnes is a historical musicologist who studies sounding expressions of identity, difference, and belonging in Central and Eastern Europe from the eighteenth century through the present. His recent work includes *Arvo Pärt's Tabula Rasa* (Oxford University Press, 2017) and "Inventing Eastern Europe in the Ear of the Enlightenment," published in the *Journal of the American Musicological Society* in 2018. He is currently working on a new book on underground festivals, discotheques, and student culture in the 1970s USSR. His research has been supported with fellowships and awards from the ACLS, the NEH, and the Association for the Advancement of Baltic Studies.

Peter McMurray is a Lecturer in Ethnomusicology at University of Cambridge and is a Junior Fellow at the Harvard Society of Fellows. He is currently completing a book and film project, *Pathways to God: The Islamic Acoustics of Turkish Berlin.* Other research interests include oral poetry and the history of tape recording. He holds a PhD from Harvard in ethnomusicology, with a secondary field in critical media practice.

Hillel Schwartz is a cultural historian, poet, and translator. His current research on the changing notions and experience of "emergency" led to a 2014 Berlin Prize in Cultural History at the American Academy in Berlin. As a historian, he is best known for studies of sound (*Making Noise: From Babel to the Big Bang, and Beyond,* 2011), of replication (*The Culture of the Copy: Striking Likenesses, Unreasonable Facsimiles,* 1996 and 2014) and of dieting (*Never Satisfied: A Cultural History of Diets, Fantasies, and Fat,* 1986). From 2009 to 2014, he cofounded and directed a medical case-management company that helped patients deal with complex issues requiring coordination of care among physicians, pharmacists, social workers, family members, and friends. Out of this experience came *Long Days Last Days: A Down-to-Earth Guide for Those at the Bedside* (2012). A graduate of Brandeis University, UC Berkeley, and Yale, Hillel Schwartz is an independent scholar but has taught occasionally in departments of history, literature, religious studies, communication, and visual arts, most recently at UC San Diego, where he was founding director of the core curriculum for Sixth College: Culture, Art and Technology.

Maria Sonevytsky is an Assistant Professor of Music (Ethnomusicology) at the University of California, Berkeley. Recent works include an article in the *Journal of Popular Music Studies* titled "The Freak-Cabaret on the Revolution Stage: On the Ambivalent Politics of Femininity, Rurality, and Nationalism in Ukrainian Popular Music" (2016), and the "Chornobyl Songs Project" (Smithsonian Folkways Recordings, 2015). Forthcoming works include an article on Crimean Tatar radio and liminal sovereignty in *Public Culture* and her first book, titled *Wild Music: Sound and Sovereignty in Ukraine* (Wesleyan University Press).

Alyson Tapp teaches Russian literature at the University of Cambridge and is a Fellow of Clare College. She has published articles on nineteenth- and early-twentieth-century Russian literature and culture, including on Tolstoy's *Anna Karenina,* on the Russian Formalist and Tolstoy scholar Boris Eikhenbaum, on Dostoevsky's *The Idiot,* and on the tram in the Russian literary imagination.

Gavin Williams is a musicologist and Leverhulme Early Career Research Fellow at King's College London. He wrote a PhD dissertation at Harvard University on sound and media in Milan ca. 1900, and was then a postdoctoral fellow at Jesus College, Cambridge. He has published articles and book chapters on Futurist music, Italian opera and ballet, and soundscapes in nineteenth-century London, and is currently writing a book on the imperial geographies of recorded sound during the first half of the twentieth century.

Flora Willson is a Lecturer in Music at King's College London, having previously held a Research Fellowship at King's College Cambridge. Her work concerns the place of opera in nineteenth-century urban history and culture. She has published in *19th-Century Music, Cambridge Opera Journal, Music & Letters, Opera Quarterly,* and elsewhere; and she is writing a book about networks in operatic culture in 1890s Paris, London, and New York.

INTRODUCTION: SOUND UNMADE

GAVIN WILLIAMS

28 February 2014

Foreign soldiers, wearing plain green battle fatigues and brandishing automatic firearms, turned up in cities across Crimea, which was, at the time, part of Ukraine. When questioned by locals, the men said they were there to protect the people and maintain public order, but refused to say where they were from or who they were working for. Yet these "little green men," as they came to be known, were understood by virtually everyone, both inside and outside Crimea, to be Russian forces. By way of indirect confirmation, Ukrainian and Crimean Tatar TV stations were blocked around the time of the soldiers' arrival, and then, with no explanation, replaced by Russian ones.[1] After a standoff with the soldiers, and with Russia's enormous Baltic Fleet looming in Sevastopol's harbor, Ukrainians surrendered government buildings without a shot being fired.[2] A tense silence was the soundtrack for Russia's annexation of

[1] John Biersack and Shannon O'Lear, "The Geopolitics of Russia's Annexation of Crimea: Narratives, Identity, Silences, and Energy," *Eurasian Geography and Economics* 55/3 (2014), 247–69; here 249.

[2] The silence of the unidentified soldiers, and of their weapons, became a journalistic trope. The militia arrived in Crimea from 28 February 2014; Russia's president Vladimir Putin acknowledged them as Russian a few weeks later on 18 March in a speech made at the Kremlin. During this speech, which was followed by a ceremony marking Crimea's (and Sevastopol's) unification with Russia, Putin rejected claims of Russian aggression by calling attention to the fact that no shot had been fired. However, a few hours after his speech, two Ukrainian soldiers were shot by Russian snipers, one fatally; see Shaun Walker and Ian Traynor, "Putin Confirms Crimea Annexation as Ukraine Soldier Becomes First Casualty," *The Guardian* (19 Mar. 2014), <https://www.theguardian.com/world/2014/mar/18/putin-confirms-annexation-crimea-ukrainian-soldier-casualty>, accessed 5 Apr. 2017.

Crimea.[3] As is well known, a referendum followed shortly afterward in which Crimeans voted (apparently overwhelmingly) to become part of Russia—although at least one group, the Crimean Tatars, were variously prevented from and pressured into voting.[4] Interpreting for baffled foreign onlookers, anthropologist Alexei Yurchak described the annexation as enacting a double vision: because Russia implicitly acknowledged Crimea as Ukrainian, it had to stake its claim to the peninsula through use of "non-Russian" militia.[5] It was an "occupation staged as a non-occupation," an original piece of political theater in which the apparently stateless soldiers, with their stubbornly taciturn behavior, took leading roles. Their silence signaled not the absence of sound, but was the means for a violent operation, subtly buffered against a likely backlash from local and international powers.

20 March 2003

Another twenty-first-century act of war, one altogether louder in execution, calls our attention to sound: in this case to a continuous rumble, punctuated by ripples of machine-gun fire and the thud of missiles (so-called smart bombs) falling from the night sky. As many as 3,000 of these bombs were launched over Baghdad in a single morning as the Shock and Awe campaign of Iraq began. The first salvos, fired before dawn, were heard around the world as news stations relayed real-time sounds and pictures. In anticipation of the event, TV broadcasts flitted between newsrooms and static, long-shot views of dimly lit Iraqi cityscapes in which the only sign of human presence was the flow of traffic along highways. The eyes and ears of the global media were thus trained, permitting spectators everywhere to witness the official beginning of the war.[6] For most

[3] Legal scholar Monica Eppinger characterized unfolding events as "quiet horror" (using the Russian idiom *tikhii uzhas*); see her article "Silencing and Backtalk: Scenes from the Crimean Occupation," *Anthropoliteia*, published online on 16 Mar. 2014, <http://anthropoliteia.net>, accessed 10 Feb. 2016.

[4] As Eppinger explained at the time, Crimean Tatars were intimidated in the days before the referendum: they were singled out by having their doors marked; some Tatars had their passports taken from them by the Russian forces until after the ballot, thus depriving them of a means of participation in the vote. At the same time, and perhaps in response to intimidation tactics—which also included the abduction of community leaders—many Tatars publicly boycotted the referendum, in order to cast doubt on the validity of its outcome. See Ibid. and United Nations, 7144th Meeting of the Security Council, Agenda: "Letter dated 28 Feb. 2014 from the Permanent Representative of Ukraine [. . .]" S/PV.7144 (19 Mar. 2014), 6.

[5] Alexei Yurchak, "Little Green Men: Russia, Ukraine and Post-Soviet Sovereignty," *Anthropoliteia*, published online on 31 Mar. 2014, <http://anthropoliteia.net>, accessed 10 Feb. 2016.

[6] The events described here took place in the early hours of the morning in Baghdad on 20 March 2003 and marked the beginning of the Iraq War in the popular imagination; but the US-led coalition's military campaign began the day before, with the dropping of "bunker buster" missiles over an industrial complex on the outskirts of Baghdad. On media coverage of the opening of the

of the survivors—as many as 7,183 Iraqis were killed by the US-led coalition during the six-week invasion—the event was, of course, never to be forgotten.[7] Yet the invasion created abiding memories for others around the world, memories sealed by the war's real-time visibility and audibility. Then-live commentaries are now available online, furnishing an archive of history in the making. Available for endless rewatching, these broadcasts draw attention to images and sounds being synchronous with unfolding events. Typical in this respect was the voice-over provided by CNN, which supplied eager reminders of the liveness of rolling pictures. Yet when the first bombs were launched, the station's anchor fell silent, just after he had encouraged viewers to "listen in."[8]

10 October 1854

"Every instant in the darkness was broken by a flash which had all the effect of summer lightning—then came darkness again, and in a few seconds a fainter flash denoted the bursting of a shell."[9] Thus William Howard Russell described one night at the height of the mid-nineteenth-century Crimean War: a night during which British soldiers dug trenches near Sevastopol to defend themselves against an imminent Russian onslaught. Conjuring the scene for readers of *The Times*, Russell appealed to audiovisual conditions of uncertainty and suspense. Seen just before they were heard, the artillery flashes projected noise into the distance; they created an uneasy silence in the British camp, affording "a strange contrast to the constant roar of the Russian batteries, [and] to the music and trumpet calls and the lively noises of the encampment of our allies." These nearby allies—within earshot, but out of communicative reach—were the French: their military bands played into the night, as though unaware of the British army's predicament. The intermittent flashes eventually revealed the enemy's infantry "moving silently towards our works"; yet the moment of mutual recognition, when the Russians finally "ascertained that we had discovered their approach," was delayed, prolonging the anxious silence. At 1:25 a.m. the moment came. Russell marked it with a sudden increase in textual amplitude: "The batteries behind them were livid with incessant flashes, and the roar of shot and shell filled the air, mingled with the constant 'ping-pinging' of rifle and musket balls."

Iraq War, see Andrew Hoskins and Ben O'Loughlin, *War and Media: The Emergence of Diffused War* (Cambridge: Polity, 2010), 31–34.

[7] <https://www.iraqbodycount.org/database>, accessed 5 Apr. 2016.

[8] "'Shock and Awe' The Beginning of the 2003 Invasion of Iraq (CNN Live Coverage)," posted by user "ytykg" on 19 Mar. 2013, <https://youtu.be/f7iorfwcmeY>, accessed 10 Feb. 2016; the "listen in" comment comes at 40:38.

[9] William Howard Russell, "The War; The British Expedition; The Siege of Sebastopol (From Our Special Correspondent)," *The Times* (London, 28 Oct. 1854), 7, https://www.thetimes.co.uk/archive, accessed 18 Mar. 2016.

This action finally stirred the French, who "on our left got under arms, and the rattle of drums and the shrill blast of trumpets were heard amid the roar of cannon and small arms. For nearly half-an-hour this din lasted, till all of a sudden a ringing cheer was audible on our right, rising through the turmoil."[10]

Wartime Sound

The sounds of war inhabit a vast, though not always clamorous, domain. The three scenes just described outline only some of the ways in which war and sound might interact. But they begin to suggest a wider point: that human experiences of war and its acoustic realities inevitably vary according to place, time, and, most importantly, political situation. This book explores one such node of attention, the sounds of the Crimean War (1853–56); in particular, it surveys the breadth and complexity of the historical experiences those sounds can recall for us now. The Crimean War was an international conflict, involving the clash of the Russian Empire with the British, French, and Ottoman Empires, backed up by forces representing would-be nations such as Italy and Poland. The allies' geopolitical aim was to reduce Russian influence over Ottoman territories. One origin for the war can be found in a violent struggle between Catholic and Orthodox Christians for control over the Church of the Holy Sepulchre in Jerusalem, then part of the Ottoman Empire.[11] Britain, Russia, and France were all in the habit of using the presence of Christians in the Holy Lands to assert political influence over the Ottomans, whose future as an empire had become a pressing concern for many powers by the mid-century. The so-called Eastern Question was everywhere debated among the elites of Western Europe: its essential concern was to forestall Russian domination over trade routes across the Black Sea.[12] The Crimean War suggested one solution to the problem. Taking impetus from the Ottomans' latest war against Russia, which began in late 1853—there had been periodic conflicts between them since the late seventeenth century—Britain, France and others weighed in on the Ottomans' side in early 1854.

Why investigate the sounds of this particular war? One reason can be found in the historical lineages that connect war, sound, and our present-day

[10] For all short quotations in this paragraph and the previous one, see ibid.

[11] There are many different explanations for the outbreak of the Crimean War; the religious interpretation presented here has recently been explored by Orlando Figes, *The Crimean War: A History* (London: Penguin, 2010), 1–22. See also Stefanie Markovits, *The Crimean War in the British Imagination* (Cambridge: Cambridge University Press, 2009), 6–12.

[12] For a history of the issue, see Lucien J. Frary and Mara Kozelsky, eds., *Russian-Ottoman Borderlands: The Eastern Question Reconsidered* (Madison: University of Wisconsin Press, 2014).

involvement with media culture. The Crimean War is an ancestor to recent global conflicts—a precedent for latter-day, post-imperial interventionist campaigns—and is thus one available historical vehicle for thinking through relationships between war, sound, and geopolitical determinations. Nowadays it has become commonplace to recognize some aspects of the war that herald "modern" warfare: vast armies fought lengthy battles of attrition on multiple, widely separated fronts (of which the Crimean peninsula was the most active); some battles involved early trench warfare; the war saw the first military use of a railroad and steam engines.[13] Even more pertinent is the war's "climate of representation" (to borrow Lisa Gitelman's phrase), as people in Britain and France, and to a lesser extent in Russia and Turkey, received news of battles at hitherto unknown speeds, thanks to photography, telegraphy, and the new, though still limited, deployment of war correspondents for the newspaper press.[14] This host of new technologies allowing distant spectatorship has prompted twenty-first-century historians to propose that the Crimean War was the "first media war."[15] The sheer proliferation of the war's media output, and the abundance of archival traces left in their wake, provide the contributors to this book with an opportunity: to scrutinize the role of media technology in the historical and geopolitical construal of wartime sound.

This opportunity in turn brings up another question that readers of this book might ask: why study war's sounds at all? Our collective aim, which in part depends on our common focus on the Crimean War's sonic archive, is to interrogate the political nature of histories of sound. In other words, the contributors—musicologists, ethnomusicologists, historians, and literary scholars—address a broad set of problems involved in constructing knowledge about the sounds of

[13] See Brian Cooke, *The Grand Crimean Central Railway. The Story of the Railway Built by the British in the Crimea during the War of 1854–1856* (Knutsford: Cavalier House, 1997).

[14] London's *Times* newspaper had used foreign correspondents on battles since the Napoleonic campaigns, but William Howard Russell reinvented the role through critical reporting on battlefield events; see Martin Conboy, *Journalism: A Critical History* (London: Sage, 2004), 117–19. Lisa Gitelman uses the phrase "climate of representation" in *Scripts, Grooves and Writing Machines: Representing Technology in the Edison Era* (Stanford, CA: Stanford University Press, 1999), 69–70.

[15] Ulrich Keller has argued the case: by focusing on visual technologies, he claims the Crimean War was unprecedentedly mediatized for people in Britain and France. In metropolitan centers such as London and Paris, Crimean battles were rendered visible through the public exhibition of photographs, paintings, and panoramas; spreading outward from urban centers, battle scenes were further transmitted to imperial audiences through the domestic consumption of commemorative maps, prints, and sheet music front covers, all of which were being sold only weeks after battles had taken place. See his *The Ultimate Spectacle: A Visual History of the Crimean War* (Australia: Gordon & Breach, 2001). For a reappraisal of the book by media scholars, see Georg Maag and Martin Windisch, eds., *Der Krimkrieg als erster europäischer Medienkrieg* (Berlin: Lit Verlag, 2010), 7–15.

the past. The wager behind this historiographical experiment is that wars may provide a fertile ground in which to explore the politics of sensory experience, not least because wars—the Crimean War is not unique in this respect, but can nevertheless be exemplary—tend to spawn prolific and diverse archives. There are numerous scholarly explanations in circulation about why this might be so; I will return to some of them in what follows. Before doing that, though, it makes sense to outline the ways in which sonic experience was technologically mediated, and so preserved for our attention. Under certain mediatic conditions, the sounds of the Crimean War did not vanish without trace; and those that have endured sometimes reveal intimate connections between what Jonathan Sterne has termed audile technique—the means, subtly and variously deployed, for negotiating aural experience—and the conditions that register their traces and thus sustain them as things in the world.[16]

Bandwidth

2014, 2003, 1854. My opening samples hint at the necessity of being selective—and the impossibility of paying equal attention to everything—when summoning up the sounds of political events. In this respect wars are not exceptional. This book strives to embrace the contingency inevitably involved. As the chapters bear out, sound history challenges us to create "lines of flight" across what remains in scattered imperial archives: to rewrite the sounds of the past in ways that not only describe but also challenge the political orders from which they emerge.[17] This introduction attempts one such line of flight, charting its own idiosyncratic itinerary through the territories explored by the book, so opening out onto multiple aural perspectives. By weaving together sonic traces left in the wake of the Crimean War, I hope to introduce larger themes to do with sound and geopolitics during wartime: themes that echo through the chapters that follow.

[16] Sterne used the introduced the term "audile technique" referring to institutionalized listening practices of doctors in detecting symptoms of the body and telegraphists in decoding messages, practices that were in turn influential on emerging cultures of sound reproduction in the late nineteenth and early twentieth centuries; see Jonathan Sterne, *The Audible Past: Cultural Origins of Sound Reproduction* (Durham, NC: Duke University Press, 2003), 87–178. Since then, uses of the term have expanded to include technical means for negotiating aural knowledge and experience in a broader sense; see, for example, Ana María Ochoa Gautier, *Aurality: Listening and Knowledge in Nineteenth-Century Columbia* (Durham, NC: Duke University Press, 2014), 3.

[17] The term "line of flight" has been widely used; it comes from Gilles Deleuze and Felix Guattari, *A Thousand Plateaus: Capitalism and Schizophrenia*, trans. Brian Massumi (Minneapolis: University of Minneapolis Press, 1987), 8–9.

A persistent theme will be the complex action of technical media, including musical media, in both producing and perpetuating the sounds of wartime. These sounds fall within an epoch spanned by Reuters, the first international news agency, founded in 1851. As business model, Reuters is more important now than it was then, when global news was little more than a pipe dream.[18] Yet Reuters and other news media received a significant boost from the Crimean War: they defined the sensory channels made available to global audiences, acting as a major gateway for wartime sound. Vice versa, and as several contributors show, news media remained ever fluid in their response to unfolding events.[19] More contemporary examples of this close interplay between media and sensory experience might be the digital satellite media that rendered Iraqi wartime audio-visibly synchronous, or the Russian signal jammers that silenced TV stations in Crimea, placing familiar news sources beyond the reach of the peninsula's inhabitants. As many readers will recall, this blackout prompted a moment in 2014 when reporting on Crimea became the site of an all-out media war between Russian and "Western" news channels.[20] In short, the mediascape of the early twenty-first century is filled with noises and silences, furnishing ever productive metaphors by which to structure the experiences of post-imperial campaigns of occupation and annexation.

Crimea was also the object of media scrutiny and international concern during the mid-1850s. Russell's report cited earlier typifies the perspective made available to British elites by newsprint. He plunged his readers into a media environment fundamentally different from our own. To understand how sound works here, we need to immerse ourselves in long-forgotten details that once occupied an implicit background for the sonic imaginings of the British public. In the passage cited previously, Russell relays incidents leading up to the siege of Sevastopol, one of the war's most prolonged and bloodiest episodes. The excerpt precedes his discussion of battlefield action, and follows on from a lengthy digression on the effects of military bands and their music on soldiers' wellbeing. Russell compares the constant presence of music within the French camp with the "gloom" that pervades British troops, whose instruments had been placed in store. Britain's regimental bands had been "broken up and disorganized, the men being devoted to the performance of duties for which the ambulance corps was formed."[21] Russell thereby informs his readers that instrumentalists have been

[18] Donald Read, *The Power of News: The History of Reuters* (Oxford: Oxford University Press, 1999), 17–18.

[19] Deborah Esch has pointed out this fluidity; see her *In the Event: Reading Journalism, Reading Theory* (Stanford, CA: Stanford University Press, 1999), 1–8.

[20] See Mikhail D. Suslov, "'Crimea Is Ours!' Russian Popular Geopolitics in the New Media Age," *Eurasian Geography and Economics* 55/6 (2014), 588–609.

[21] Russell, "War; British Expedition," 7.

reassigned to deal with sickness and injury, giving the silence around camp grim implications. As part of a larger critique of the army, he hints at the role that music ought to play in life around the British camp (the role it does play for the French):

> Every military man knows how regiments, when fatigued on the march, cheer up at strains of their band, and dress up, keep step, and walk on with animation and vigour when it is playing. At camp, I always observed with pleasure the attentive auditory who gathered every evening at the first taps of the drum to listen to the music. At Aladyn and Devno the men used to wander off to the lines of the 77th, because it had the best band in the division; and when the bands were silenced because of the prevalence of sickness and cholera, out of a humane regard for the feelings of the sick, the soldiers were wont to get up singing parties in their tents in lieu of their ordinary entertainment. It seemed to be an error to deprive them of a cheering band at the very time they needed it most. The military band was not meant alone for the delectation of garrison towns, or for the pleasure of officers in quarters, and the men were fairly entitled to its inspiration during the long and weary march in the enemy's country, and in the monotony of a standing camp ere the beginning of a siege.[22]

This passage offers Russell's contemporaries journalistic scene setting—comparisons with previous British encampments at Aladyn and Devno (along the Varna river in present-day Bulgaria) are thrown in for good measure—and establish a lugubrious mood for what follows. His words are clearly calculated to draw attention to the plight of common soldiers. At the same time, the absence of the military band forces him to reflect on its uses. As an appeal to musical authority, "every military man knows" may seem unpromising, but the numerous practical functions Russell attributes to the band are borne out by historical accounts: it raised morale, kept soldiers in lockstep, provided entertainment (and mitigated the endemic boredom), and gave soldiers and officers the chance to interact.[23] Deeply embedded within army life, the military band also provided an important connection between the army and society at large—serving among other things as a conspicuous tool for recruitment (a function nowadays fulfilled by khaki-clad representatives assigned to shopping malls and by the extensive PR machines of national armies). In other words, the military band

[22] Ibid.

[23] See Trevor Herbert and Helen Barlow, *Music and the British Military in the Long Nineteenth Century* (Oxford: Oxford University Press, 2013), 240–53.

fulfilled several functions within society during the Crimean War: not only in Britain, but, as chapters in this book demonstrate, in many other places besides.

This brief dip into Russell's journalism can suggest many things. First, wartime around 1850 fundamentally resists straightforward comparison with our own wartimes; and that such comparisons, when they are made, ought to take historical media into account. The long-forgotten interplay between military bands and newsprint might encourage us to think afresh about the mediatic conditions that underwrite our own experiences of wartime sound. Secondly, and more obviously, Russell draws our attention to the importance of the military band, both in print and in the flesh. One undeniable (if not unexpected) observation made by this book is that the military band was central to representations of the Crimean War. Yet the contributors go further, considering the band's role as a medium for channeling sounds and shaping sensory experience on a transnational scale.

Some of the milestones in the internationalization of the band are well known. For example, in 1828 Giuseppe Donizetti (brother of Gaetano, about to become world famous as an opera composer) was recruited by the Ottoman imperial court. He was tasked with instituting a European-style military band to replace Janissary marching bands—which had once struck fear into the hearts of opposing armies, and had until more recently served as an exotic musical sideline for listeners to European operas, symphonies, and dance music. Donizetti trained a generation of Ottoman musicians to read and play from Western notation.[24] The Ottoman acquisition of such a figure reveals a wider point about the military band up to and beyond the mid-century. As the lavish possessions of national and imperial courts, bands were a resonant and mobile means for the projection of geopolitical power, both within and beyond the spaces of empire.

Although the band was an important medium of display, it could be feeble on occasion. This much has already been observed in Russell's report, in which the band created a jarring impression in the context of battlefield action: the lively music of the French camp had a demoralizing effect on the British as they prepared to fight the Russians.[25] Accounts of everyday experiences of bands in

[24] See contributions by John Morgan O'Connell and Ruhi Ayangil in *Giuseppe Donizetti pascià: Traiettorie musicali e storiche tra Italia e Turchia*, ed. Federico Spinetti (Bergamo, Italy: Fondazione Donizetti, 2010).

[25] The incongruity of music on the battlefield became a literary trope, one that can also be found in Tolstoy's writings on the Crimean War; see Dina Gusejnova's chapter, "Sympathy and Synesthesia: Tolstoy's Place in the Intellectual History of Cosmopolitan Spectatorship," in this book, 7–9, 14–15. The incongruity of music on—or indifference of music to—the battlefield may suggest a nineteenth-century precursor for Michel Chion's much discussed notion of "anempathetic sound" in film; see his *Audio-Vision: Sound on Screen*, trans. Claudia Gorbman (New York: Columbia University Press, 1994), 8–9.

wartime are hard to come by, but for the most part appear at moments when sounds break down, revealing a chasm between imperial aspirations and mundane realities. Typical in this respect was the cacophony that ensued when several British bands joined forces at Scutari (Üsküdar), near Istanbul, in a rendition of "God Save the Queen" for Victoria's birthday on 24 May 1854. The bands were evenly spread across the valley, yet no attempt was made to coordinate tuning.[26] The international embarrassment cued by this event, which became known as the "Scutari incident," led to a series of institutional reforms within the army, including the inauguration of the first school dedicated to the training of military musicians.[27]

We can continue to sketch the background for inter-imperial contact through the band by briefly considering some instrumental reforms pursued in France. In some ways, the technological solutions can be considered as an extension of the country's imperial aims of domination and expansion. General anxiety over the state of the empire's bands had surfaced during the 1840s, around the time Adolphe Sax conceived a series of acoustically improved instruments.[28] His revamped musical outfit included the saxhorn, saxtromba, saxtuba, and saxophone, each name proudly displaying his personal brand. Best-known now, the saxophone was intended to meld timbres of the trumpet and clarinet, while being powerful and versatile enough to be used in both indoor and outdoor spaces with ease.[29] In April 1854, a month after France declared war on Russia, Sax emerged victorious from a band competition involving direct comparison between a traditional military band and his own, technically enhanced collection of instruments. The event took place at the Champs de Mars before the eyes and ears of thousands of military and civilian spectators, including General de Rumigny, France's minister of war. Sax's victory led, later that year, to his becoming the official supplier of musical instruments to the French Army.[30]

While Sax's takeover came too late for his instruments to see service in Crimea—it took time to manufacture the quantity of instruments the army

[26] Barlow and Herbert, *Music and the British Military*, 140–41.

[27] Ibid., 140–46.

[28] In 1848, music theorist and composer George Kastner was calling for the "amélioration complète de nos musiques militaires" (complete improvement of our military music), writing in support of a government commission by France's ministry of war for reform in contemporary military music. See his *Manuel général de musique militaire à l'usage des armées françaises* (Paris: Didot Frères, 1848), xiii.

[29] See Sax's 1846 patent, which is included and translated in Stephen Cottrell, *The Saxophone* (New Haven, CT: Yale University Press, 2012), 343.

[30] As Cottrell points out, the new sonority, tethered to the Sax trademark, catered for a culture that increasingly fed on the promise of novelty extended by the ever renewing commodity form. Cottrell, *Saxophone*, 15–22.

required—they nevertheless formed part of the international wartime soundscape. To put this point slightly differently: Sax's inventions encourage us to pay attention not only to the war's sounds as experienced by its participants, but also transformations in technical means that gave rise to wartime sound.[31] In this sense, Sax's proto-industrial workshop in Paris might be considered the crucible within which French military music achieved a newly forceful register during the second half of the nineteenth century—a register advertised early on, and worldwide, by saxophone virtuosi such as Charles Jean-Baptiste Soualle, who gave concerts in China and Hong Kong on the "turcophone," one of the saxophone's many early appellations.[32] Soualle and others offered elite colonial audiences in Asia and Australia exoticized samples of the latest sounds to emanate from the metropolitan West—an orientalism achieved at the expense of the Ottomans, who had recently become Paris's ally in the war.

While Sax's instruments were not part of the mid-century military band, we would miss something important about an evolving medium if they were left out of account. They encourage us to embrace sounds that were not (yet) heard, or at least not widely so—potential sounds that became widespread only later on. Not all technologies were emergent and innovative, of course: the majority of those to be considered in this book were old, sewn into the fabric of the everyday long before the Crimean War began. And yet, if war is not only productive of sounds and sonic experiences, but is also a process by which technical means of perception, inscription, and dissemination are transformed, then the sounds of its nineteenth-century Crimean eruption should also include technologies that were conceived and discussed, even if they were not widely sensed or perceived. The saxophone, for example, brings closer inaudible connections between technological innovation and France's capitalist war machine, connections that would be missed if we were to focus too narrowly on the "sonic" environment. Sax's invention provides a material counterpart to the railway tracks laid down

[31] R. Murray Schafer's notion of the soundscape (*The Soundscape: Our Sonic Environment and the Tuning of the World* [New York: Knopf, 1977])—which he has also called "the sonic environment"—has come in for criticism in recent years, not least because soundscape suggests an objective reality that precedes the experiencing subject (albeit an evolving reality, open to modification through composition). By contrast, those who have used the term more recently have stressed the affordances between environment and the political subjectivities of listeners in fashioning sonic experience; see Bruce Smith, *The Acoustic World of Early Modern England: Attending to the O-Factor* (Chicago: University of Chicago Press, 1999), 44; and Emily Thompson, *The Soundscape of Modernity: Architectural Acoustics and the Culture of Listening in America, 1900–1933* (Cambridge: MIT Press, 2002), 1; Sophia Rosenfeld, "On Being Heard: A Case for Paying Attention to the Historical Ear," *American Historical Review* 116/2 (2011), 316–34.

[32] Cottrell, *Saxophone*, 109–18.

by the British in Crimea in 1854 to transport people and goods between the nearby towns of Inkermann and Balaklava.[33] As a nascent commodity, and as a tool, the instrument projects mid-nineteenth-century empires at war.

Throughout this book, musical instruments emerge across international milieux as key technologies in the construal of wartime sound. Individual chapters show the many types, uses, and plans for instruments, and observe them intended for diverse ends—violent, symbolic, mundane. The readiness with which weapons and instruments, such as cannons and church bells, melt into each other during wartime is a long-standing historical theme.[34] This book shows how such insights can be extended, as we chart the ways in which instruments come to be multiply deployed and imaginatively weaponized.

Voice/Writing

Organology provides just one way to reimagine the sounds of the Crimean War, and instruments were only one means by which the war's sounds were mediated. More prolific were those that involved explicit foregrounding of linguistic modes of inscription, storage, and transmission. Voice, paper, handwriting, movable type, telegraphy: these are the technical means that the chapters gathered here most often encounter, and so most often employ, in charting the war's sounds. Taking impetus from media theory, some contributors hazard a more ambitious claim: that these verbal and graphic incarnations of sound comprise nodes within a larger network that had a broader impact on sonic experience around the midcentury. Here I am invoking Friedrich Kittler's idea of "discourse networks," which undergirded his well-known though widely contested thesis that "media determine our situation."[35] Less technologically weighted, and more responsive to historical events, is the notion of a "climate of representation," something that I loosely introduced in a previous discussion.[36] The term

[33] See Cooke, *The Grand Crimean Central Railway*.

[34] On continuity between weapons and media, see Paul Virilio, *War and Cinema: The Logistics of Perception*, trans. Patrick Camiller (London: Verso, [1984] 1989); Friedrich Kittler, *Gramophone, Film, Typewriter*, trans and ed. Geoffrey Winthrop-Young and Michael Wutz ([1986] Stanford, CA: Stanford University Press, 1999). On bells and cannon, see Edward V. Williams, *The Bells of Russia: History and Technology* (Princeton, NJ: Princeton University Press, 1985), 58–60; and for a more recent musical treatment of this topic, see James Q. Davies, "Instruments of Empire," in *Sound Knowledge: Music and Science in London, 1789–1851*, ed. James Q. Davies and Ellen Lockhart (Chicago: Chicago University Press, 2016), 145–74.

[35] Kittler, *Gramophone, Film, Typewriter*, xxxix; see also Friedrich Kittler, *Discourse Networks 1800/1900*, trans. Michael Metteer and Chris Cullens (Stanford, CA: Stanford University Press, [1985] 1990), 369–72.

[36] Gitelman, *Scripts, Grooves, and Writing Machines*, 69–70.

is borrowed from media scholar Lisa Gitelman, who uses it to characterize a later moment in the nineteenth century, when an ensemble of technologies for linguistic inscription—notably shorthand and typewriting—came to channel broader experiences of sound. According to Gitelman, these related representational techniques provided the conditions for the emergence of phonography, widely (if not exclusively) understood as the writing of the voice.[37]

Phonography was still in the future in the 1850s, but can provide a way of taking the measure of the Crimean War's climate of representation. For example, the cultural energies that shuttled between voice, sound, and paper in the 1850s were to leave their mark on early phonography when three wax cylinders were made in London in 1890. The purpose of these cylinders, produced by the Edison Phonograph Corporation, was to raise funds for British veterans of the Light Brigade, those already mythic warriors of the Battle of Balaklava, whose dwindling number were living in destitution in the last decade of the nineteenth century. Their squalid living conditions outraged public morality.[38] To champion their cause, and drum up sales for the charitable initiative, the war's most famous personalities were recruited to speak, among them Florence Nightingale, the aristocratic nurse who had gone to Scutari to tend to British soldiers returning from Crimea.[39]

In her phonograph message, Nightingale adopted a role with which she had become long familiar, as a paragon of female service to the nation.[40] Before the horn, she delivered her words with queenly pacing and precision: "When I am no longer even a memory, just a name, I hope my voice may perpetuate the great work of my life. God bless my dear old comrades of Balaklava and bring them safe to shore. Florence Nightingale." Her declamation is impressive and was no doubt thoroughly rehearsed; it may even appear sung to twenty-first-century hearers. In two aborted takes, Nightingale tripped over her words, giving lie to the notion that her elaborate diction represented her normal speech.[41] Beneath the overtly Edisonian values invoked by Nightingale's recording—a fulfillment of his intended use of the phonograph for "preserving the sayings, the voices,

[37] Ibid., 1–20. See also Lisa Gitelman, *Always Already New: Media, History and the Data of Culture* (Cambridge. MA: MIT Press, 2008), 25–88.

[38] Mark Bostridge, *Florence Nightingale: The Woman and Her Legend* (London: Penguin, 2009), 516.

[39] Nightingale implemented reforms in military hospitals, in the process becoming an international celebrity—although nowhere more so than in Britain, where she attained something like cult status. Ibid., 508–10.

[40] Markovits, *The Crimean War in British Imagination*, 98–120.

[41] The authenticity of the recording has been subject to debate. Nightingale's recordings, along with the wax cylinders discussed in what follows, are preserved at the British Library; see "Florence Nightingale Cylinder 1890," catalogue number C1693/1.

and the last words of the dying members of the family, as of great men"—we can detect older vocal mediations.[42] The recording session supposedly took place in Nightingale's London residence, before a team of expectant technicians, the scenario itself recalling the stenographic lineages of early phonograph cultures that Gitelman and others have traced. Nightingale seems to dictate, to use the mechanism as she would a secretary; she wields her voice as a means of writing down its sound. This may seem a circuitous way of putting things, but as classicist Shane Butler has shown, since antiquity alphabetic writing has recurrently served as a conduit for the preservation of vocal sound.[43] While innovative in its means, in its approach to vocal writing Nightingale's cylinder directs our attention toward longer-standing connections between sound and writing.

Equaling Nightingale in celebrity and prestige, Alfred Tennyson was also persuaded to have his voice recorded. The poet had not taken an active part in the Crimean campaign, but became inextricably associated with it through his poem "The Charge of the Light Brigade," written in response to breaking news from Balaklava. This was one of the war's many infamous episodes, and became so partly through the dissemination of his eponymous poem.[44] To approach the wartime climate of representation within which Tennyson's poem appeared, we might begin with the battlefield miscommunications that precipitated the charge. On 25 October 1854, the Light Brigade received an order from Lord Raglan, commander of the British troops. The order was ambiguous and misinterpreted (perhaps willfully) by an officer, as requesting the immediate deployment of troops. The result was fatal: rather than pursue a retreating battalion, the Light Brigade undertook a frontal assault on a well-prepared line of Russian guns. A bugle sounded the advance. Not long afterward, more than 156 men were dead, missing, or mortally wounded; fewer than 200 (out of around 670) returned to the British camp with themselves and their horses intact.[45] Despite this disaster, Britain and France proceeded to victory at Balaklava. Thanks to telegraphy, the outcome of the battle relayed to London and Paris in the hours that followed. However, newspaper readers had to wait several days for corroborating reports, such as the one by William Howard Russell, who, in more than 10,000 handwritten words, sent by international post, conjured distant events for newspaper readers. Transformed through typesetting, Russell's

[42] For a discussion of Edison's "library of voices," see John M. Picker, *Victorian Soundscapes* (Oxford: Oxford University Press, 2003), 114.

[43] Shane Butler, *The Ancient Phonograph* (New York: Zone, 2015), 1–29. Along similar lines, Friedrich Kittler argued that one of the functions of handwriting in German romanticism was to elicit the imagined sounds of voices; see Kittler, *Discourse Networks 1800/1900*, 77–84.

[44] Markovits, *The Crimean War in the British Imagination*, 148–66

[45] See Lara Kriegel, "Who Blew the Balaclava Bugle? The Charge of the Light Brigade and the Afterlife of the Crimean War," *Interdisciplinary Studies in the Long Nineteenth Century* 20 (2015), 1–17.

handwriting—and by extension his first-person "voice"—rebounded throughout London's mid-century public sphere.

I have already tried to evoke the immensely detailed sounds conveyed by Russell's journalism; Tennyson's poem can be understood as a further mediation of the wartime news network, transmuting newspaper reports through heavy rhymes.[46] In turn, his poem recirculated through the pages of the British broadsheets in a quasi-official tribute to the men who had fallen. When he penned the words, Tennyson was already well known in Britain as the poet laureate. By the time he delivered the poem into the phonograph's horn in 1890, then in his eighties, the wartime news media that facilitated the poem's early career had been entirely effaced. The poem now fulfilled a new set of purposes, as both a charitable gesture and an authentic, collectible historical document, capturing the swan song of the wizened sage.[47] Given these changed circumstances of transmission and reception between 1854 and 1890, it would be a conceit to say we can "hear" the media networks of the Crimean War inside the fizz of the wax cylinder. Yet as this brief reconstruction of mid-century news suggests, those networks played an important role in sustaining Tennyson's voice as an audible trace—whether we encounter it in its carefully preserved archival afterlife at the British Library Sound Archive, or much more readily in one of its many digital reincarnations online. Tennyson's cylinder makes clear that phonography is often, if not always, remediating older sonic media, such as, in this case, those of 1850s wartime news.

The third wax cylinder provides another case of sonic remediation—as with the saxophone, by instrumental means—in showcasing Balaklava's fateful bugle call. The cylinder comes packaged with its own historical context, its protagonist introducing himself (and the record) as follows: "I am Trumpeter Landfried, one of the surviving trumpeters at the charge of the Light Brigade at Balaklava. I am now going to sound the bugle that was sounded at Waterloo"—then, following a noticeable pause—"and sound the charge as was sounded at Balaklava on that very same bugle." Then follows another, shorter pause; and then a voice (one similar to Landfried's own, and perhaps intended not to be noticeably different) adds, "the twenty-fifth of October, Eighteen-Fifty Four." An unidentified female voice contributes a date, time, and location—"Record made at Edison House on Northumberland Avenue, August the Second, Eighteen-Hundred and Ninety"—giving the document a final seal of authenticity before the bugle itself finally sounds. As he informs us, Landfried was indeed present on the morning of 25 October 1854; he may even have witnessed the Light Brigade's charge.[48] But

[46] Markovits, *The Crimean War in the British Imagination,* 158–60.

[47] On the wax cylinder as a historical document, see Lisa Gitelman, *Paper Knowledge: Towards a Media History of Documents* (Durham, NC: Duke University Press, 2014).

[48] Kriegel, "Who Blew the Balaclava Bugle?" 3.

the record misleads through omission, for Landfried did not sound the charge at Balaklava; nor was he part of the Light Brigade (exploiting the more capacious sense of the preposition in the phrase "*at* the charge of the Light Brigade"), but instead attached to the regiment of the 17th Lancers. Speaking more hesitantly than Nightingale and Tennyson, Landfried tells us the bugle we are about to hear was "sounded at Waterloo," but then, after a long pause, that he will "sound the charge as it was sounded at Balaklava." Landfried's awkward doubling of active and passive constructions ("I will sound . . . as it was sounded") gestures toward the bugle's multiple players over the course of its long history. And this, together with the accretion of historical time implied by "Waterloo," "Balaklava," and the present ("1890"), frames the instrument as a medium that can hold on to sounds of the past. It becomes a medium able faithfully to recover sanctified sounds: in this case, sounds that precipitated the charge of hundreds of men and horses plummeting toward their destruction.

Wartime

Sounds can be made to carry far beyond war zones and so take on an urgency of communication. During 1850s wartime, particular sounds became a sustained topic in soldiers' letters, newspapers, literary and musical works, and theatrical productions, picking up charges that were both aesthetic and ethical.[49] Along with instruments and occasionally musical notation, textually mediated sounds helped make tangible the experience of living through a war, both for combatants and for those far removed from battlefields. The duality of this experience structures the modern condition known as wartime, which, as literary scholar Mary Favret has shown, took on recognizable contours during the European experience of the Napoleonic Wars. What she calls "wartime" has a particular meaning: it was "the experience of war mediated, of time and times unmoored, of feeling intensified but adrift."[50] Even several decades later, telegrams from Sevastopol could take many hours to arrive in London; detailed reports still took days, if not weeks, to arrive by post. As several chapters in this book suggest, this experience of wartime seems to have been an international phenomenon by the

[49] Indicative of these representational dynamics is the disclaimer for Russell's column: "The letters of our special correspondent from the scene of war, although naturally a few days in arrear of those leading communications which reach us through the agency of the telegraph, are always replete with interest, and are calculated indeed to serve far more important purposes than those of momentary amusement." "The Letters of Our Special Correspondent," *The Times* (London, 21 Oct. 1854), 6.

[50] Mary Favret, *War at a Distance: Romanticism and the Making of Modern Wartime* (Princeton, NJ: Princeton University Press, 2010), 9.

mid-nineteenth century. Sights and sounds of the battlefield were usually out of date, always out of time, and often felt to be so.

That circumstance is one reason for this book's interest in sound in Favret's "wartime," rather than, say, sound in warfare, or even sound in war. Our focus is meant to shift attention away from battlefields and much-studied (elite, male) military actors, toward the temporalities established by sounds in motion: temporalities that embrace civilian actors, and, crucially, help to make up for the conspicuous absence of women in discussion of war's sounds. As outlined in this book, a wider social purview is granted by our focus on the time lapse between battlefields and elsewhere: the relation between those who claimed to hear Crimea and those for whom they claimed to hear it. This book is, accordingly, organized around Crimea's manifold elsewheres. Contributors cast their auditory coordinates widely across territories and cultures, attempting to rehear the war through the ears of elites in Petersburg and London, British operagoers in Constantinople, religious leaders in the Caucasian Imamate, Latvian troops stationed in Riga, soldiers from Italy and Poland stationed in manifold theaters of war, and Tatar communities in Crimea itself. In each case we are dealing with a particular construction of wartime: an experience of temporality that, to a greater or lesser extent, permeated everyday life in territories far removed from battles.

This experience of wartime for media publics in Istanbul, or London, or Paris, or Petersburg, was clearly not the same everywhere. It may be worth pointing out here some of the differences that emerge between, say, readers of literary journals, operagoers, and sheet-music publics on the one hand; and, on the other, those whose wartime experience was "mediated" in completely different ways. Among the latter, we might include people at the periphery of metropolitan centers, whose access to information was precarious; or, in Maria Sonevytsky's chapter, the indigenous communities of Crimea, for whom the destructive impact of warfare was an intimate reality; or those Ottoman subjects mentioned by Peter McMurray, people who beheld telegraph wires not only as a means of information, but also as a way to venerate the Sultan; or the Russian soldiers who waited on the Baltic coast, in appalling conditions, for an attack that never came. As Kevin Karnes's chapter relates, many of this last group did not return from the war—they either died from illness or starvation, or were endlessly redeployed as lifetime conscripts—and so seem to drift outside "wartime" altogether. The variegated wartimes that emerge from Karnes's and other chapters unfold a complex transnational soundscape, deeply enmeshed in geopolitical inequalities sustained by imperial power. These imbalances manifested themselves in many guises—in the overwhelming noises attributed to particular armies; in the supremacy often attached to Western military music; in the "unknowable silences" variously imposed on political Others. It soon becomes clear that sounds often serve as highly mobile metaphors for geopolitical might.

Perhaps this much could have been anticipated. As the obverse of cosmopolitanism, and as its accompanying shadow, we could have expected war to reveal an unequal soundtrack to intercultural conflict.[51] Celebrated studies of orientalism and imperialism during the nineteenth century have long stressed the role of music as a cipher for political power.[52] What is more, the foundational texts of what subsequently took shape as sound studies have repeatedly underscored territorialization as a basic function of sounds of all kinds.[53] A more novel theme to emerge from the following pages, however, is the role of sound, and of voice in particular, in fashioning mental geographies. Thus Andrea F. Bohlman approaches Polish legion songs about Crimea as a political technology for redrawing national boundaries; Karnes interprets a Latvian album as "exploding associations" between the peoples and spaces of Eastern Europe; and Delia Casadei explains that the war gave Italians the opportunity to hear (and mishear) voices of the would-be nation from the outside, and so conjure them in proto-national form. A now remote but important scholarly model for these chapters is Steven Feld's ethnographic studies, which homed in on the interconnection of voice, sensing, and place.[54] We should also flag up more recent precedents in Katherine Bergeron's discussion of phonetics in the "mouth's complex geography" in mid-nineteenth-century French *mélodie*; and Mary Ann Smart's insight, worked into a proposition by Martha Feldman, that "voice is nothing if not relational, always situated at boundaries."[55] Compatible with this notion are chapters in this book that illustrate ways in which voice articulates relations between self and other, inside and outside, here and there, while also possessing the power to disrupt and remake these fundamental spatial and political distinctions.

[51] On the relation between war and cosmopolitanism, see Srinivas Aravamudan, "Introduction: Perpetual War," *Periodical of the Modern Language Association* 124/5 (2009), 1505–14.

[52] Edward Said, *Orientalism* (New York: Penguin, [1978] 2003); Ralph P. Locke, *Musical Exoticism: Images and Reflections* (Cambridge: Cambridge University Press, 2009), 176–79; Damien Mahiet, Mark Ferraguto, and Rebekah Ahrendt, "Introduction," *Music and Diplomacy from the Early Modern Era to the Present* (New York: Palgrave MacMillan, 2014), 1–18.

[53] Jacques Attali, *Noise: The Political Economy of Music*, trans. Brian Massumi (Manchester, UK: Manchester University Press, 1985), 6–9; Schafer, *The Soundscape*, 73–78.

[54] Steven Feld, "Waterfalls of Song: An Acoustemology of Place Resounding in Bosavi, Papua New Guinea," in *Senses of Place*, ed. Steven Feld and Keith H. Basso (Santa Fe, NM: School of American Research Press, 1996), 91–136.

[55] Katherine Bergeron, *Voice Lessons: French Mélodie in the Belle Epoque* (New York: Oxford University Press, 2009), 94; Mary-Ann Smart, "Callas Listening," *Women and Music: A Journal of Gender and Culture* 9 (2005), 106–10; Martha Feldman, "The Interstitial Voice: An Opening," *Journal of the American Musicological Society* 68/3 (2015), 658.

Silence . . .

Bruce Smith's *The Acoustic World of Early Modern England* (1999) excavated the sonic environment of Shakespeare's England and stands out as an early model for the kind of scholarly endeavor this book pursues.[56] More closely related to the topic at hand, though, is Mark M. Smith's interrogation of the soundscapes of the American Civil War. His *Listening to Nineteenth-Century America* (2001) reconstructed experiences of war by way of the sonic traces left by written accounts:

> Time and again the imagery of how each section [i.e., the American North and South] sounded was recorded first in the ear, then in a print version that stripped the sounds of their nuance and replaced them with a clumsy, written representation, thus giving readers access to a captured record of sectional aurality that they in turn could repeat with their voices to other ears.[57]

Smith outlined the dynamic interplay between ear, voice, and page, only one of which the historian may access. In their denuded form as writing, sonic experiences become metaphors, "crystallized [. . .] in remarkably clear and candid [aural images]," Smith wrote.[58] In this way, he managed to tease out the threat of industrial progress perceived by slaveholders in the American South within the noises of war, while those same sounds were welcomed by southern slaves as "the melody of victory."[59] In other words, he demonstrated that the boundaries between sound on and off the battlefield are always blurred, and there is an ever present, heightened parallax introduced to wartime sound because of conflicting perspectives and political purposes.

Writing in 2001, in advance of a wave of scholarship on the history of sound, Smith likened the task to switching on several lights at once in a dark room: the sources of sound are manifold and overwhelming in their cumulative effect. Smith's recommendation was that sound scholars choose their switches carefully, proceeding methodically and selectively. Nearly two decades on, his

[56] Bruce Smith, *The Acoustic World of Early Modern England: Attending to the O-Factor* (Chicago: University of Chicago Press, 1999); for another widely cited example of exploring historical soundscapes, in another context, see Alain Corbin, *Village Bells: Sound and Meaning in the Nineteenth-Century French Countryside*, trans. Martin Thom (London: Macmillan, [1995] 1998).

[57] Mark M. Smith, *Listening to Nineteenth-Century America* (Chapel Hill: University of North Carolina Press, 2001), 8.

[58] Ibid., 8.

[59] Ibid., 196.

advice speaks to a continued feeling of unease about doing sound history: that sound's ubiquity makes it damagingly promiscuous as an object of study, too yielding to the ideas the scholar may wish to apply, too responsive to the light switches she may choose to flip. As though to circumvent this danger, various theoretical advances have been proposed over the intervening years, amounting to a sea change in how scholars approach sounds of the past: not as an objective totality that precedes the perceiver (the sound-flooded space of Murray Schafer's "soundscape," for example), but as the co-production of perceiver and perceived, involving countless affordances between humans, technology, and environment. Emily Thompson's *The Soundscape of Modernity* (2002) examines early-twentieth-century America by way of technologies for sound, showing how "modern" sonic experience was negotiated by musical patterns of attention and innovations in acoustic science.[60] Along similar lines, Jonathan Sterne investigates the progressive externalization of human ears during the nineteenth century in *The Audible Past* (2003), and the use of audile technique to transduce sound by way of various forms of writing.[61] These means of inscription themselves transformed the nature of sound, according to Sterne, perhaps most conspicuously through creating a division between original sounds (sounding presence) and their traces—and later in the nineteenth century, between originals and their reproductions.

This focus on technology has, among other things, denaturalized sound—it is not simply "out there" to be retrieved—and has made the complex and fluid media of perception central to conceptions of the historical soundscape. And yet Smith's methodological quandary persists. It may be impossible to know whether historical sounds are in some way representative of the past, or to a greater or lesser extent random in their persistence in archives and cultural memory. To put it another way, if we seek out the sounds of the past, how can we be sure to notice significant *absences* of sound? It seems that the historian must always leave open the possibility of finding silence less metaphorical in kind. Beyond evocations of silence by historical actors, there is an infinitely vaster silence left by sounds that have disappeared altogether, and that may be much more difficult to notice than those which, for one reason or another, have left a more permanent trace. We might call this second-order absence an archival silence, to refer to historical experiences of sound (this would include now vanished evocations and impositions of silence) that have since been forgotten and become otherwise irretrievable. While the frailty of memory usually takes the

60 Thompson, *The Soundscape of Modernity*, 33–50.

61 Sterne, *The Audible Past*, 32–35. Animal ears were also investigated and increasingly conceived as detachable; see Jonathan Sterne, *MP3: The Meaning of a Format* (Durham, NC: Duke University Press, 2012), 61–91.

position of an explanatory default when it comes to this kind of silence, there may be other reasons why historians encounter silence where there ought to be, or where they might expect, sonic traces. This kind of archival silence corresponds to what Jacques Derrida called the archiviolithic: the destructive force that brings about loss within the archive, through inciting "forgetfulness, amnesia, and the annihilation of memory."[62]

This silencing force delineates complex historical itineraries across the present book. It seems that, depending on where and when in the world we choose to direct our ears, we encounter a greater or lesser heft and/or availability of historical materials pertaining to wartime sound. This effect is no less true of traditional venues of historical research, such as libraries and archives, than of the digital repositories that have partially and unevenly reproduced them.[63] There are, self-evidently, fault lines in memory according to different national and imperial histories, as well as diverse cultural engagements with archival and media technologies.[64] In the case of the Crimean War, we find rehearsed in sonic terms the truism that history is often told, and so now frequently heard, from the victor's perspective: that of Western Europe, particularly the perspectives of Britain and France, in whose archives countless materials are stored, and in whose tongues the history of the war has overwhelmingly been written and read.[65] My introduction has doubtless betrayed this perspectival bent. However, as I have drawn on French musical instruments and British phonograph records, I have nevertheless been attempting to unsettle the dominance of British and French accounts. Other contributors to this volume, seeking to explore different aural vantage points, have also been obliged to read imperial accounts against the grain.

Yet the question of interpretation—what to do with archival silence—remains contentious, and in fact provides a point of divergence between contributors to this volume. Once we have established that certain sounds are absent from the archive, itself no straightforward task, we are left with scholarly and ethical choices. We can, for example, choose to understand archival silence in historical terms, as the ruptures and absence produced by acts of silencing. The prosecution of war may itself efface sonic experiences (along with many other,

[62] Jacques Derrida, *Archive Fever: A Freudian Impression*, trans. Eric Prenowitz (Chicago: University of Chicago Press, 1996), 11.

[63] On this topic, see Benjamin Walton, "Quirk Shame," *Representations* 132/1 (2015), 121–29.

[64] For a skeptical view of the power of archives, see Richard Thomas, *The Imperial Archive: Knowledge and the Fantasy of Empire* (London: Verso, 1993), 1–10; on the dispersal of sonic archive in colonial contexts, see Anna María Ochoa Gautier, *Aurality: Listening and Knowledge in Nineteenth-Century Columbia* (Durham, NC: Duke University Press, 2014), 3–4.

[65] Figes also remarks on the traditional dominance of British and French accounts, and on the need to consider Russian and Ottoman accounts; see his *Crimea*, xxiv.

more obviously important things) of certain groups, such as the Tatars, who underwent mass displacement during the Crimean War.[66] Another response is to take archival silences as an opportunity for advocacy, in attempting a scholarly act of recovery, or unsilencing. As ethnomusicologist Ana María Ochoa Gautier has argued, and as Sonevytsky discusses in her chapter in this volume, music may be particularly useful for such acts of unsilencing because its complex social ontology provides multiple opportunities to recover aspects of under-heard historical cultures and experiences.[67] Reading people from the sounds that remain tends to put the scholar in the position of doubting herself: endemic uncertainty as to whether she might be over-interpreting, thus giving the idea of "overhearing" a double meaning, as Sonevytsky's chapter points out. The problem of overhearing (and the hermeneutics of suspicion it unleashes) becomes inevitable when sonic evidence is lacking, but is a pervasive concern. As witnessed throughout this book, historical actors and communities are frequently encoded and obscured by the making of "sounds"—the reification of aural experiences performed by particular people at particular times and places—which often serve as metonyms for the social bodies from which they emanate. Archival silence, the absence of sounds and silences, becomes immediately political. It has forced contributors in this volume to develop their own ways of negotiating an unbounded and inscrutable realm.

. . . And Archives

That these problems should arise in a book on sound history may not come as a surprise. Such conundrums have beset historical writing on music, and historical writing more generally, for many decades. The illusions sponsored by imperial archives will be familiar to readers versed in subaltern studies and feminist theory. Familiar too will be some of the strategies we have pursued in attempting to correct, or at least draw attention to, institutional biases in what gets preserved, archived, and narrated—but that, ultimately, create an unresolvable tension across the present book: between our focus on nineteenth-century wartime sound and the recovery of aural perspectives of women and others. The idea that archives might yield up counter-histories, or that we can read them "against the grain," has a long pedigree, in particular the postcolonial problem of giving density to the subaltern in historical accounts. Writing in the 1980s, Ranajit

[66] Mara Kozelsky, "Casualties of Conflict: Crimean Tatars during the Crimean War," *The Slavic Review* 67/4 (2008), 866–91.

[67] See Ochoa Gautier, "Silence," in *Keywords in Sound*, ed. David Novak and Matt Sakakeeny (Durham, NC: Duke University Press, 2015), 183–92.

Guha notably argued that an opportunity for archival recovery was inherent in a dialectical relationship between oppressor and oppressed. For him, ripples in routines of power could be read as signs of colonial violence.[68] Also advocating an askew perspective on archives, although taking a different approach, Gayatari Spivak called attention to the (necessarily strange) ways in which the subaltern can make herself understood. In a celebrated essay, "Can the Subaltern Speak?" (1988), Spivak described the problematic listening engendered by subaltern narratives and invented a mode of archival reading on the lookout for (at least partly) unreadable gestures, especially those of subaltern women.[69]

Within the sprawling interdisciplinary field of sound studies, the legacies of postcolonial theory might be acknowledged as providing a link to enduring problems. As the title of Spivak's essay suggests, and as postcolonial theory bears out, listening and silence have long been master tropes for critics of imperialism and patriarchy, both in diagnosing operations of violence (as in the cliché of the silence of the archive) and in articulating modes of redress. There are, for example, countless contemporary academic projects enjoining us to listen to unheard or under-heard voices of the past.[70] Much more could be said about the complex aurality of feminism and postcolonial theory, requiring exertions beyond the remit of this book. However, the contributions gathered here suggest that histories of sound are well placed to inherit from postcolonial theory and to participate in its broader re-evaluation across the humanities.

Prescient in this regard, or now it seems, was Antoinette Burton's *Dwelling in the Archive* (2003). Although her book did not deal with sound directly—her focus was rather the exclusion of women's stories from public archives in late-nineteenth-century colonial India—Burton's approach to the problem of archives can be usefully revisited here. The disciplinary stakes she outlined may seem all too familiar to historical musicologists and historically inclined practitioners of sound studies. Faced with growing, widespread impatience

[68] Ranajit Guha, *Elementary Aspects of Peasant Insurgency in Colonial India* (Durham, NC: Duke University Press, 1999).

[69] Reproduced in Rosalind Morris, ed., *Can the Subaltern Speak? Reflections on the History of an Idea* (New York: Columbia University Press, 2010).

[70] There are have been some striking reactions against this trend, however. A group of historians of slavery and the Black Atlantic have called time on what they call the "Question of Recovery." They argue that, while the building of black archives has been vital to twentieth-century campaigns for liberation, the recovery of such voices as a political project should be called into question. Rather than strain against archival silence, we would do better to accept the generative tension between recovery as an imperative "fundamental to historical writing and research" and the "impossibility of recovery when engaged with archives whose very assembly and organization occlude certain historical subjects." See Laura Helton et al., "The Question of Recovery: An Introduction," *Social Text 125* 33/4 (2015), 1–18.

with postcolonial theory among contemporary historians, along with widespread demands for retrenchment within long-furrowed archives, Burton declared that "[s]cholars like Said and Spivak have not dispensed with the archive. More unsettling by far to traditions of imperial history (especially in Britain), they have insisted on it as the unstable ground of imperial desire and colonial power [. . .]."[71] Shifting focus on to sound, a supposedly more evanescent category of inquiry, does not avoid the problem of engaging the archive. On the contrary, the presumed omnipresence of sound can make it a much more imperious term of historical investigation, stimulating fresh illusions of incontrovertible knowledge and historical completeness. Yet there is another aspiration that may be discerned amid the turn to sound, which, as an area of knowledge and experience can call into question what an archive is in the first place: the forms of power and desire that attend its constitution. As others have pointed out, there is no one place to begin looking for sounds of the past. In *The Singing of the New World* (2007), Gary Tomlinson has described the scattered nature of Aztec sounds in the following terms: "The European domestication of Mexican speech, song, and writing exemplifies a broadly dispersed discursive adjunct to European conquest, colonization, enslavement, and even extermination of native Americans."[72] Tomlinson is dealing with an extreme example, of course. However, the dispersive and often incidental quality of sonic traces across and between archives—which may be official repositories, though are often more heterogeneous and personal in nature—is a widespread phenomenon, which often calls our attention to the political forces that attend archival preservations.

A case in point here, among many that could be singled out, are the sounds captured "incidentally" through nurses' accounts of the Crimean War. The names of some of the authors are well known. Already mentioned above, Florence Nightingale penned her proposals to modernize nursing practice in documents that turn out to be, beyond their stated purpose, a landmark in creating silence at the bedside of hospital convalescents (as Hillel Schwartz has pointed out elsewhere; see also his chapter in this volume).[73] There are countless travelogues and autobiographies, providing abundant if indirect resources of sounds, such as the memoirs of Mary Seacole, the Jamaican nurse and

[71] Annette Burton, *Dwelling in the Archive: Women Writing House and Home in Late Colonial India* (Oxford: Oxford University Press, 2003), 141.

[72] Gary Tomlinson, *The Singing of the New World: Indigenous Voice in the Era of European Contact* (Cambridge: Cambridge University Press, 2007), 18.

[73] Hillel Schwartz, "Inner and Outer Sancta: Ear Plugs and Hospitals," in *The Oxford Handbook of Sound Studies*, ed. Trevor Pinch and Karin Bijsterveld (Oxford: Oxford University Press, 2010), 273–97.

hotelier who established a restaurant in Balaklava during the Crimean War. Seacole unfortunately receives zero coverage in this book, although the unusualness of an autobiography advancing the perspective of a Jamaican-British woman of the mid-nineteenth century has been discussed elsewhere.[74] No less remarkable, though less often mentioned, is the *Autobiography of Elizabeth Davis* (1857): the life story of a domestic servant who grew up in Merioneth in North Wales. Her birth name was Betsi Cadwaladyr and her mother tongue was Welsh, although she learned English in her teens, following her dramatic escape to Liverpool around the turn of the nineteenth century. Toward the end of a remarkable life, during which she crisscrossed continents on board merchant ships, she traveled to Crimea to work as a nurse and found herself in charge of an enormous hospital kitchen. Her autobiography reaches its climax in Crimea, and in doing so arrives at a critique of Nightingale's failure to distribute vast stockpiles of charitable supplies to wounded and sick British soldiers.

A unique vista onto mid-nineteenth-century migrant labor, Cadwaladyr's autobiography can also tell us something important about the incidental yet thoroughly political nature of sonic inscription, storage, and preservation. Dwelling a moment longer over her book can, more generally, help explain the structural precariousness of women's sounds in the present volume. As sensational as Cadwaladyr's accusations about Nightingale were in 1857, what strikes us now are the conditions of possibility that conspired to make her life writeable. After collapsing with exhaustion in Crimea, and following her return to the Britain in 1856 at age sixty-seven, Cadwaladyr met the pioneering historian of Wales Jane Williams, who conducted a series of interviews that enabled Williams to write Cadwaladyr's biography, relayed throughout in the latter's first-person perspective, probably in Welsh. In a preface, Williams described her method of reassembling Cadwaladyr's story:

> A cursory reader may suppose that the writer had merely to listen and record, but the task of preparing the narrative has really involved much care and labour. To seize the first floating end of each subject that chanced to present itself, to draw it out, to disentangle it, to piece it, to set the warp straight and firmly in the loom, and to cast the woof aright so as to produce the true and original pattern of tapestry, has required

[74] See Lynn McDonald, *Mary Seacole: The Making of a Myth* (London: Iguana, 2014). See also Francine Fernandes, "Injustice Anywhere Is a Threat to Justice Everywhere," *Political (Dis) Engagement: The Changing Nature of the 'Political,'* ed. Nathan Manning (Bristol: Policy Press, 2017), 213–15.

> sedulous application. The winding of silk worms' cocoons without a reel, is scarcely a task of more difficult manipulation.[75]

A modern-day ethnographer would take issue with Williams' metaphor, which cast Cadwaladyr in a passive role; the autobiography emerged through interaction, including mutual listening and speaking. Yet, one measure of Williams' commitment to the task of reproducing Cadwaladyr's voice, as she heard it, is the difference in register between her usual literary style (in evidence in the preceding extract) and the choppy, relatively unadorned prose that relates the events of Cadwaladyr's life. Another metric is the role of particular "sounds," which feature copiously in Williams's ethnographic introduction (and characterize Cadwaladyr's national and religious formation as the distinctive product of Welsh Methodist culture), but only rarely appear in Cadwaladyr's own life history.[76] And yet, "sound" as a medium of communication features vividly, if more indirectly, in the latter, through the polyphony of voices Cadwaladyr remembers, and Williams conveys, in snippets of conversation.

At this point, I could comb through the few examples in which Cadwaladyr, via Williams, mentions particular sonic experiences: an exercise that might ultimately reveal something about the aurality of working-class women's culture in the nineteenth century. However, it is perhaps more important to notice that, while Williams expends literary and ethnographic energies on writing the "sounds" of Cadwaladyr's past, Cadwaladyr's own attention is mainly directed elsewhere. The desire of the historian and the ethnographer clearly revolves around isolating sounds and voices—unwinding silken cocoons, to borrow Williams's words—and such scholarly desires make the act of listening to voices of the archive both problematic and, more fundamentally, possible. We can never hear Cadwaladyr's voice, however much we strain, but the unusual archive established by the cooperation of these women does something even more astonishing. It allows us to witness sounds and voices taking shape through a

[75] Deirdre Beddoe, "Introduction," *The Autobiography of Elizabeth Davis: A Balaclava Nurse, Daughter of Dafydd Cadwaladyr*, by Elizabeth Davis and Jane Williams (Cardiff, UK: Honno, 1987), xiii.

[76] Here is a sample of Jane Williams' quasi-ethnographic approach to sounds: "Few persons who have chanced to travel through the Principality can forget the sight of Welshwomen knitting with unremitting industry while walking along the roads carrying heavy burdens upon their heads. [. . .] matrons and maidens would assemble together in some pleasant nook [. . . and] ply their work with busy fingers, and sing together the sweet national airs of their country, pausing at times to relate to each other some wild legendary tale, connected by tradition either with the place or with the tune. On such occasions Cadwaladyr's daughter [i.e., Betsi] was often a listener, though forbidden to attend their regular meetings." Williams, *Elizabeth Davis*, 2 vols. (London: Hurst and Blackett, 1857), 26.

tense interaction: the messy intersubjectivity of listening as it makes itself manifest on the page.

Cadwaladyr's/Williams' *Autobiography* also reminds us of a challenge bequeathed to sound history by feminist and postcolonial thought: not to succumb to the fantasy that, by including Others (women, the subaltern, and so on) in the stories we tell about the past, we will one day arrive at an all-encompassing History. As Burton pointed out, "[t]riumphalism about the capacity of history—including feminist history—to see all its subjects effectively reproduces the discourses of surveillance and total vision that underwrote colonial modernity and its political manifestations, history prime among them."[77] If we substitute hearing for seeing, the relevance becomes clear. "Embracing [such wide-angle history] would require us to participate in the hubris of the panopticon rather than face the ultimate fragmentation and ghostliness of all archives [. . .]."[78] This warning from the not-so-distant scholarly past could, with small adjustment, be made for the current turn to sound, as new archives are everywhere mined and written into being.

Humanism

This book is no exception, of course. It simultaneously addresses and brings into existence a sonic archive of the Crimean War. Yet the sounds that we mine may also undermine scholarly protocols of collection and comparison, and not only for the reasons just outlined. Our focus on sound in wartime faces another basic challenge, calling into question an habitual association between sound and human presence, often conjured in the figure of a listener of some kind. This representational logic has been called into question in the wake of the 2003 Iraq War: such "phenomenological anthropocentrism," to borrow Steve Goodman's term, can break down under wartime conditions.[79] The metonymy that connects sound to the human may snap, as sounds become detachable from, even inimical to, humans and their experience of the world. Although this issue has received attention largely in relation to wars of the present day and recent past, we might briefly outline its contours. In an article published in 2006, Suzanne Cusick

[77] Burton, *Dwelling in the Archive*, 143.

[78] Ibid., 143–44.

[79] ". . . [T]he phenomenological anthropocentrism of almost all musical and sonic analysis, obsessed with individualized, subjective feeling, denigrates the vibrational nexus at the altar of human audition, thereby neglecting the agency distributed around the vibrational encounter and ignoring the nonhuman participants of the nexus of experience." Steve Goodman, *Sonic Warfare: Sound, Affect and the Ecology of Fear* (Cambridge: MIT Press, 2009), 82.

has explored music's use as an instrument of torture in Guantánamo Bay: in an attempt to erode resistance, prisoners of war and alleged terrorists were routinely subjected by their US captors to heavily amplified music, blasted loudly enough to deprive them of sleep and cause pain.[80] Cusick considered the cultural meanings generated by violent, masculinist rock as it was (and elsewhere no doubt still is) channeled for the purposes of torture; but she also indicated a particular fold within musical representation, as sounds become inseparable from the harm they do, and harming others becomes perversely musical. Along parallel tracks, J. Martin Daughtry has considered the functions of sound in Iraqi wartime.[81] On the basis of interviews with returning American soldiers and Iraqi non-combatants, Daughtry has signaled the ways that "[w]artime violence besieges the sensorium, introducing empathic and corporeal pain into bodies and forcing subjects to confront their radical finitude."[82] He has even invented a term, thanatosonics, to mark the threshold at which sound no longer affects bodies positively, but attends their destruction.

The most philosophically trenchant account of this destructive dimension of sound is Steve Goodman's *Sonic Warfare: Sound, Affect, and the Ecology of Fear* (Cambridge, MA: MIT Press, 2009), which examines the use of sound in military conflicts to generate fear. Goodman takes his cue from the Israeli army's use of sound bombs over the Gaza Strip. The use of overwhelming vibrations, decoupled from the explosions that would normally accompany them, has led to virtualization of fear, sound's use as a weapon in its own right.[83] As Goodman argues, sonic assaults (including sound bombs) induce fearful feelings, intercepting us at a pre-individual level. Detectable across a range of contemporary cultural phenomena, not only in war, sonic warfare is, Goodman defines it,

> the use of force, both seductive and violent, abstract and physical, via a range of acoustic machines (biotechnical, social, cultural, artistic, conceptual), to modulate the physical, affective, and libidinal dynamics of populations, of bodies, of crowds.[84]

[80] Suzanne G. Cusick, "Music as Torture/Music as Weapon," *Trans: Revista Transcultural de Música* 10 (2006), <http://www.sibetrans.com/trans/articulo/152/music-as-torture-music-as-weapon>, accessed 4 Apr. 2016; in a subsequent article Cusick pointed out that the use of sound in torture can be deployed *musically*, thereby calling into question cherished beliefs about the innocence of musicology's object of study; see her "Musicology, Torture, Repair," *Radical Musicology* 3 (2008): 24 paragraphs, <http://www.radical-musicology.org.uk>, accessed 5 Sept. 2015, §11.

[81] J. Martin Daughtry, *Listening to War: Sound, Music, Trauma, and Survival in Wartime Iraq* (Oxford: Oxford University Press, 2015); J. Martin Daughtry, "Thanatosonics: Ontologies of Acoustic Violence," *Social Text* 32/2 (2014), 25–51.

[82] Daughtry, "Thanatosonics," 25.

[83] Goodman, *Sonic Warfare*, xiii.

[84] Ibid., 10.

It is worth underscoring the eccentricity of Goodman's position, which stands out within the broader discussion on music, sound, and war. Unlike Cusick and Daughtry, he avoids discussion of experience and all supposedly anthropocentric understandings of listening and hearing. Coming from the Deleuzian tradition, he takes it as read that sound affects us, and other entities, in ways that both precede and go beyond conscious knowing. He outlines instead an ontology of "vibrational force," in which inaudible sound, even potential sound, is on par with the narrow strip of sounds that humans may claim to hear.

A philosopher by training, Goodman has thrown down a gauntlet to musicology and sound studies, which remain broadly humanist in their (inter)disciplinary orientations. *Sonic Warfare* challenges us to expand our remit beyond music and sound, to include sounds that were never heard, or perhaps never sounded, but may nevertheless have had vital consequences. I have already described one scenario from the Crimean War that might stand as an example of what Goodman describes, in the form of Adolphe Sax's enhanced musical armory; I have suggested that, while saxophones remained largely beyond the realm of perception in 1850s wartime, they nonetheless belong to transnational context of empires at war. This point might seem, at first glance, a decidedly twenty-first-century academic formulation. But consider Figure I.1, which shows one "use of the saxophone during wartime": to blast an enemy soldier in the face.[85] It is telling that the instrument shown here is not in fact a saxophone. This cartoon is probably based on a descriptive account, the reed mouthpiece and the finger holes along the shaft hinting at an imaginative extrapolation, so demonstrating the instrument's discursive presence before it became an audible and visible phenomenon. Appearing in *Le Charivari*, Paris's major satirical newspaper, and following a string of decisive victories for France, the image calls attention to the pomp of Sax's modernized military band, an institution by now deemed a needless presence in battle.

The noisiness of Figure I.1 sticks in the mind. It imagines the overwhelming effect that sound can have over the enemy. This fictional saxophone gestures beyond humanistic listening to an idea of music as force, pointing to deeper complicities between military action and instrument building—complicities that were aural in kind, even as they evaded audible experience. Although never realized, two further inventions were dreamed up by Sax during the Crimean War. One was the Saxotonnerre: a mammoth organ whose pipes were to be driven by a locomotive engine; Sax claimed it would have been loud enough to broadcast Meyerbeer's overtures throughout Paris.[86] Another, also

[85] Charles Vernier, lithograph, "De l'utilité du saxophone en temps de guerre," *Le Charivari* (15 Nov. 1854).

[86] According to the *Revue Gazette Musicale*, the instrument would be "operated by vibrating blades, submitted to pressure of four or five atmospheres. The blades are huge steel bars vibrating

Figure I.1 Charles Vernier (illustrator), "De l'utilité du saxophone en temps de guerre" [On the use of the saxophone during wartime]. *Le Charivari* (Paris: 15 Nov. 1854), 20. Reproduced from the personal copy of Frédéric Maillard, and accessible at his website <http://www.lacaveavents.com/docs.php?>; used with permission.

citywide in intended address, was a cannon with a shaft ten meters in diameter, capable of firing round shot weighing half a ton over a mile in distance.[87]

under high pressure." Michael Segell, *The Devil's Horn: The Story of the Saxophone; From Noisy Novelty to the King of Cool* (New York: Picador, 2005), 27. See also Leon Kochnitzky, *Adolphe Sax and His Saxophone* (New York: Belgian Government Information Center, 1949), 40.

[87] Cottrell, *Saxophone*, 17.

Thereby, Sax envisioned a technical solution to protracted sieges such as the one seemingly never ending at Sevastopol, ongoing for nearly a year at the time he designed the weapon. Just a few missiles from the Saxocannon would have been enough to raze an average-sized city to the ground. The Saxocannon and Saxotonnere were never built, but might have been. They belong to a genealogy of ever more powerful instruments, at once imaginary and conceivable, that also include latter-day innovations such as electric loudspeakers, sound bombs, and perhaps even bunker-buster missiles. Such an organology would gather sound machines that transcend humanist concerns, pointing toward a history in which vibration would figure less as the basis for experience, more as its annihilating double.

The connection, through the activities of Adolphe Sax, between the invention of musical instruments and the design of weapons might give us pause. It suggests, among other things, that proximity between real and imaginary objects forms an important vector to the politics of music and sound: a dimension that cannot be recovered solely from the perspectives of listeners, but must be inferred from the residues of affective phenomena that inhabit cultural archives in unpredictable ways. Sound, in this sense, is not always heard, and so sonic histories should not necessarily be restricted to the audible realm. This lack of restriction is, of course, easier imagined than it is to trace historically. In the present book, which is more concerned with acts of listening, affective experience nevertheless hovers at the edges. Fuller descriptions of the chapters will shortly follow, but first a word on submerged connections between them in this regard.

For writers in this volume, affect is understood as inseparable from the media epistemologies that wartime engenders.[88] Alyson Tapp, for example, discusses Tolstoy's writings on the Crimean War, showing how sensory partitions (epistemology) are intimately connected with their affect (ontology). On Tolstoy's battlefield, cannon fire "shakes not just the ear organ, but your whole being"—his vibrational ontology is prepared and sustained by nineteenth-century assumptions about the behavior of sound. By the same token, Flora Willson considers the opera house in Constantinople as a mediating site through which wartime violence was perceived and structured. Taking in a panoply of wartime sounds, Hillel Schwartz juxtaposes the noises of the Crimean front alongside whistling in Russia and popping champagne corks in Britain, and many other resonances besides. Emphasizing the non-signifying elements of these sounds, Schwartz posits their common basis in potential energy: their culturally determined preexistence as sounds about to explode.

[88] Favret, *War at a Distance*, 12.

Unmaking Sense

In charting a zigzag course through the themes of this book, this introduction has tried to give a sense of the scholarly orbits within which the chapters move. There are still more ways to navigate through the chapters, which are of course linked by their focus on the Crimean War as a historical event, but also share common themes and approaches. In the first part, "Sound, Technology, Sense," the book's first four chapters deal with the technical mediation of sound in different political and cultural contexts. Dina Gusejnova opens with "Sympathy and Synesthesia: Tolstoy's Place in the Intellectual History of Cosmopolitan Spectatorship," a wide-ranging assessment of humanist interpretations of war in the European sentimental tradition (and in its aftermath); she aims to tease out Tolstoy's contribution to war writing in terms of the literary mediation of sensations such as sounds, notably through the novel. Peter McMurray's chapter, "The Revolution Will Not Be Telegraphed: Shari'a Law as Mediascape," explores the war as a sonically and telegraphically mediated event in the south and northwest Black Sea region, in particular in the Caucasian Imamate and the Ottomans Empire. My chapter, "Gunfire and London's Media Reality: Listening to Distance between Piano, Newspaper, and Theater," homes in on the pervasive representation of gunfire across different media forms in London in late 1854. Bringing this part to a close, Maria Sonevytsky in "Overhearing Indigenous Silence: Crimean Tatars during the Crimean War" submits to scrutiny the relative lack of historical sources pertaining to Crimean Tatar experiences of the war.

The linking concepts that shape this first part are broad and return throughout the book; they include the historical production of sensory experience, cultural memory, and technologies of the archive. This part opens with explorations of sonic mediation, then broaches mediation's steady erasure and the reality of archival loss. Hence Gusejnova surveys the literary evidence of the slow and uneven encroachment of cosmopolitan tendencies in late-eighteenth- and nineteenth-century European societies. She argues that multichannel sensory experiences generated through literary montage, and later on through film montage, were a historical and cognitive emergence. In other words, she aims to show that there is a fundamental connection, at the level of the senses, between war and nineteenth-century forms of cosmopolitanism: greater understanding between people was driven, in part, by violent wartime encounters, as well as their literary, visual, and sonic mediation. Similarly concerned with the political uses of sounds, McMurray makes innovative use of legal sources to recover aspects of the sonic past, by comparing contemporaneous societal reforms stimulated by the war in Turkey, Daghestan, and Chechnya. He argues that sound, voice, and telegraphy played a foundational role in anchoring new laws as issuing from authoritative bodies: ultimately those of the Sultan and the Imam. By positing these figures as unattainable acoustic origins, McMurray shows how sound and communication media became mutually reinforcing practices, simultaneously transmitting the content of legal reforms and the furnishing the cultural codes by which they were understood. My chapter is also concerned with the operations of historical media. It argues that theaters, newspapers, and printed music were mutually inflecting domains in wartime London: areas of sonic knowledge and experience that gave particular significance to musical and sonic simulations of the battlefield both at home and in the urban public sphere. I consider the implications of this historical mediation of wartime sound, and try to show that the macabre fascination produced by gunfire was linked to the invisibility of low-ranking soldiers. Whereas I consider the politics of representation, Sonevytsky tilts her investigation toward the politics of a lack of representation. By probing the slender archive left by Crimean Tatar songs, she attempts to recover experiences that have disappeared from cultural archives, in large part because the Tatars were subject to Russian imperial powers. Sonevytsky asks what can be done by historians, and by activists, with memories preserved through sound, ultimately turning her attention toward Russia's present-day annexation of Crimea and contemporary efforts to use musical memory as a means of political resistance.

In the next part, "Voice at the Border," Andrea F. Bohlman's chapter, "Orienting the Martial: Polish Legion Songs on the Map," also foregrounds the question of archival loss; but the common focus between her contribution and those in the book's second part is the role of voices, both spoken and sung, in defining geographical boundaries during wartime. In exploring the fragmented archive represented by Polish military involvement in the Crimean War, Bohlman mulls evocations of military might in legion songs. She argues that these songs were a political technology for preserving and promoting Polish nationhood and creating a virtual landscape for the cultivation of a future homeland, while also, more pragmatically, stimulating nationalist sentiment both at home and abroad. In "Who Sings the Song of the Russian Soldier? Listening for the Sounds and Silence of War in Baltic Russia," Kevin Karnes examines the war's less eventful and often overlooked Baltic theater. He considers conscription, encampment, combat, and mourning as defining events that structured the experiences of hundreds of thousands of Russian soldiers stationed on the Baltic Coast, homing in on the broader effects of militarization in Latvia in particular. As Karnes shows, the mass mobilization of Russian troops occasioned many first encounters: between culturally heterogeneous Romanov subjects; between "Russians" and Europeans from the West; between Europeans both Eastern and Western as well as non-European others. Finally, Delia Casadei's chapter, "A voice that carries," likewise addresses the geopolitical uses of aurality, by sketching a history of Italy "as heard from the outside" during the war. She charts the ideology of the *bella voce* as means of projecting and disrupting national boundaries, both in the years before national unification and, even more so, by way of literary accounts that came later in the nineteenth century.

Such experiences served to refashion mental geographies of Europe, altering its ever shifting boundaries at the mid-century. Karnes argues that listening to voices in wartime shattered associations between peoples and spaces within so-called Eastern Europe. Along similar lines, Bohlman argues that poems and songs served to sing a nation into being, redrawing a constantly shifting imaginary border between Poland and the imperial forces that kept it splintered. Similarly noting the capacity for voices to make (new) sense of geographical distinctions, Casadei asks what was at stake in the Sardinian troops' ability to organize themselves, even to understand themselves, amid countless regional dialects. This problem was thrown into relief as the army set sail from the Italian peninsula. By following the voyage in literary accounts, Casadei uncovers a telling episode in the history of attending to Italian sounds: one in which voice and the capacity for language are fashioned into politicized and even oppositional terms.

In the third part, "Wartime as Heard," the final chapters contemplate the ways aural perception was structured during the Crimean War. In "Operatic Battlefields, Theater of War," Flora Willson explores how (mainly Italian) opera inflected listening for British officers and tourists on the move. At home and on the battlefield, and particularly in the spaces in between, opera became enmeshed in cycles of transport and mediation. Willson concentrates on operatic perceptions in and around the Pera district of Constantinople, the site of the city's first opera house (a must-see for elite Britons en route to Crimea), as well as those associated with traveling military bands connected with the Ottoman imperial court. In the next chapter, "Earwitness: Sound and Sense-Making in Tolstoy's *Sevastopol Stories*," Alyson Tapp revisits one of the Crimean War's most celebrated literary productions, the *Sevastopol Stories*, written while Tolstoy was serving as an officer in Crimea. Considering this key text (placed in a larger context by Gusejnova in "Sympathy and Synesthesia") in more detail, Tapp interrogates the workings of sound in the *Stories*, elucidating the different valences of battlefield sound at degrees of remove from the war zone. From afar, battlefield sound in Tolstoy is both meaningless and often figured as musical; yet with increasing proximity it becomes a cipher for unmediated reality, and ultimately for truth, becoming a means to gesture toward authentic experiences of combat. Also concerned with aurality, but in a different vein, Hillel Schwartz's chapter, "InConsequence: 1853–56," deploys historical listening as critical and creative method. Beginning at the Crimean front, he discusses the pervasive whistling of dying horses—untold in number and often ignored in the face of the war's many human miseries—and goes on to pursue human and nonhuman whistlers in mid-century Russia and industrial Britain. Whistling thus generates a panoply of war-related homophones and connotations, inviting us to rehear, among other things, the decline of shrilling round shot on the battlefield and the ascendency of spiraled, hissing bullets.

The three chapters in the final part can be read as a whirlwind tour of aesthetic distinctions across the Crimean War's sprawling territories. Beginning with the perceptions of Britons in Constantinople, it proceeds to Tolstoy's documentation of experiences on the Russian side of the front, and culminates in a comprehensive rehearing of the Crimean battlefield. Schwartz's tour de force ultimately transports us back to London, with an extended analysis of "Pop Goes the Weasel," one of the most whistled tunes in wartime Britain. Venturing an explanation for the refrain's ubiquity, Schwartz writes that its "lexical potential energy could be enlisted to do highly kinetic, cultural work in dozens of contexts"—contexts that were military, political, literary, and scientific, in which the very idea of potential energy was itself emergent. Vast shifts in the history of listening are subjected to interrogation in Tapp's and Willson's chapters, too. Willson examines elite perceptions of foreign battlefields and cityscapes to probe a grand, oft noted, and complicatedly global shift in the history of listening: that of middle-class audiences falling silent in theatrical spaces during the nineteenth century, supposedly with the intention of devoting concentrated attention to elite music. Willson argues that these listening habits, formed in part in the opera house, persisted well beyond its hallowed enclosures as war came to extend the complex geographies of attentive listening at the midcentury. Meanwhile, Tapp shows how Tolstoy, in his attempts to represent the ultimate truth of battlefield experience, cordoned off audible reality from its supposedly less immediate visual counterpart. He thus reproduced an audiovisual split already pervasive in nineteenth-century culture, and, as Tapp suggests, further deepens the rift through his literary refractions of war's overwhelming sounds.

This outline suggests one route through this book. The organizing themes drawn out above—organology, technologies of inscription, wartime, archival loss and silence, humanism—suggest others. When read together, the chapters combine to form a variegated geopolitical picture of the sounds of the Crimean War. But they may also prompt reflection on the sonic turn in musicology, literary studies, and the historical disciplines, as that turn accrues a history of its own. As we cycle through the themes of sonic mediation, a need grows to account for mediation's undoing: whether in considering sounds that have been lost over time, becoming un-mediated, absent, silent; or by heeding sounds that play an active part in their own demise, in destroying their own medium; or by encouraging listeners to forget through methods more or less coercive and violent. If sounds are made through complex relations between biological systems and cultural processes involving technological media, then they can be unmade by the same means. It becomes a priority that we should to be able to account for the breakdown of sonic experience under particular historical, social, and political conditions.

This point touches on the aims and aspirations of sound studies as a historical discipline, whose agenda was set by Jonathan Sterne more than a decade ago. As he put it then, motivating his own focus on nineteenth-century sound technology, "the history of sound must move beyond recovering experiences to interrogating the conditions under which that experience became possible in the first place."[89] Yet, as his manifesto wound to a close, he wrote, "the question of experience still lingers"—thus gesturing to the vexed puzzle of writing about sonic experience without succumbing to the illusion that such experience is separate from, or alien to, writing and other inscriptional acts and products. To think of sound with and as writing has proved an intractable and generative problem in sound studies in the twenty-first century.[90] Expanding notions of technology and "audile technique" to embrace writing of all kinds—as does Ana María Ochoa Gautier in *Aurality* (2014)—may allow us to appreciate the complexity of the issue. Ochoa Gautier seeks to extrapolate from nineteenth-century archival traces a means to explore "ontologies and epistemologies of the acoustic, particularly the voice, produced by and enmeshed in different audile techniques, in which sound appears simultaneously as a force that constitutes the world and a medium for constructing knowledge about it."[91] She aims to combine the inscriptional technologies that provide conditions of possibility for experience with a reconstruction of sound as a force and an agent within a particular historical and cultural order.

In contemporary histories of sound, the question of experience still lingers—to repeat Sterne's melancholic, closing remark—as it must in all historical inquiry, not only in sound studies.[92] And yet this observation should not give rise to regret. The impossible need of recovering experience has long been productive for thinking about sound and many other things besides. Reading sound as a force and an agent from the impressions it has left behind—inscriptions

[89] Sterne, *The Audible Past*, 28.

[90] On this problem, see the volume edited by Deborah Kapchan, *Theorizing Sound Writing* (Middletown, CT: Wesleyan, 2017).

[91] Ochoa Gautier, *Aurality*, 3.

[92] In his classic study, Hayden White understood the representation of historical experience in the nineteenth century as a problem of realism; see his *The Historical Imagination in Nineteenth-Century Europe* (Baltimore: Johns Hopkins University Press, 1973), n13, 33. Meanwhile, Michel de Certeau famously characterized the historian's acts of inscription as a way of marking the difference between present and past, and so "calming the dead who still haunt the present; see *The Writing of History*, trans. Tom Conley ([1975] New York: Columbia University Press, 1988), 2.

in wax, handwriting on paper, earth upturned on the battlefield—is a fraught scholarly endeavor, caught between more or less convincing appeals to plausibility, and more or less successful creative acts of synthesis and imagination. Yet it may be worth asking as much of technologies and inscriptions: that they tell us not only about conditions of possibility for sounds, but also the shaping of historical experiences, human and otherwise. Approached in this way, technologies and inscriptions can lead us away from "sounds" and toward what people and others have been able to make of them in different times and places: the technical, cultural, and social resources that go into producing sense in both its literal and figurative guises.[93] This connection between sound and making sense will be a constant theme, and a recurrent problem, throughout the present book. It can also serve as a transcoder for the chapters that follow, each of which suggests, in its own way, that we can observe the making of sound through the unmaking of sense that wartime brought about.

[93] Gilles Deleuze describes sense as the "frontier" that runs between sensation and sense-making; see his *The Logic of Sense*, trans. Constantin V. Boundas ([1969] London: Continuum, 2003), 35.

Hearing the Crimean War

PART I

SOUND, TECHNOLOGY, SENSE

1

Sympathy and Synesthesia

Tolstoy's Place in the Intellectual History of Cosmopolitanism [1]

DINA GUSEJNOVA

The importance of sympathy on a global scale seems to be a distinctly twentieth- and twenty-first-century cosmopolitan ideal.[2] Yet the connection between the passions and politics had, of course, weathered long-standing philosophical debate before the age of economic globalization.[3] The issue acquired new pertinence with the geographical expansion of the communications industry in the context of modern warfare.[4] The increasingly global and widely mediated wars provided a new challenge to the viability of impartial spectatorship, to use a concept which Adam Smith had chosen to describe the human capacity for

[1] I would like to thank fellow members of the UCL-based Passionate Politics group, Axel Körner, Uta Staiger, and Tim Beasley-Murray, for providing me with opportunities for discussion of some core themes connected with emotions and politics. The Leverhulme Trust has enabled me to work on this chapter in conjunction with my project on Cosmopolitanism and War. More thanks are due to the anonymous reviewers for constructive suggestions.

[2] For a review on the growing literature on empathy and political theory, see Samuel Moyn, "Empathy in History, Empathizing with Humanity," *History and Theory* 45/3 (Oct. 2006), 397–415. See also Michael Freeden, "Emotions, Ideology and Politics," *Journal of Political Ideologies* 18/1 (2013), 1–10.

[3] Carl von Clausewitz, *On War*, ed. and trans. Michael Howard and Peter Paret (Princeton, NJ: Princeton University Press, 1970). See also Elisabeth Krimmer, *The Representation of War in German Literature: From 1800 to the Present* (Cambridge: Cambridge University Press, 2010), and Elisabeth Krimmer and Patricia Anne Simpson, eds., *Enlightened War: German Theories and Cultures of Warfare from Frederick the Great to Clausewitz* (London: Camden, 2011). On the link between emotions, ethics, and politics, especially in the political thought of the Scottish Enlightenment, see Thomas Dixon, *From Passions to Emotions: The Creation of a Secular Psychological Category* (Cambridge: Cambridge University Press, 2002).

[4] Jürgen Osterhammel, *The Transformation of the World: A Global History of the Nineteenth Century*, trans. Patrick Camiller (Princeton, NJ: Princeton University Press, 2014), 37–39.

morality.[5] New technologies such as the telegraph made participants and observers increasingly implicated even in distant affairs. But technological change was not the only factor. As I shall argue in this chapter, a similar effect of sympathetic imbrication was achieved by new narrative techniques, which were used by some of the leading European writers of the time. The speed of war reporting and the changing character of war depiction in documentary writing as well as literary fiction conveyed a more multi-perspectival perception of war. This augmented the sense of a shared humanity among participants and observers of war, a sense which chimed with cosmopolitan ideals articulated by eighteenth-century philosophers.

Wars involving multiple empires gave new visibility and audibility to previously unrecognized populations and their vernacular speech. They lent new identities to previously disparate political communities and thus, arguably, gave rise to new conceptions of the state and the political. The Prussian military theorist Carl von Clausewitz drew on his experiences gained in the Napoleonic wars to argue that wars could have a positive effect, building ethical communities of nations by creating specific rules of conduct within and between them.[6] At the same time as shaping communities along national lines, military conflicts also entailed "cosmopolitan moments" by providing zones of contact between people of different classes and ethnic backgrounds.[7] The Crimean War, a conflict between four major empires (the Russian, British, French, and Ottoman), was one such moment.[8] In this encounter, different political communities competed not only for territory but also for the authority to make universal claims about

[5] On spectatorship, and the broader theme of "unsocial sociability" it implies, see Istvan Hont, *Politics in Commercial Society: Jean-Jacques Rousseau and Adam Smith*, ed. Michael Sonenscher and Béla Kapossy (Cambridge, MA: Harvard University Press, 2015).

[6] On Clausewitz and the concept of the state, see Peter Paret, *Clausewitz and the State: The Man, His Theories and His Times* (Princeton: Princeton University Press, [1976] 2007), 173 and passim. For the link between global conflicts and the emergence of cosmopolitan theories, see Dina Gusejnova, ed., *Cosmopolitanism in Conflict: Imperial Encounters from the Seven Years' War to the Cold War* (London: Palgrave Macmillan, 2018).

[7] Ulrich Beck, *World at Risk* (Cambridge: Polity Press, 2007), 47–66. On war and ethnicity, see Anthony Smith, "War and Ethnicity: The Role of Warfare in the Formation, Self-Images and Cohesion of Ethnic Communities," *Ethnic and Racial Studies* 4 (1981), 375–97.

[8] Twenty-first-century scholarship on the war has emphasized this aspect. See Lucien J. Frary and Mara Kozelsky, eds., *Russian-Ottoman Borderlands: The Eastern Question Reconsidered* (Madison: University of Wisconsin Press, 2014), esp. Mara Kozelsky, "The Crimean War and the Tatar Exodus," 165–93; Candan Badem, *The Ottoman Crimean War, 1853–1856* (Leiden and Boston: Brill, 2010); and the republication of a contemporary source, Adolphus Slade, *Turkey and the Crimean War: A Narrative of Historical Events* (Cambridge: Cambridge University Press, [1867] 2012).

frontiers of faith and ethnicity.[9] Thus, in one sense, this conflict exposed the negative traits of Europe as a cosmopolitan society of "monarchs, feudal lords, moneyed men, however they may be differently assorted in different countries," as Marx and Engels put it in their journalism of the time.[10]

But this war also fleshed out a different, if no less elite-driven, vision of community among Europeans, which turned the critique of conflict into the foundation for a new cosmopolitan sensibility. The rise of world news reporting had the effect of making distant conflicts more present. And yet, this immediacy cannot be ascribed to the development of new technologies such as the telegraph alone. As important was the evolution in the techniques of narrative and war reporting, which evoked human sympathy across boundaries of class, nation, and empire, also contributing to a change of attitudes toward war.[11] It gave rise to a new type of sentimental writing, one that encouraged a critical perspective on war itself. Stretching the limits of propriety to which readers of European literary works were accustomed, some writers involved in these nineteenth-century wars provided intimate, firsthand accounts that deliberately resisted judgment of events in terms of military logic. Yuval Harari has described this transformation through the term "flesh-witnessing"—a way of thinking about war as a sublimely incommunicable experience of the type that had become fashionable in Europe since the mid-eighteenth century.[12] A new language of humanitarianism evolved by the mid-nineteenth century, influenced not only by wars but also by causes such as abolitionism.[13] In British society in particular, the cult developing around a nurse, Florence Nightingale, eclipsed that of even the most famous generals.[14]

[9] The religious dimension is foregrounded in Orlando Figes, *Crimea* (London: Penguin, 2010).

[10] They go on, "The Turks are fit for sudden starts of offensive action, and stubborn resistance on the defensive, but seem not to be fit for large combined manoeuvres with great armies. Thus everything is reduced to a degree of impuissance and a reciprocal confession of weakness, which appears to be as reciprocally expected by all parties. With governments such as they are at present, this Eastern War may be carried on for thirty years, and yet come to no conclusion." Karl Marx and Friedrich Engels, *New York Tribune,* 7 Apr. 1853, reprinted in Karl Marx and Frederick Engels, *The Russian Menace to Europe,* eds. Paul Blackstock and Bert Hoselitz (London: George Allen and Unwin, 1953), 121–22.

[11] Stefanie Markovits, *The Crimean War in the British Imagination* (Cambridge: Cambridge University Press, 2009).

[12] Yuval Noah Harari, *The Ultimate Experience: Battlefield Revelations and the Making of Modern War Culture, 1450–2000* (Basingstoke, UK: Palgrave Macmillan, 2008), 231.

[13] Michael Barnett, *Empire of Humanity: A History of Humanitarianism* (Ithaca, NY: Cornell University Press, 2011); Moyn, "Empathy in History."

[14] For an early account of the Nightingale cult, see Lytton Strachey, *Eminent Victorians: Cardinal Manning, Florence Nightingale, Dr. Arnold, General Gordon* (London: Chatto & Windus, 1918). I am grateful to Rüdiger Görner for alerting me to this reference.

According to many historians, Crimea was the first global war on the European stage—although there are some other contenders for the status, too, such as the Seven Years' War.[15] What makes this label meaningful for the Crimean War is that it featured new forms of long-distance communication, such as telegraphy and war correspondents, establishing connections between theaters of war and European communities at home. These media were not just means of communicating descriptions of battles post factum: some leading parties in the conflict found out about the outbreak of hostilities from the media before the news had time to reach them through diplomatic channels. One of the striking outcomes of the war, at least for the British side, was the emergence of a new fashion—one might even say, an ideology—of humanitarianism, which tied conventional patriotism to a cosmopolitan perspective.[16] War not only became, increasingly, a contact zone for distant cultures and communities; combatants also became more connected to noncombatants, turning the theater of war into an ever more immersive experience.[17] In this sense, global warfare complicated and reconfigured familiar distinctions between friends and enemies.[18]

Within this evolving trend, Leo Tolstoy acquired a distinctive tone of voice. The war inspired him to fragment his experience into a series of impressions of visual, auditory, and textual encounter. As a technique of imagination that involved the reader in co-producing new associations, his approach was to have a great influence on a variety of other art forms, especially cinema, in which and through which military conflicts were represented, and it is through a cinematographic lens that a twenty-first-century reader of Tolstoy tends to experience his writing. As the great film director Sergei Eisenstein recalled, Tolstoy mastered the art of montage long before the emergence of photographic or cinematic techniques.[19] For Tolstoy, as for Eisenstein, montage was principally an

[15] Cf. Daniel Baugh, *The Global Seven Years War, 1754–1763: Britain and France in a Great Power Contest* (Harlow, UK: Longman, 2011); see also Frans de Bruyn and Shaun Regan (eds.), *The Culture of the Seven Years' War: Empire, Identity, and the Arts in the Eighteenth-Century Atlantic World* (Toronto: University of Toronto Press, 2014).

[16] The transnational, interimperial imaginary of a Christian world defending its frontiers from Muslim or other kinds of "barbarism," for instance, in the so-called Bulgarian atrocities, formed part of what Leonard Woolf described as the growth of an "international" morality; see Leonard Woolf, "International Morality," *Essays on Literature, History, Politics, Etc.*, by Leonard Woolf (London: Hogarth, 1927), 153–70.

[17] See Stephen Conway, *A Short History of the American Revolutionary War* (London: I. B. Tauris, 2013).

[18] On the link between cosmopolitanism and the practice of war, see, for instance, Cécile Fabre, *Cosmopolitan War* (Oxford: Oxford University Press, 2012); and Mary Kaldor, *New and Old Wars* (Cambridge: Polity, 2006).

[19] On montage, see the classic essay by André Bazin, *Qu'est-ce que le cinéma? Cinéma et sociologie*, vol. 3 (Paris: Cerf, 1961).

intellectual intervention, which made the reader complicit in a common work of imagination by the creator's use of words, images, and sounds to trigger multiple associations. These new associations, as well as dissociations, between images and concepts, and indeed the representation of time itself, or the discussion of war in conjunction with the concept of peace, could be used either to evoke a new idea, or to make the viewer or reader discard an old one. According to Eisenstein, with his mental montages Tolstoy taught his readers how to look at a clock without imagining time, or how to look at a soldier in the war without thinking about war. The overpowering effect of his writing of war was to impress upon readers the experience of suffering, both for individuals and for the people as a mass.[20] By representing events through a combination of multiple sensory experiences, Tolstoy enjoined his readers to witness war synesthetically, through combining seemingly unrelated dimensions of reality in a kind of montage.[21] Tolstoy's eye for detail, such as the description of pain in a particular part of a soldier's body, or the sound of a waltz or mazurka as it mingled with thoughts about war, frustrated those who expected him to provide a conventional war report that would reconstruct strategic and tactical aspects of the events.[22] The account of the Crimean experience served as the foundation for *War and Peace* (1869)—whose Russian name is also translatable as *War and World,* or even *War and Community,* because of the multiple meanings of the Russian word *mir*—an epic novel that consciously eschewed the logic of national and imperial interests to foreground more universally human concerns.[23] As is well known, and exhaustively discussed, Tolstoy's writings are suffused with ideas drawn from Enlightenment philosophical literature as well as strands of sentimentalism in English and French fiction.[24] In this respect, not just the content but also the

[20] On Tolstoy's mastery of montage, see Sergei Eisenstein, *Montazh* (1938, new ed. Moscow: Muzey kino, 2000).

[21] On the Napoleonic wars and the hyperreal sense of war, see Jan Mieszkowski, *Watching War* (Stanford, CA: Stanford University Press, 2012), 4.

[22] On Tolstoy's frustrations with military history, see Donna Tussing Orwin, "War and Peace from the Military Point of View," *Tolstoy on War: Narrative Art and Historical Truth in "War and Peace"* ed. Rick McPeak and Donna Tussing Orwin (Ithaca, NY: Cornell University Press, 2012), 98–111.

[23] Katheryn B. Feuer, *Tolstoy and the Genesis of "War and Peace,"* ed. Robyn Feuer Miller and Donna Tussing Orwin (Ithaca, NY: Cornell University Press, 1996). See also Leo Tolstoy, "Neskol'ko slov po povodu knigi "Voina i mir'," *Polnoe sobranie sochinenii,* 90 vols. (Moscow: Gosudarstvennoe izdatel'stvo Khudozhestvennaia literatura, 1928–58), vol. 16, 9–13.

[24] See Boris Eikhenbaum, *The Young Tolstoy* (1922), trans. Gary Kern (Ann Arbor, MI: Ardis Press, 1972). Russian edition: Berlin-Petersburg: Grzhebin, 1922. Available at <http://feb-web.ru/feb/tolstoy/critics/emt/emt-001-.htm>, accessed 15 Aug. 2015. Among other sources, the book is based on the diary of Tolstoy, *Dnevnik L'va Nikolaevicha Tolstogo,* ed. V. Chertkov, vol. 1 (1895–99) (Moscow, 1916). Boris Eikhenbaum, *Molodoi Tolstoi* (Berlin: Grhebin, 1922). On Sterne, 30–31, 37. On Rousseau, 33. On Rousseau and dissolution of canonical form, 36. On war revealing bad sides of an "entire class of people," 41. On Tolstoy's library, see L. N. Tolstoy, *Biblioteka L'va Nikolaevicha*

form of his works on war, which included the *Sevastopol Stories* (1854–55)—produced while Tolstoy was serving as an officer in the Crimean War—and *War and Peace,* constituted an intervention in contemporary debates about social justice and political thought.[25] Both works reproduced experience though different registers of speech and the allusion to different levels of detail. For example, the *Sevastopol Stories*—which multiply fused the genres of the novel, soldiers' letters, and literary journalism—expose the fact that the language and sentiment of the elite formed part of the same culture as that of a purported enemy, France, while being audibly detached from that of their own people. Meanwhile, one of the most impressive characters of *War and Peace,* Platon Karataev, is a footsoldier whose wisdom about life exceeds that of the intellectuals in Russian society. Tolstoy also found a way of communicating his sensory experience of the war without attaching his characters' passions to national or strategic interest, something that greatly irritated a number of his contemporaries.[26]

Not only in Tolstoy's work, but in Russian society at large (as in other places such as Britain) the experience of the Crimean War brought national and class differences to the fore. Such disparities were immediately visible in the design of uniforms, of course, but also made themselves felt in subtler ways, such as in the interactions between people of different social standing. New technological equipment and new types of tasks, such as digging trenches, made the disparities seem wider, and made class differences newly tangible. Against this larger context, it is perhaps less surprising that the Russian formalists, in interpreting Tolstoy's narrative method during the 1920s, suggested that Tolstoy's wartime writings put into practice the kind of critical estrangement that they were trying to develop on the basis of Marxist critiques of ideology.[27] As Marx notably put it, invoking a visual metaphor, "if ideology as a whole makes peoples' condition appear upside down before their eyes, as though in a camera obscura, then this phenomenon arises from their historical life process, just as the turning of objects upside down on their retina arises immediately from their physiological

Tolstogo v Iasnoi Poliane, Vol. I, Knigi na russkom iazyke (Moscow: Izdatel'stvo Sovetskaia Rossiia, 1958). On other sources of Tolstoy's writings on the sentiments, see "Izbrannye aforiszmy i maksimy Laroshfuko," Tolstoy, *Polnoe sobranie sochinenii,* vol. 40, 283–310.

[25] Leo Tolstoy, *Sebastopol,* trans. Frank D. Millet (New York: Harper & Brothers, 1887), first published in the periodical *Sovremennik;* and Leo Tolstoy, *War and Peace,* first published in four volumes in 1867–68. Tolstoy himself wrote about his writing process in "Neskol'ko slov po povodu knigi 'Voina i mir,'" *Russkii arkhiv* 3 (1868). On the circumstances of writing, see Mstislav Zvialovski, "Kak pisalsia i pechatalsia roman 'Voina i mir'," *Tolstoi o Tolstom: Novye materialy,* vol. 3, ed. V. G. Chertkov and N. N. Gusev (Moscow: Tolstoy Museum, 1927).

[26] See Dominic Lieven, "Tolstoy on War, Russia, and Empire," *Tolstoy on War,* 12–25.

[27] On estrangement as a device, see Viktor Shklovsky, *Iskusstvo kak priem* (1919), in *O teorii prozy* (Moscow: Krug, 1925).

process."[28] The war worked like Marx's camera obscura, rendering the contours of social oppression sensible—both visible and audible. Tolstoy clearly registered this kind of visual and (as we will see) aural identification of an ideology of class. Yet he also went further, developing a narrative technique that the formalist theorists later came to call "estrangement." It is the intellectual formation of this technique that chiefly interests me in what follows.

Compassion through Montage

The first of the *Sevastopol Stories*, "Sevastopol in December," consists of a walk behind the front line, where everyone, including each noncombatant, suffers from different levels of pain. To initiate conversation with a wounded solider, Tolstoy (as narrator) asks, "Where have you been wounded" (*Ty kuda ranen?*)—note, not, "Which battle did you fight in?," "How did you get wounded?," or the even more basic "Did you win?"[29] The soldier dutifully points to his leg. It is only after a few seconds that Tolstoy (and the reader) realizes that the soldier is experiencing this pain in an amputated limb. Best not to think about it, the soldier advises himself. "When we don't think we don't feel." This knowledge clearly belongs to the soldier; but it also encapsulates a wartime ethics of reading: Tolstoy's readers, many of them far from the battlefield, might have read this passage as an implicit reproach for their own unthinking, unfeeling attitude toward the events unfolding in Crimea.

As an officer, Tolstoy enjoyed access to all levels of cultural production in the war, both high and low. As in his other works, music played a central role in his perception of events. During breaks from fighting, Tolstoy resorted to playing the piano in his fashionable rooms. Tolstoy even composed a song himself, dedicated to the Russian losses at the Battle of the Tchernaya River of 16 August 1855.[30] He also mentioned incidental music which he overheard, including Tatar songs and gypsy romances (the latter sung by a Russian

[28] Karl Marx and Friedrich Engels, *Die deutsche Ideologie*, 1: "Wenn in der ganzen Ideologie die Menschen und ihre Verhältnisse wie in einer Camera obscura auf den Kopf gestellt erscheinen, so geht dies Phänomen ebensosehr aus ihrem historischen Lebensprozeß hervor, wie die Umdrehung der Gegenstände auf der Netzhaut aus ihrem unmittelbar physischen." *Marx-Engles-Werke*, vol. 3 (Berlin: Dietz, 1958), 26.

[29] Frank Millet's translation—from the French—which strikes me as inaccurate, is "Where do you feel badly now?" Tolstoy, *Sebastopol* (1887), 23.

[30] Anon, "Dve pesni krymskoi voiny," *Poliarnaya zvezda* 3 (London, 1857). According to Tolstoy's biographer P. I. Biriukov, this publication severely damaged Tolstoy's military career. In Biriukov, *Tolstoy*, vol. 1, 4 vols. (Moscow: Posrednika, 1906–1922). However, Tolstoy himself never acknowledged authorship, and it was published anonymously.

aristocrat, Prince Galitsyn). While presented as quotidian, such musical occurrences are rarely neutral within the wartime milieu. Instead, they usually serve to tune the reader's attention to the purity of lower-class cultures. The auditory leakage of lower-class music into Tolstoy's text presents a threshold of attention that, while accessible to the reader, remains inaudible to aristocratic combatants and noncombatants alike. The latter class remain garrulously wedded to notions of heroism and aristocracy—outmoded ideals, which in practice constantly elude them.

In addition to highlighting *levels* of culture and forms of social distance, Tolstoy's *Sevastopol Stories* contain scenes produced through what we might call a montage of sensory experiences.[31] "Natural" sounds, such as the roar of the sea, the humming of bees, or the sound of human voices, are routinely combined and compared with the "unnatural" sounds of war, particularly the whistling of the cannonballs and bullets. In a memorable passage, the strains of a waltz mingle with the far-off noise of firing from the bastions, as heard from town of Sevastopol.[32] Into this acoustic backcloth, Tolstoy weaves further sights and sounds that furnish information about the wartime cityscape. We hear and so are made aware of the presence of officers, cadets, soldiers, and young women promenading in festive costume. It is important to note, however, that while sound is crucial, all senses are in play in wartime Sevastopol: as swarms of militia and civilians glide through the town, he notices, "by the bright light of a fire which burned behind a fence, a row of dark-leaved acacias."[33] Yet these sensory observations typically contain an element of social and political commentary. For example, Tolstoy delights in describing partial inversions of social roles, listing, for instance, how Prince Galitsyn performed a "gypsy song in a magnificent style." He also relishes describing how soldiers spend (copious) idle time in the camp in learning new songs and romances, while the officers more elevated in rank learned folk songs from itinerant villagers.

Tolstoy blends a range of senses and experiences to flesh out the social iniquities he witnessed, not only wartime noise and musical sound. He also uses the multilingual character of the social interactions he observed to a similar effect. Like Tolstoy himself, aristocratic officers seek to articulate their patriotic intent in the language of their enemy—namely, French—and fail. Indeed, they are at their most comic when attempting to display to all concerned that they are comfortable speaking French. To recreate verisimilitude of these linguistic pretensions at the front, Tolstoy uses French in the *Sevastopol Stories* to convey

[31] Alyson Tapp discusses Tolstoy's literary mediation of sounds in more depth in her chapter in this volume, 204–12.

[32] Tolstoy, *Sebastopol* (1887), 44.

[33] Tolstoy, *Sebastopol* (1888), 171.

everyday aristocratic platitudes among Russian officers—as well as, of course, the speech of French soldiers—while Polish (also rendered directly) is used to convey emotions typical of the nobility. In stark contrast, Russian soldiers barely speak any French, of course; yet, significantly, this deficit does not prevent them from feeling solidarity with French prisoners of war.

Tolstoy understood this multi-sensory polyphony of voices in explicitly moral terms, as information that would convey firsthand experience of the front to home audiences. He did so not only through counterpoint, but also through jarring effects created by juxtaposition, evoking multiple, often contradictory sensory experiences. Increasingly, as already mentioned, these sensory dissonances worked to generate sympathy for the ordinary soldier in contrast to higher-ranking officers. At the same time, though, literary montage contributes to a loss of what could be described as the rational description of war as a conflict of interests in which he, officer Tolstoy, represents one interested party. Tolstoy never states his own views on narrative technique outright. Instead, he presents himself in writing as an aristocratic officer, but in a detached way, by splitting himself into multiple subjects of war time experience whose simultaneous perceptions contradict one another. It is in this sense significant that all the officers in his account suffer from a kind of alienated experience of events. Their frequently rather distant experience of wartime, suffused with the consumption of cultural pastimes such as card games or chamber music, also distracts them from having to register the pain felt by ordinary soldiers.

The cosmopolitan feelings produced in conflict zones such as Crimea were characteristically normative and determined. The war concentrated within a specific geographic setting what had previously been dispersed over the globe. In this way, it brought to the surface what David Harvey has termed "geographical evils"—a term borrowed from Edmund Burke—without necessarily undoing them. Indeed, some deeply uncosmopolitan sentiments were widely reinforced by the Crimean experience.[34] What seems to have been at stake for Tolstoy was a cosmopolitan commitment that both transcended belonging to any particular empire and emphasized widespread injustices in the military sphere—injustices sustained by radical differences between soldiers and officers (people who were subjects of the same empire but did not share the same class) as well as between empires, insofar as these were expressed in the institutional peculiarities of various imperial armies.

[34] See David Harvey, "Cosmopolitanism and the Banality of Geographical Evils," *Public Culture* 12/2 (2000), 529–64. On Edmund Burke's original use of the term, see Richard Bourke, *Empire and Revolution: The Political Life of Edmund Burke* (Princeton: Princeton University Press, 2015), 220. For an exploration of the ways in which post-Enlightenment mental geographies were refashioned during the Crimean War, see Kevin Karnes's chapter in this volume, pp. 131–34.

To put this another way, the Crimean War transformed the experiential basis of subjective experience, and Tolstoy's frustration of his readers' expectations from the reportage genre enacted this transformation at the level of representation.[35] The social experience of nineteenth-century wars, along with new communication media, provided a challenge for eighteenth-century theories of proximity and distance in relation to the human capacity for sympathy also in terms of social class.[36] Tolstoy used the technique of literary montage in order to extend sympathy's reach to a maximum, well beyond conventional boundaries of social and geographical proximity and distance. He thus performed what we might call a cognitive intervention—one of countless similar ideological rewirings that took place around this time—that contributed to a new spirit of social reform in European society.[37] In Russia, six years after the Crimean War, serfdom was abolished and the wide-ranging reforms of the 1860s began to shape public opinion among the intelligentsia to which Tolstoy belonged. In Britain, a new system of public honors was symbolic of a new appreciation of the lower classes: the Victoria Cross—first introduced in the Crimean War—recognized the heroic achievements of ordinary infantrymen for the first time.

Tolstoy's reflection of these trends challenges not only established attitudes to his novelistic activities, but the very idea of the political effects of the novel in modern culture. In Benedict Anderson's celebrated argument, the genre of the novel—Tolstoy was its most prominent exponent in Russia—was particularly conducive to nation building and the construction of national imagined communities. The novel showed society as a "bounded intrahistorical entity," a set of "large, cross-generation, sharply delimited communities"—ultimately, an abstraction that, Anderson argues, lent itself to national uses and boundary lines.[38] But the case of Tolstoy can inspire other readings, readings that take his literary output not as a monument of national literature, but as a microcosm of worldliness. It provided a mirror of a wider European elite sociability, and as

[35] See Arthur Kleinman and Erin Fitz-Henry, "The Experiential Basis of Subjectivity: How Individuals Change in the Context of Societal Transformation," *Subjectivity*. ed. Joao Biehl, Byron Good, and Arthur Kleinman (Berkeley: University of California Press, 2007), 52–66.

[36] Cf. James Chandler, *An Archaeology of Sympathy: The Sentimental Mode in Literature and Cinema* (Chicago: University of Chicago Press, 2013).

[37] On the Russian reforms of the 1860s (especially the abolition of serfdom) in transnational perspective, see Bruce Lincoln, *The Great Reforms: Autocracy, Bureaucracy, and the Politics of Change in Imperial Russia* (Dekalb: Northern Illinois University Press, 1990); and Robert R. Franklin, "Tsar Alexander II and President Abraham Lincoln: Unlikely Bedfellows?" *University of Hawai'i at Hilo: Hawaii Community College* 10 (2012), 74–84.

[38] See, for instance, an argument Benedict Anderson has reiterated since *Imagined Communities: Reflections on the Origins and Spread of Nationalism* (London: Verso, 1991), in "El Maladado País," *The Spectre of Comparisons: Nationalism, Southeast Asia and the World* (London: Verso, 1998), 333–59, esp. 334.

such serves as an indictment both of aristocracy and of nationalism.[39] By distancing himself from this debauched elite, while simultaneously introducing the reader to more demotic characters, Tolstoy used the wartime setting to achieve a kind of disalienation of his readers. He thereby contributed to a production of moral sentiments, consciously exceeding the limitations of human capacity for sympathy with communities that were marked by their geographical or social distance.

Literary Sentimentalism and Panorama

As we know from his letters and interviews, Tolstoy's account of the Crimean War was greatly influenced by reading he had done prior to enlistment.[40] It was Jean-Jacques Rousseau who triggered Tolstoy's turn toward sentimentalism. Prompted to join the Rousseau society in Geneva in 1905, Tolstoy wrote to its directors

> Rousseau has been my master since the age of fifteen. [. . .] Rousseau does not age. Quite recently I happened to reread several of his works, and I experienced the same feeling of spiritual uplift and of admiration that I experienced reading him in my first youth.[41]

Tolstoy's thinking was influenced by two expressions of sentimentalism: the philosophy of moral sentiments associated with Adam Smith and others, and the literary mode of sentimental writing, with which he had become familiar principally through the writings of Laurence Sterne and Stendhal.[42] The works of Samuel Richardson, Adam Smith, David Hume, and Rousseau were well represented in his library at Yasnaya Polyana, and his own account of his reading in *Confessions* testifies to their extended use.[43] Tolstoy was also a subscriber to the

[39] See, for instance, Stefan Zweig, *Casanova, Stendhal, Tolstoy: Adepts in Self-Portraiture* (German original 1929) (New Brunswick, NJ: Transaction, 2012); John Burt Foster Jr., *Transnational Tolstoy: Between the West and the World* (London: Bloomsbury, 2013), 107ff; and Charlotte Alston, *Tolstoy and His Disciples: The History of a Radical International Movement* (London: I. B. Tauris, 2014).

[40] According to a biography by P. I. Biriukov, *Biografiia* (Moscow: Posrednik, 1911), 279–80.

[41] Tolstoy to Bernard Bouvier, President of the Societé de Jean-Jacques Rousseau in Geneva, 20 Mar. 1905; cited in Hugh McLean, *In Quest of Tolstoy* (Boston: Academic Studies Press, 2008), 143–44.

[42] Feuer, *Tolstoy and the Genesis of "War and Peace"*; Foster, *Transnational Tolstoy*, 107ff.

[43] G. Alekseeva, "Iz istorii yasnopolyanskoi biblioteki," *Istochniki po istorii russkoi usadebnoi kul'tury* (Yasnaya Polyana, Russia: RGGU, 1997), <http://oiru.archeologia.ru/biblio050.htm>, accessed 31 Mar. 2017.

"Collection of British Authors" series, an editorial project by a Saxon publisher that distributed the works of authors such as Wilkie Collins, Bulwer-Lytton, the Bronte sisters, and, of course, Sir Walter Scott. What philosophical and literary sentimentalism shared was a focus on the large, panoramic overview combined with detailed knowledge of particular human experiences—experiences that might seem insignificant in the grand scheme of things, but acquire importance through their cumulative effect.

Stendhal's description of the Battle of Waterloo in *La chartreuse de Parme* (1839) served as Tolstoy's greatest literary inspiration for the *Sevastopol Stories* and, later, for *War and Peace*. He was fascinated by the possibilities of Stendhal's specifically human panorama as it unfolded in the Waterloo battle scene:

> Has anyone before been able to describe war as it really is? Do you remember Fabrice as he crossed the battlefield and understood "nothing"? [. . .] In Crimea, I could see it all with my own eyes. But I repeat, all I know about war I learned from Stendhal.[44]

Continuations of this nineteenth-century tradition of literary realism can be found in Nabokov and Proust. As war becomes an object of esthetic contemplation, this literary tradition turns increasingly inward to deal with cosmopolitan themes through the multi-sensory experience of a particular subject. For Tolstoy and Stendhal, subjective optics were a means for conveying a political argument about war; yet for Proust and Nabokov, synesthetic experience is no longer the means to an end (imbuing a story about war with moral and cosmopolitan meaning) but becomes the chief object of narration, as well as being a desirable challenge for performers of a fashionable sort of literary genius.

In *La chartreuse de Parme,* Stendhal introduces the reader to the twilight of the Napoleonic era from the point of view of a young Italian prince, Fabrice. Brought up by Jesuits, he is driven to join Napoleon's army by two simultaneous motivations: the desire to be a hero and a profound dislike of the Enlightenment—or, as he puts it, "any work published after 1720, with the possible exception of the novels of Walter Scott."[45] Fabrice ends up at Waterloo, where he wants to "kill one of the enemy" and experience a "real battle."[46] His disappointment is considerable

[44] According to Biriukov, *Biografiia,* 279–80. Cited by Eikhenbaum in *Young Tolstoy.* On Stendhal and Tolstoy, see also Leonid Grossman, "Stendhal i Tolstoi," *Sobranie sochinenii v piati tomakh,* vol. 14 (Moscow: N. A. Stollar, 1928), <http://az.lib.ru/g/grossman_l_p/text_1914_stendal_i_tolstoy.shtml>, accessed 19 Sept. 2015.

[45] Stendhal, *The Charterhouse of Parma* (1839), trans. Margaret Mauldon (Oxford: Oxford University Press, 1997), 95.

[46] Ibid, 59.

when he sees clear lines of conflict vanish before his eyes. Stendhal's panoramic description itself shatters Fabrice's fanciful ideas (which the author could have purloined from Walter Scott) as "le bruit de cannon" grumbles on "comme une basse continue."[47] Fabrice later considers joining the republican war effort in America before becoming embroiled in struggles for power in his native Parma. During this time, he is briefly imprisoned on account of his political intrigues. As it turns out, his army training—notably his skills in wartime communication techniques such as a Morse-like semaphore alphabet—is more useful to him in exchanging secret messages with his aunt and his lover than on the battlefield.

As with Tolstoy at Sevastopol, Stendhal had experienced war firsthand and was, in fact, one of the few French soldiers who returned to tell the tale following Napoleon's catastrophic Russian campaign. Echoing Stendhal's account of Waterloo, Tolstoy's *Stories* open with a two-paragraph panorama of Sevastopol during the siege. Despite a large number of possible agents—officers, sailors, merchants, and several women are mentioned in this sweep across the city—the real focus of attention is on sensations. Music is treated much like the smells evoked in this panorama: both are employed for their affective qualities, their ways of working on human minds unable to resist them; the music of the military bands is interlaced with the stench of rotting human flesh. We then return to Russian folk music, which an anonymous soldier admits he has never heard sung in a village, along with a waltz that "aristocrats" (like Tolstoy) recognize as belonging to the past, but sounds new to many of their subordinates as well as to the ordinary residents of the city. It is the accumulation of these pungent details that most strongly recalls the sentimental tradition, especially the travel writings of Sterne, which were enjoying great enthusiasm in Russia around this moment.[48]

The panorama was not only an imaginary edifice for Stendhal and Tolstoy. Invented in the late eighteenth century, panoramas were spaces of entertainment for Europe's educated and well-traveled audiences. In them, the metaphorical "theater of war" could become a three-dimensional, immersive experience.[49] Unlike ordinary theaters, and certainly unlike the battlefield itself, the panorama placed the spectator at its perspectival center. Thus it not only made the experience of war overwhelming for an otherwise unaccustomed spectator; the panorama also changed the nature of the relationship between the viewer and what was seen. War was no longer just a process that was distantly imagined away

[47] Ibid, 59.

[48] On the connection between Western European and Russian sentimentalism, see Grossman, "Stendhal i Tolstoi"; Hilde Hoogenboom, "Sentimental Novels and Pushkin: European Literary Markets and Russian Readers," *Slavic Review* 74/3 (Fall 2015), 553–74.

[49] Bernard Comment, *The Panorama* (London: Reaktion Books, 1999), 8.

from the urban centers. By way of the panorama it could be brought near to oneself, to one's body, and its representation, painted on walls, could mingle with the inner life of the visitor herself.

Sympathy and the Little Finger

Reading Adam Smith was fashionable in Russian "educated society" throughout the nineteenth century.[50] Setting the trend in a novel of 1825, Alexander Pushkin informs us that his hero, Eugene Onegin—a man typical of his time among the social elite—had Adam Smith on his reading list.[51] Following suit, many aristocratic libraries (including Tolstoy's own) contained editions of Adam Smith's works, particularly his *Theory of Moral Sentiments* (1759). In that book, Smith notoriously expressed his skepticism regarding the possibility of a truly global love of mankind. If, Smith mused, a European "man of humanity" were to hear that the "great empire of China, with all its myriads of inhabitants, was suddenly swallowed up by an earthquake," he would soon overcome any "melancholy reflections upon the precariousness of human life, and the vanity of all the labours of man." His thoughts would instead wander to those effects of distant calamities, which had an impact on his own affairs: settling upon these affairs, he would soon cease to contemplate the calamity altogether. Even the conjecture of a much more "frivolous disaster" such as the loss of his "little finger to-morrow" would deprive him of his "sleep to-night," whereas he will "snore with the most profound security over the ruin of a hundred millions of his brethren, and the destruction of that immense multitude seems plainly an object less interesting to him, than this paltry misfortune of his own."[52] Even though Smith could envisage, twenty years later, the existence of a global market, his understanding of modern commercial society, characterized by a "sympathetic exchange of positions," still imagined the practical existence of boundaries of individual and national interest, which were in turn defined by criteria such as

[50] Nikolai Leskov, in drawing the portrait of a self-taught man from the people, emphasized his diligence in reading Adam Smith to compensate for his lack of inborn gentility. Nikolai Leskov, "Nesmertel'nyi Golovan," *Istoricheskii vestnik* 12 (1880), 641–78.

[51] Aleksandr Pushkin, *Evgeny Onegin*, chap. 7 (St. Petersburg: Smirdin, 1833). On Vasily Zhukovsky, Afanasy Fet and others in Tolstoy's circle, who were also steeped themselves in the ideas of the Scottish Enlightenment, see Natal'ia Samover, "Literatur und Psychologie: Vasilij Zukovskijs Ausarbeitungen zu einer philosophisch-pädagogischen Sprache der Personalität," *Diskurse der Personalität: Die Begriffsgeschichte der "Person" aus deutscher und russischer Perspektive*, ed. Alexander Hardt and Nikolaj Plotnikov (Paderborn, Germany: Wilhelm Fink, 2008), 299–321.

[52] Adam Smith, *Theory of Moral Sentiments* (Oxford: Oxford University Press, [1759] 1976), 134.

proximity.[53] The little finger of one's own nation, in other words, still tended to matter more than a large mass of suffering in another's.

Despite thinking of human economy as potentially global, Smith was, in this sense at least, no cosmopolitan thinker. In contrast to Smith, Immanuel Kant—the greatest advocate of cosmopolitanism in the eighteenth century—thought the idea of global humanity transcended any practical limitations of human sensibility. Citing the fragility as well as the relativity of sentiment, Kant insisted that the normative ideal of "perpetual peace" had to be built on the solid foundations of reason, which included a categorical imperative to think of principles of humanity as an "end in itself."[54] This cosmopolitan order did not require validation by experience: even the existence of societies mutually at war with one another posed no risk to the theory. As he argued in his work on anthropology and geography, human developments were in practice spread unevenly across the globe, and some existing human societies and cultures were inherently inferior to others.[55] Likewise, further limits to Kant's cosmopolitanism can be found in his belief that war might provide a resource for imagining new normative orders, if only by expanding the domain of regulated ethical behavior.[56]

In the course of the nineteenth century, it was not just the speed of access to information that changed the correlation between distance, difference, and sympathy. Equally important was the way in which the *small details* of large calamities were conveyed to distant audiences in order to produce cosmopolitan fellow feeling. Tolstoy's experiences of war enabled him to combine Kant's cosmopolitan vision with a different approach to moral sentiments in Smith's sense. Challenging the seemingly immutable law of the distant earthquake, Tolstoy demonstrated that the capacity for sympathy depended not so much on

[53] James Chandler, *An Archaeology of Sympathy: The Sentimental Mode in Literature and Cinema* (Chicago: University of Chicago Press, 2013), 302.

[54] Immanuel Kant, "On the Common Saying: 'This May Be True in Theory, but It Does Not Apply in Practice,'" *Political Writings*, ed. Hans Reiss (Cambridge: Cambridge University Press, 1991), 61–92.

[55] Immanuel Kant, *Von den verschiedenen Racen der Menschen* (Königsberg: H. Hartung, 1755); and *Immanuel Kants physische Geographie*, ed. Friedrich Theodor Rink (Mainz and Hamburg: Gottfried Vollmer, 1804). See also Immanuel Kant, *Anthropology from a Pragmatic Point of View*, ed. and trans. Robert B. Louden (Cambridge: Cambridge University Press, 2006). For critiques of Kant's anthropology and geography, see Harvey, "Cosmopolitanism and the Banality of Geographical Evils"; John H. Zammito, *Kant, Herder, and the Birth of Anthropology* (Chicago: University of Chicago Press, 2002); and Stuart Elden and Eduardo Mendieta, eds., *Reading Kant's Geography* (New York: State University of New York Press, 2011).

[56] Immanuel Kant, *Kritik der Urteilskraft*, ed. Karl Vorländer (Hamburg: Meiner, 1968), 263. See also the discussion in Elisabeth Krimmer, "Transcendental Soldiers: Warfare in Schiller's *Wallenstein* and *Die Jungfrau von Orleans*," *Eighteenth-Century Fiction* 19/1 (2006–7), 99–121. For the wider context, see Krimmer, *The Representation of War in German Literature*.

the subject's real distance from suffering, but on his *perception* of that distance. Such an internalized reading of distance would also correspond to another key example in Smith's theory, that of the "impartial spectator" who tempers what would otherwise be rampant self-interest. With the growth of the global economy—and especially of the global media—in the nineteenth and twentieth centuries, there has also been a globalization of fellow feeling. Wars in particular generate new institutional and individual practices devoted solely to global forms of humanitarianism such as the International Red Cross—activities that were previously restricted mostly to religious organizations.[57]

Yet the Crimean War also became a cosmopolitan moment thanks to the production of moral sentiments by the increasingly global literary public sphere, authors like Tolstoy could tap into. In Franco Moretti's *Atlas of the European Novel,* we can see the diffusion of the initially Francophone and Anglophone genre across a European readership from as early as 1810.[58] As Margaret Cohen has argued, the rise of the novel from the late seventeenth to the nineteenth century was accompanied by a process of shaping a readership that was increasingly plugged into a sentimental code of a trans-imperial humanity.[59] Because of such narrative interventions, the European "man of humanity" could now be aware not only of his little finger, but also empathize with, say, a distant man's wife in a remote theater of war. In other words, nineteenth-century cosmopolitanism was not just a technological but a cognitive and narrative achievement. The spirit of reform, into which Tolstoy's narrative dissolution of war reporting slotted well, was eminently suited to the support shown to Russia by the United States during the Crimean War.[60]

Later in the nineteenth century, the legacies of wartime cosmopolitanism were more securely in place. For example, when the Russo Turkish War came in 1877–78, the Tolstoyan Russian publisher Ivan Sytin began mass publication of popular literature featuring many texts sourced from continental Europe.[61] Another late-nineteenth-century Tolstoyan, the painter Vasilii Vereshchagin, toured Europe and North America with his graphic, hyperrealist depictions of

[57] Barnett, *Empire of Humanity.*

[58] Franco Moretti, *Atlas of the European Novel, 1800–1900* (London: Verso, 1998), 179.

[59] On sentimentalism and the shaping of communities, see Margaret Cohen, "Sentimental Communities," *The Literary Channel: The Inter-National Invention of the Novel,* ed. Margaret Cohen and Carolyn Diever (Princeton, NJ: Princeton University Press, 2002).

[60] Aside from personal affinities between President Lincoln and Tsar Alexander II, their abolitionist projects against slavery and serfdom in their respective countries also received an impulse from their joint intentions to spite Britain and France. See Frank A. Golder, "Russian-American Relations during the Crimean War," *American Historical Review* 31 (1925–26), 462–76.

[61] Cf. Sytin's own memoirs, in Ivan Sytin, *Zhizn' dlia knigi* (Moscow: Gosudarstvennoe izdatel'stvo politicheskoi literatury, 1960).

Russian and British wars. The Baltic Russian lawyer Mikhail von Taube became one of the team of architects behind the Hague conventions of international law. Nor should we forget the most famous Tolstoyan of them all: Mahatma Gandhi.[62] All of these figures justified their internationalist orientation with reference to Tolstoy's example.

From Phantom Pain to Cosmopolitan Spectatorship

The overlaying of sensory experiences through montage, along with the multilingual multiperspectivism that this technique implies, is the principal literary construction through which Tolstoy resists the dilemma expressed in Smith's old parable of the little finger. We cannot sympathize with the plight of those we cannot sense. In his version of the Crimean War, we see, hear, and feel more than it was customary for European readers to bear. But with whose pain should we sympathize? For Mikhail Bakhtin, Tolstoy epitomized a shift to a new type of narrator and a new type of hero, one who was of interest for his feelings and failings rather than for acts and achievements.[63] To find evidence of this shift, we might recall Tolstoy's pointed question: where does it hurt? Perhaps we can never tell; even the wounded soldier was unsure, as he feels pain in a leg that is missing. This scene describing phantom pain is in one sense a key to Tolstoy's vision of war in the *Sevastopol Stories*, one that remains unsurpassed even in *War and Peace*. Tolstoy himself found theoretical expression for this technique in the writings of authors such as Ralph Waldo Emerson, whose 1841 essay on self-reliance was in his library.[64] The transcendentalist idea that only searching the soul and individual perception could bring enlightenment resonated with Tolstoy's confessional style of witnessing.

For Tolstoy, the medical jargon of war, noticeable through such expressions as *perforatio pectoris*—puncture of the chest—was a central element in eliciting a reflexive form of compassion.[65] The term "phantom pain" gained currency in

[62] On Tolstoy as Gandhi's favorite author and his appearance in *Indian Opinion*, see Isabel Hofmayr, *Gandhi's Printing Press: Experiments in Slow Reading* (Cambridge, MA: Harvard University Press, 2014), 74 and passim.

[63] On literary sentimentalism from Stendhal to Brecht in this sense, see Mikhail Bakhtin, "The Problem of Sentimentalism," *Sobranie sochinenii v semi tomakh*, ed. S. G. Bocharev (Moscow: Russkie slovari, 1996), 3: 304–5.

[64] Ralph Waldo Emerson, "Self-Reliance," *Essays: First Series* (Cambridge, MA: Harvard, 1841). On the influence of American transcendentalism on Tolstoy and its presence in his library, see Aleksandr N. Nikoliukin, *Vzaimosviazi literatur Rossii i SShA* (Moscow: Nauka, 1987).

[65] Tolstoy, *Sebastopol*, 83.

American medical journals ten years after the Crimean War, based on studies of the suffering of war veterans from the American Civil War.[66] Yet phantom pain was hardly a new medical phenomenon: Descartes had already described it in the seventeenth century.[67] In his *Sevastopol Stories*, Tolstoy used his narration of a soldier's phantom pain to elicit a particular kind of sympathy, and also a particular kind of shame, from the reader—feelings that he too, as an unscathed witness of the war, experienced. By connecting pain with the sounds of war and waltzes, he created not just a simultaneity of pain and pleasure, but also made them emblematic of the reader's own phantasmatic injury. Ideally, his *Stories* made it impossible for the reader to hear a waltz again without feeling the pain of the combatants in Sevastopol.

Yuval Harari and James Chandler have shown how the sentimental tradition, affected by such forms as letter writing, travel literature, and war literature, boosted the capacity for sympathy in the eighteenth-century mind.[68] Increasingly during the nineteenth century, world-historical events were no longer the exclusive domain of the (near or distant) past, but were potentially always ongoing. In this context, the processing of information became an important business opportunity in its own right, as witnessed by the inexorable expansion of the newspaper press. Smith's erstwhile European gentleman no longer had the option of thinking or not thinking about an "earthquake in China" and then proceeding calmly about his business; thinking and talking about the earthquake had increasingly become the object of business.

In response to this situation, writers like Tolstoy realized that, if sympathy was held to be proportionate to the degree of exposure to another person's suffering, then levels of sympathy could be increased (and even manipulated) in concert with levels of exposure to the *detail* of that suffering. These authors not only conveyed distant pain but also progressively focused on the particular locations in which suffering might occur. Authors and artists with immediate experiences of nineteenth-century wars—Stendhal, Tolstoy, and the painter Vasilii Vereshchagin, but also journalists such as Karl Marx and Friedrich Engels

[66] On phantom limbs, see Stanley Finger and Meredith P. Hustwit, "Five Early Accounts of Phantom Limb in Context: Paré, Descartes, Lemos, Bell, and Mitchell," *Neurosurgery* 52/3 (2003), 675–86; on the origins of the modern science of phantom limbs in the American Civil War, see Stanley W. Mitchell, "Phantom Limbs," *Lippincott's Magazine of Popular Literature and Science* 8 (1871), 563–69. In connection with empire, see the case of post-Soviet mourning in Sergei Oushakine, *The Patriotism of Despair: Nation, War, and Loss in Russia* (Ithaca, NY: Cornell University Press, 2009), 84 and passim; also Katherine Verdery, *The Political Lives of Dead Bodies: Reburial and Postsocialist Change* (New York: Columbia University Press, 1999), 26.

[67] See Tommy L. Lott, "Descartes on Phantom Limbs," *Mind and Language* 1/3 (Autumn 1986), 243–71.

[68] Harari, *The Ultimate Experience*; Chandler, *An Archaeology of Sympathy*.

(writing for the *New York Tribune* and not for the European press)—created a set of narrative and pictorial devices that enabled a leap of imagination beyond the front line. The cultivation of a social conscience, added to an extended capacity to listen to one's inner "impartial spectator," were calculated to make it impossible to be indifferent even to distant suffering. This is the achievement of a literary tradition on war writing and its pictorial equivalents in the nineteenth century: it made it possible to speak of local suffering from multiple perspectives and thus construct a view of the war from the perspective of a "man of humanity."

In this context, harrowing journalistic detail enabled a shift of perspective toward a decidedly more cosmopolitan point of view. This is perhaps the Crimean War's most conspicuous legacy—one that lingered on in the transformation of sites of war into sites of memory, to use Jay Winter's expression here.[69] Years after the end of the war, the American Samuel Clemens (better known as Mark Twain) visited Sevastopol on a tour of Europe. He remarked how keenly other tourists were buying various paraphernalia from the ruined city:

> They have brought cannon balls, broken ramrods, fragments of shell—iron enough to freight a sloop. Some have even brought bones, including one labelled "Fragment of a Russian General," even though it later turned out to be a piece from a horse's jawbone.[70]

By focusing on the material details of war memory, Clemens noticed something essential about this war. Unlike generals, in death, horses are no longer "Russian" or "French." The suffering of horses points to a cosmopolitan dimension of war. It is the nonhuman, paradoxically, that puts humanity in perspective.

Memories of the Crimean War remained within close reach in twentieth-century mass culture, to such an extent that even histories of the two World Wars are sometimes narrated with reference to Crimea. It survived most clearly in British and French public memory: in France through the naming of public places after major battles (Pont de l'Alma, café Malakoff, and, perhaps even Proust's fictional Balbec, which echoes Belbek); in Britain it entered the popular imagination through clothes—the cardigan (after the British general James Thomas Brudenell, the 7th earl of Cardigan, who was in charge of the Light Brigade) and the balaclava, after the eponymous battle of 1854. To a twenty-first-century

[69] Jay Winter, *Sites of Memory, Sites of Mourning: The Great War in European Cultural History* (Cambridge: Cambridge University Press, 2007).

[70] Mark Twain, *The Innocents Abroad; or, The New Pilgrims' Progress: Being Some Account of the Steamship Quaker City's Pleasure Excursion to Europe and the Holy Land with Descriptions of Countries, Nations, Incidents, and Adventures as they Appeared to the Author* (Hartford, CT: H. H. Bancroft, 1869), 385.

audience, the "thin red line" is more likely to evoke associations with the 1998 Terence Malik film, which deals with the Second World War, than with the Battle of Balaklava. Likewise, *The Charge of the Light Brigade* (originally a poem by Tennyson) became the subject of a 1936 Warner Brothers film, whose plot shuttles between two battles, one in the Crimea and the other in India; the script initially confused the order of the battles, picturing the 1857 Sepoy rebellion as having happened before the Battle of Balaklava. Between 2006 and 2014, the contorted memory of this event not only made a comeback in Russian popular historical writing and the state-owned public media, but also provided the ideological fuel for the enforced incorporation of the peninsula into the Russian Federation in 2014.[71] In the new post–Cold War Russia, the memory of the Crimean War epitomizes the many humiliations Russia had once supposedly suffered at the hands of the "West." It is one of the historical ironies that this idea of the "West" is most clearly associated with the United States, and yet, during the Crimean War, it was from Civil War–ravaged America that imperial Russia had received its greatest informal support in military and economic terms.[72]

Today, the Crimean War lives on in mediated memory in a variety of ways, some of which ought to be worked through archaeologically in order to disentangle them from subsequent or preceding wars and conflicts. One outcome of such an excavation might be to conclude that in embracing the multimedia age, modern war reporting has lost the element of multi-sensory "counterpoint," as well as the sense of cognitive dissonance, that was so characteristic of nineteenth-century cosmopolitanism as espoused by Stendhal and Tolstoy. In the age of global media and YouTube montage, it seems to be more rather than less difficult to recreate a cosmopolitan point of view by means of synesthetic or ekphrastic connections and allusions. Smith's parable of the Chinese earthquake would seem to apply again, after all, since the media has achieved new ways of distancing us from "humanity"—by making the sights and sounds of war a banal feature of the everyday that washes over us as part of our regular news consumption.

In Tolstoy's writings, the intervention of the narrator as a master of montage was central to attuning his readers' moral sentiments to a cosmopolitan point of view. This ethos was carried further by the international community of Tolstoyans, in an extension of the very elite sociability that he criticized.[73]

[71] See N. V. Skritskii, *Krymskaia voina: 1853–1856 gody* (Moscow: Veche, 2006). For a TV version of this report, see this contribution on the Crimean War from a Russian state-owned TV channel in 2014: <http://www.youtube.com/watch?v=HNNUpoZarkA>. For the earlier book publication of the official Russian view of Crimea in post-Soviet history, see N. V. Skritskii, *Krymskaia voina*, in *Voennye tainy Rossii* (Moscow: Veche, 2006).

[72] Golder, "Russian-American Relations"; Alexandre Tarsaïdzé, *Czars and Presidents: The Story of a Forgotten Friendship* (New York: McDowell, Obolensky, 1958); Albert A. Woldman, *Lincoln and the Russians* (Cleveland: World Publishing, 1952).

[73] See Alston, *Tolstoy and His Disciples*.

The Tolstoyan wager was this: provided that we trust combatants not to distort their sentiments—a huge assumption, it now seems—a society based on mutual sympathy would be compatible with a cosmopolitan world order. One could also argue that by synthesizing multiple sensory experiences—the remembered action of war, the memory of a waltz—Tolstoy reached another extreme of Smith's parable. By giving up on the anesthetized, rational assessment of war, he made his readers more distant. When you see war as another reflection of the human condition, the loss of a finger can hurt more than the defeat of one's empire.

2

The Revolution Will Not Be Telegraphed

Shari'a Law as Mediascape

PETER MCMURRAY

> . . . [T]here is no law without the voice. It seems that the voice, as a senseless remainder of the letter, is what endows the letter with authority, making it not just a signifier, but an act.
>
> —Mladen Dolar

In September 1859, Imam Shamil, an Avar revolutionary and Islamic fighter, surrendered himself to Russian forces at the Battle of Ghunib, marking the end of the Imamate of the North Caucasus.[1] The Tsar allowed Shamil, after a decade of imprisonment in Russia, to make the *hajj* pilgrimage to the holy cities of Mecca and Medina. Shamil's departure from the Russian Black Sea port of Anapa on 4 June 1869 was described by a contemporary as follows:

> Two hours after the noon prayer on that day, Shamil boarded a ship leaving from Anapa. News of this event was sent on the long thread known as "the telegraph." It reached Shura about four hours after the noon prayer . . . What is stranger, the arrival of that news from Anapa to Shura in two hours, while the journey takes two months by trade

[1] I am grateful to Carolyn Abbate, Steven Connor, and Benjamin McMurray for conversations about voice, sound, and law, and their intersections during the early stages of this project. Later conversations with James Parker, as well as his 2017 workshop, "Acoustic Justice," were also invaluable. My epigraph comes from Mladen Dolar, *A Voice and Nothing More* (Cambridge, MA: MIT Press, 2006), 54–55.

> caravan, or the fact that the Tsar sent Shamil just like that on the *hajj* pilgrimage?[2]

This account by Shamil's official chronicler not only highlights Shamil's miraculous release, but also marks its miraculousness through comparison with an equally astonishing technology: the telegraph.[3] Coming over a decade after the Crimean War, this event offers a way into multifaceted developments in cultural life that took place during, around, and in the midst of that war, which entailed political reform and revolution, Islamic holy war (*ghaza*), and the expansion of communication media.

Both the Ottoman Empire and the Caucasian Imamate became significant zones of wartime struggle, as they oscillated with increasing force between inward-looking legal reform (within the Empire or Imamate) and the outward-looking geopolitics of wartime combat. In this chapter, I contrast two major Islamic reform movements in the Black Sea region during the Crimean War: Ottoman *tanzimat* and North Caucasian *ghazavat*. Both date back several decades earlier, to the Ottoman Edict of Gülhane of 1839; and, in 1827, the initial attacks on Daghestani khans by the first Imam of the eponymous Imamate, Ghazi Muhammad, along with his chief aide Shamil. For both of these movements, Islamic law, or *shari'a*, played a central role, but in divergent ways: the Ottomans sought to diminish the importance of *shari'a* law within their revamped state apparatus, while the leadership of the Caucasus Imamate sought to make it the law of the land. While both efforts lasted for decades, the years of the Crimean War mark a particular flurry of activity in these efforts. Yet much more generally, and well beyond the context of the Crimean War, the divergences in these self-fashioned modernities raise larger questions about the inevitability of a singular, Western Modernity, while also highlighting the deeply sonic underpinnings of these modernities by different forms of *shari'a* law.

[2] Muhammad Tahir al-Qarakhi, "The Shining of Daghestani Swords in Certain Campaigns of Shamil (Selected Passages)," trans. Ernest Tucker and Thomas Sanders, *Russian-Muslim Confrontation in the Caucasus: Alternative Visions of the Conflict between Imam Shamil and the Russians, 1830–1859*, ed. Sanders et al. (London: Routledge 2004), 67.

[3] Although the miraculousness of Shamil's life extends beyond the scope of this chapter, it is one of the central tropes in the chronicle by al-Qarakhi and warrants close attention. For example, as he travels to Mecca by ship, the following is reported to take place: "When the ship came one night to a place where a pharaoh had once drowned, the waves of the sea rose up. The ship's captain complained to Shaykh Shamil. [Shamil] gave a note to one of his comrades and told him to throw it in the sea without having it touch the ship. He threw it in and [the sea] calmed down. We found that news of this [event] had reached Alexandria and the *sharif* [governor] of Mecca before Shamil left the ship. This is the end of what his associate told us." Ibid., 69.

In tandem with these reform efforts, I want to consider a series of mediations of wartime events through writing (most famously by Marx and Tolstoy, but also by later revolutionaries like the Young Turks and Lenin) and to a more limited degree through telegraphy, a then-recent invention that (especially for the Ottomans) was both part of the reforms and part of the mediation of reforms. In considering these mediations, I am interested in the sensory valence of law and legal reform/revolution. I will ask the following: How is law generally—and especially Islamic *shari'a* law—produced through sound and what is its relation to voice? How does law inhabit place—whether the mountain ranges of the Caucasus or the hills, towers, and spires of nineteenth-century Istanbul? And how is it communicated within Islamic contexts like those in and around the Crimean War? In short, I will argue that both *tanzimat* and *ghazavat* elucidate the media logics of Islamic law—its encoding, transmission, and communication—within imperial contexts.

In exploring these connections, I consider first the sonic-juridical nexus of the Naqshbandi Imamate of the Caucasus in the mid-nineteenth century, followed by that of the Ottoman Empire. Again, from the perspective of general jurisprudence and religious law, *shari'a* increased in importance in the Imamate while Ottoman legal reforms actively diminished its privileged position. But in both cases, sound performed crucial functions as both the *what* and the *how* of law: that is, as both what was regulated and as the medium in which law itself was generated, pronounced, and circulated. This view of sound's central but multifunctional role in law shares much with James Parker's notion of "acoustic jurisprudence," which he distills from the proceedings of Simon Bikindi's trial for inciting genocide in Rwanda through music.[4] Thus, after comparing these two cases (the Imamate and the Ottoman Empire), I reflect briefly on the more trenchant general connections between sound (especially voice) and law. Having done so, I then turn to the introduction of telegraphy as an example of a particular (and new) communications medium that emerged in this same period; its geographies had significant juridical consequence as both an extension and a displacement of the physical voice. These technical redistributions of voice via telegraphy offer a useful reminder that technologies do not simply come into being, nor do they merely ripple from some starting point (e.g., Western Europe) outward to the rest of the world; rather, they quickly become entangled with other media and cultural techniques such as laws, alphabets, and language more generally. I conclude then by briefly considering some of the conjunctions and frictions between telegraphy and Islamic law.

[4] James E. K. Parker, *Acoustic Jurisprudence: Listening to the Trial of Simon Bikindi* (Oxford: Oxford University Press, 2015).

Holy War in/about/against Sound: The Naqshbandi Imamate of the Caucasus

The Imamate and the Ottoman Empire have long been understood to be in close contact throughout the Crimean War, with Imam Shamil, one of the namesakes of the Imamate, serving as an important intermediary for Ottoman officials and their allies in the fight against Russia. In the period leading up to the war, Shamil (who had become Imam of the Northern Caucasus in 1834) sought aid from the Ottoman Sultan in fighting the Russians—but until 1853 was rebuffed because of treaty stipulations. Once war had broken out, the alliance strengthened somewhat, but for some time the Ottomans seemed generally disinterested in the Caucasus, to judge from the leisurely pace of official communications between them.[5] As with much of the war (as evidenced, for example, by Karl Marx's writings), the speed of information dissemination was a critical issue. For example, in summer 1854, Shamil's son, named Ghazi Muhammad after the Imam, led a raid and took a number of prisoners, among them a French national working as a governess attached to the French royal family. Western Europeans were scandalized—in part because of exaggerated reports—by this "barbarous atrocity."[6] A flurry of news reports and diplomatic letters followed, with the Allies leaning on the Ottomans to convince Shamil to release his prisoners. Shamil shrewdly opted for an exchange, trading hostages for another one of his sons (held in captivity by the Russians)—but did so, crucially, before he received correspondence from wartime allies to direct his actions.

Shortly after the episode, Shamil's response to English colonel William Fenwick Williams, who was stationed nearby, highlights some of the dynamics of Islamic law and communication media in play:

> In the name of God the merciful and clement [. . .] We received your letter and understood its purport and meaning. We rejoiced to hear of the successes of our ever-victorious arms over our virtueless enemies, and the prostration of their pride in every engagement that has taken place. May the Lord be praised! [. . .] God forbid that we should do anything which might be considered disgraceful by the Mohammedan laws or by the exalted government. We had liberated the women before the arrival of your letter, and had you been acquainted with the true circumstances you would not have found fault with us; for everybody knows that we are always humane; that we expend our breath in reciting

[5] Candan Badem, *The Ottoman Crimean War, 1853–1856* (Leiden: Brill, 2010), 152–53.

[6] Lord Stratford de Redcliffe, quoted in ibid., 203.

the holy words of the Lord of the Creation, and scorn the enmity of the infidels our foes.[7]

Written in Arabic, the letter opens with an invocation of God (the *basmalah*) and immediately acknowledges its place in a broader correspondence. The only explicitly meaningful metric for Shamil is Islamic theology and law: the mark of martial success is the "prostration" of "virtueless enemies" and the standard for his own actions is "Mohammedan laws." And yet, of all the Christian nations involved, (apparently) only the Russians constitute "the infidels our foes," highlighting the strategic nature of alliances between English, Ottoman, and Daghestani forces.

Noteworthy is the emphasis Shamil places on "expend[ing] breath in reciting the holy words of the Lord," probably a reference to practices of *dhikr*, the ritual recitation of names and attributes of God, practiced among the Imamate rebels, who had aligned themselves closely with the Naqshbandi Sufi order. These recitations might have been a part of the Imamate's military tactics as well, given their use of audible recitations during a meeting between Shamil and his Russian counterpart, Kluge von Klugenau, and in the final battle depicted in Tolstoy's "Hadji Murat," in which one of Murat's companions, Jurban, does the same.[8] These audible recitations are all the more intriguing, given the Naqshbandi aversion to reciting *dhikr* aloud, as explained by Shamil's teacher (*murshid*) Jamal al-Din: "The silent *dhikr* is the most sublime form of veneration of God . . . the *dhikr* of the heart (*al-dhikr al-qalbi*) is a witnessing [of God] (*shuhud*) and it leads to the Presence of God (*hudur*) and closeness to Him (*qurba*)."[9]

Given the centrality of a silent or non-uttered *dhikr*, it should come as no surprise that the sound of breath also played an important role for these Naqshbandis, as can be observed in the following descriptions of Ghazi Muhammad (the first Imam): "Shamil said of him that he was 'silent as a stone,' others, that men's hearts were glued to his lips; with a breath he raised a storm in their souls."[10] Shamil was also known to use *dhikr*, coupled with intense fasting, to cause himself to pass out, after which he used to come to and announce that

[7] Atwell Lake, *Kars and Our Captivity in Russia* (London: Richard Bentley, 1856), 340–41. Also cited in Badem, *The Ottoman Crimean War,* 205. Badem speculates that the phrase "exalted government" refers to the Ottoman sultanate.

[8] Al-Qarakhi, "The Shining of Daghestani Swords," 37; and Tolstoy, "Hadji Murat," *The Cossack and Other Stories*, trans. David McDuff and Paul Foote (London: Penguin, 2007), 462.

[9] Anna Zelkina, *In Quest for God and Freedom: The Sufi Response to the Russian Advance in the North Caucasus* (London: Hurst, 2000), 113.

[10] John Frederick Baddeley, *The Russian Conquest of the Caucasus* (New York: Longmans, Green, 1908), 240.

while unconscious he had received instructions from the Prophet.[11] Much less ecstatic were Shamil's interactions with the Ottomans, which were largely indirect. He maintained correspondence with them through the Crimean War, but was generally disappointed by their military failures. Meanwhile, the Ottomans bestowed military honors on some of Shamil's forces. Later, on his way to Mecca, Shamil also stopped for some time in Istanbul (or "Islambol," as his chronicler calls it) and finally met the Sultan.[12]

Yet despite the relative lack of direct contact throughout this era, both the Caucasus Imamate and the Ottoman Caliphate wrestled with similar geopolitical, cultural, and religious challenges—but addressed them in nearly inverse ways, especially in their regard for Islamic legal traditions. The concurrent (if not parallel) histories of their legal reforms, situated within the larger cultural history of the Crimean War, can evoke the various sensory entanglements and the multifarious valences of sound relative to Islamic *shari'a* law in particular. *Shari'a,* literally meaning "path" or "way," is a general term for the broad frameworks of Islamic law and morality.[13] A critical question facing both the Ottomans and North Caucasus was the place of *shari'a* in their broader legal and moral systems, particularly as they moved to modernize, albeit in strikingly different ways: the Ottomans adopted a more European orientation in administration, internal affairs, and technology—an orientation less focused on *shari'a*—while the Imamate looked to the institution of *shari'a* as a step forward from blood feuds, drinking, and subjugation to imperial Russia.

As suggested previously, these divergent trajectories in legal reform vis-à-vis *shari'a,* and more broadly in projects of modernization, destabilize some of the more totalizing narratives of European and Western modernity. At the same time, these reform movements reveal the deeply sonic foundations of law—in this case, *shari'a* in particular, though parallel developments in, say, English common law also readily come to mind—and thus by extension of these legal-minded

[11] Cf. Moshe Gammer, *Muslim Resistance to the Tsar: Shamil and the Conquest of Chechnia and Daghestan* (London: Frank Cass, 1994), 239–40.

[12] Al-Qarakhi, "The Shining of Daghestani Swords," 67.

[13] I generally refer to it here as "Islamic law" for lack of better terminology, although as Wael Hallaq points out, such terminology has traditionally set up *shari'a* to fail through comparison with Western juridical standards, even though it has arguably proved itself less coercive and punitive that nation-state law; see Wael Hallaq, *Sharī'a: Theory, Practice, Transformations* (Cambridge: Cambridge University Press, 2009), 1–3. Khaled Abou al-Fadel goes further still: "The word 'Shari'a,' which many very often erroneously equate with Islamic law, means 'the way of God and the pathway of goodness,' and the objective of Shari'a is not necessarily compliance with the commands of God for their own sake"; see his article "The Islamic Legal Tradition: A Comparative Law Perspective," *The Cambridge Companion to Comparative Law,* ed. Mauro Bussani and Ugo Mattei (Cambridge: Cambridge University Press, 2012), 295–312, esp. 299.

forms of modernity, as well. At risk of oversimplifying these reformist modernities, I might formulate their sonic qualities more generally as follows: law is profoundly sonic and modernity is deeply enmeshed in law, but the particular relationships between these entities (sound and law, law and modernity) leads to a polyphony of modernities. The examples of Ottoman and North Caucasus legal reform offer two useful points of comparison for thinking through such phonojuridical permutations within such modernities.

To understand the sonic qualities of Islamic law in these instances, we must first consider how these two geopolitical entities conceptualized Islamic law and legal reform. Discussing the Islamization of law in the North Caucasus (Daghestan, Chechnya), Anna Zelkina distinguishes between two major legal systems that were simultaneously active: *shari'a,* on the one hand, and *'adat* (Arabic sing. *'āda* or *'ādah,* pl. *'ādāt;* Turkish: *âdet* or just *adet*), or "customary law," on the other.[14] She challenges the scholarly assumption that "almost all matters of civil law were settled on the basis of the *shari'a,* and criminal offences on the basis of the *'adat,'* " arguing instead that the two were closely interwoven and that a more pertinent distinction may be drawn between the majority of Daghestani areas and Chechen *tuqums* (in which *'adat* was profoundly influenced by *shari'a*) and the high-mountain regions of Daghestan and the Vaynakh *tuqums,* which "had been Islamicised only superficially and remained largely based on the pre-Islamic value system."[15] Zelkina enumerates differences in legal offices (e.g., *qadi, ulema,* mullahs, and secular judges), land ownership, legal procedure (e.g., oaths, witnessing), criminal law, and, most notoriously, blood feuds.

In rare written accounts, we can witness attempts at Islamicization in action, which already begin to highlight the centrality of sound to legal reform efforts. In the late eighteenth century, for example, Sheikh Mansour of the Naqshbandi Sufi order initiated a national-Islamic struggle with the express desire to bring people back to Islam. His first attempt to preach was a sonic spectacle atop a village mosque, where he had gathered people to pray:

> From the roof of the mosque, he addressed the people who flocked up to the mosque in great numbers, exhorting them to abandon their evil deeds, put aside enmity and fighting against each other and be reconciled, stop drinking vodka and wine, give alms and follow the Muslim law. Having finished his speech he returned home with the people following him, ordered two sheep to be slaughtered and he immediately distributed the meat among all those present.[16]

[14] Zelkina, *In Quest for God and Freedom,* xiii.

[15] Ibid., 41.

[16] Quoted in ibid., 60.

This complex call to repentance—there would be many more like it in the early nineteenth century—required the supplanting of traditional law. What is more, this transformed value system guided military confrontations between the Imam's soldiers and Russian imperial troops as early as 1775.

Half a century later, Russia annexed the entire North Caucasus, provoking a fresh round of resistance led by three imams over several decades, at the heart of which lay jurisprudential questions about the comparative value of oral and written law and their dissemination. Ghazi Muhammad took up the cause, declaring a holy war not only against the Russians, but also simultaneously against traditional *'adat*:

> Be aware that in recent times people have been governed [according to] the customs [*'adat*] of their ancestors. They promoted them to the rank of religious obligation and preferred them to the Qur'an and Sunna [. . .] Those who have reached a position of leadership and pre-eminence [. . .] boast as if they were the greatest of all the kings and there is no other way for them to exercise leadership except by skillfully imposing the customary law (*'urf*).[17]

Here one set of oral legal traditions, customs/customary law (*'adat, 'urf*), are upended by a written text deeply embedded in a tradition of recitation (the Qur'an) and oral traditions about the prophet Muhammad's actions (Sunna). Indeed, we might understand these parallel legal modalities as competing forms of orality. On the one hand, as forms of customary law, *'adat* and *'urf* effectively constituted an oral legal tradition. On the other hand, the Sunna had a long history of passing back and forth between orality and writing, from its original chains of oral transmission into written legal compendia, which were then often transmitted orally.[18] Much like the Qur'an itself, the other major source of

[17] From Ghazi Muhammad's *Bahir al-Burhan li-'Irtidat 'Urafa Daghestan*, quoted in ibid., 138–39. For more on the relation of *'urf* and *'adat* to *shari'a*, see Ayman Shabana, *Custom in Islamic Law and Legal Theory: The Development of the Concepts of 'Urf and 'Adah in the Islamic Legal Tradition* (New York: Palgrave MacMillan, 2010), 32–38.

[18] The oral transmissions of legal texts can be seen across the centuries from accounts by Ibn 'Arabī of reciting and listening to hadith traditions in Mecca to Brinkley Messick's descriptions of oral-aural transmission of such texts in the nineteenth-century Ottoman Empire and its twentieth-century successor states. Even in twenty-first-century Istanbul, one can readily find 'alim teachers in major mosques in Istanbul like Eyüp Sultan teaching such texts regularly to listeners who generally learn primarily through audition. See Muhyīddīn Ibn 'Arabī, *Divine Sayings: 101 Hadīth Qudsī*, trans. Stephen Hirtenstein and Martin Notcutt (Oxford: Anqa Publishing, 2004), 80–81; and Brinkley Messick, *The Calligraphic State: Textual Domination and History in a Muslim Society* (Berkeley: University of California Press, 1993), 75–98. I discuss Messick in greater detail later in this chapter. He argues that late Ottoman reforms and their aftermath ultimately led to a shift away

Islamic law, these Sunna texts exist in a regular movement between their written and sonic forms.

Not only was this a holy war about the (often sonic) mediations of law (i.e., through voice, writing, etc.), it was also conveyed *through* sound and sought to exert control *over* sound. Such reforms can be understood as part of a broader Naqshbandi theology. For instance, Jamal al-Din al-Ghazi-Ghumuqi, the Sufi teacher (*murshid*) of Ghazi Muhammad and of Imam Shamil, wrote the following sonocentric description of his order:

> The Naqshbandi *tariqa* is the most exalted and the best one since it has retained its original basis [*asl*] without additions and reductions [and has upheld] the way of the Companions of the Prophet and [is] free from the innovations [*bid'a*] which were introduced by some Sufi sheikhs, such as the loud *dhikr*, listening [to music—*sama*] and dancing which did not exist at the time of the Prophet or the righteous caliphs.[19]

Not only do sonic practices serve here as a kind of litmus test of orthodoxy (loud *dhikr* is inappropriate, while silent *dhikr* presumably is not), but they also frame those distinctions as being fundamentally legal ones, justified juridically through their ties (or lack thereof) to the practices of the Prophet, his Companions, and the early caliphs.

However, cultivating these broad legal and theological ideals was made possible only through individual leaders and their particular sonic capabilities. Such talents allowed figures like Ghazi Muhammad to use one form of sound to condemn another:

> [Ghazi Muhammad's] eloquence and the power of his words, combined with deep knowledge, made his listeners receive his teaching with enthusiasm, so that people gradually got used to the rules of the *shari'a* and stopped noticing the harshness of its demands, the men gradually gave up smoking and stopped drinking, women covered up and the young people started to behave well and stopped singing, apart from the hymn [of the Naqshibandi *dhikr*] "*la ilaha illa allahu,*" as not befitting a true Muslim.[20]

from an audition-based model of legal transmission—an interesting claim that similarly observes a deep wellspring of orality/aurality in Islamic legal traditions.

[19] Quoted in Zelkina, *In Quest for God and Freedom*, 110.

[20] Ibid., 143.

As seen here, Ghazi Muhammad's legend was bound to his access to power through sound, both through his own sonic presence and absence, but also through his ability to silence others. Similarly, when he first visited his own Sufi teacher, Jamal al-Din al-Ghazi-Ghumuqi, he was said to have hidden behind the door to listen to the teacher to test him—only to be greeted acousmatically by Jamal al-Din.[21] In his chronicle, Imam Shamil further emphasizes Ghazi Muhammad's sonorous power:

> He was Ghazi Muhammad al-Gimrawi (May God sanctify his secret!), the heaven-favored scholar who led them. He called on [Muslims] to obey the *shari'a*, practice it, and reject customary laws. He spoke thunderously on this subject and composed a letter denouncing the people [of Daghestan] as infidels. It was a poem titled "The Splendid Proof of How the Daghestani Authorities Rejected [True Belief]" and here is a part [of it]:
>
> > "The chronicles of customary law are collections of poetry by followers of the treacherous one who is stoned [that is, Satan . . .].
> > If the one who follows customary law were equal to the one who follows *shari'a*,
> > Then there would be no difference among us between the pious and the debauched.
> > Why [else] were messengers sent, the *shari'a* established, and the Qur'an revealed with its rules?" [. . .] [22]

Shamil highlights Ghazi Muhammad's "thunderous" speech alongside his poetic letter; those who heard it spoken aloud by him found "their tempers cooled and they accepted what he said."[23]

More generally, sound-based transmission of the divine message remained extremely important, to the detriment of customary law, which was increasingly understood as a "treacherous" form of poetry (presumably as opposed to ethical guidance). And messengers, Islamic law, and the Qur'an all began to mediate—that is, spread—that message. *Pace* McLuhan, the message did not collapse into medium here, but the message was from the outset a form of sounding—a recitation (literally, a *qur'an*)—that lent sonic power to its adherents and messengers.

The practice of enacting *shari'a* through, about, and against sound continued into the Imamate of Ghazi Muhammad's successors after his death in 1832. Not

[21] Ibid., 136.

[22] Al-Qarakhi, "The Shining of Daghestani Swords," 13.

[23] Sanders et al. *Russian-Muslim Confrontation in the Caucasus*, 14.

long after the battle in which Ghazi Muhammad was killed (and Imam Shamil almost fatally injured), Shamil became involved in a series of interactions that well encapsulate his sonic methods. According to his own chronicle, on the first day of Ramadan (1248 AH/January 1833 CE), Shamil traveled to Gimrah. After performing prayers, he came across a group of women working alongside an old man. He beat them all with a stick because of their gender mingling, then was arrested by a local judge's assistant and later flogged until his war-wounds reopened. He attended Friday prayers in the local mosque and challenged the local judge's ruling that he, Shamil, was not allowed to enforce *shari'a*. Subsequently, he joined his family for Ramadan and was again traveling just before the beginning of the festival 'Eid's celebrations at the end of the month. What followed next highlights the role of sound in his legal campaign; it is worth quoting from his chronicle at length:

> When he went out before dawn to a community mosque to perform ablutions, he came upon a group of hypocrites who had lit a fire and were banging a drum. They danced and cursed the *murids* [Sufi devotees], even calling them "those who have sex with their mothers." They said, "Tomorrow we will drink wine, party, and dance. Then we will see how humiliated those *murids* will be."
>
> Shamil was focusing on heading back [to his house at that moment] but said to himself, "If I leave here [now without doing anything], then I cannot claim to possess true faith." He pulled out his dagger and went after this group, saying literally, "God will make known who will conquer and who will be disgraced!" The group was seized with dread of him. They threw down their drum and all ran away. Some jumped in the water while others slipped through the crowd at the [mosque] gate.
>
> Imam Shamil ripped off the drumskin. He broke its rim and threw it after those who had fled, saying, "Take this donkey skin of yours." Next to his house [there lived] a hypocrite—an ally of the Russians. They had set him up as Gimrah's headman in a house that Ghazi Muhammad had bought for Sheikh Muhammad Efendi al-Yaraghi. In the morning, those hypocrites [who had been celebrating during the night] came to that vile man's house to complain about Shamil. "He destroyed the drum that the judge had ordered nine of the village musicians to beat at dawn."
>
> The village elders came to Shamil's door and yelled, "You are creating unrest. The likes of you are responsible for stirring things up and taking advantage of this!"
>
> Shamil spoke [to them] from [the flat] roof of his house and raised his voice so that those who standing in front of this vile man's house [next door] could hear him. He said, "I found them saying various things. Let them do what they will. By God, I will not give up trying

> to stop forbidden actions and I will fight them. Even if I am alone, God is sufficient [for me]. Whoever wants to, let him believe; and whoever doesn't want to, let him not believe." His words were sharp and rough. Now humiliated, the group that had gathered at that evil man's house then dispersed.
>
> Shamil spoke at the mosque in front of the congregation at a holiday prayer gathering. "Do you think that with the death of Ghazi Muhammad the *shari'a* was weakened? By God, I will not let it falter [even] by small measures, but will strengthen it by large measures with God's help. You know that I have more knowledge, strength, and followers than he [did]. Let the opponents of *shari'a* come forward in battle. 'The more honorable ones will expel the meaner ones from there.'" Upon [hearing] this, the *murids'* leaders were bolstered, the people stood up to support the *sharia*, and Shamil's opponents grew less confident. The village headman [mentioned previously] fled to the Russians on the pretext of a quarrel with [Shamil].[24]

This series of incidents illustrates a number of aspects of the Imamate's *ghazavat*. First, violence was not reserved for Russian imperial forces and there were apparently plenty of people (at least in the 1830s, though probably well beyond) who not only had reservations about *ghazavat* but also had legal sanction (and even command) to act contrarily. Also, much of the conflict is carried out in sound: not only the sound of drums, but also the accusation that Shamil was creating unrest, as well as his subsequent preaching (on a roof, at a mosque). In addition, the drummers seem to be adhering in some fashion to the Islamic calendar, with a fast during Ramadan and a subsequent festival ('Eid), suggesting that this was not simply a struggle between an orally transmitted customary law and Islam, but rather between two very different interpretations of the relative proportions of these two traditions in contemporary practice. Finally, if somewhat metaphorically, the destruction of the drum highlights the slippage between sound and violence, a theme that will emerge more fully in what follows.

The Re-Order of Things: Media Logics of Ottoman *Tanzimat* Edicts

The end of the Crimean War resonated through the Ottoman Empire—by means of sonic and other media technologies, including legal pronouncements.

[24] Al-Qarakhi, "The Shining of Daghestani Swords," 24–25.

Figure 2.1 Oil painting by Halil Say. "The first telegraphic text: The Allied Forces have entered Sevastopol. Minister of Foreign Affairs, Fuad Paşa, 9 September 1855." No. 759, Istanbul PTT (Post, Telegraph and Telephone) Museum; used with permission. © Istanbul PTT (Post, Telegraph and Telephone) Museum, No. 759; reproduced with permission.

The fall of Sevastopol was marked by a floating parade down the Bosphorus on 10 September 1855. It was "a gay and beautiful spectacle," recounted a British visitor: "The roar of guns from the different vessels was tremendous—enough to startle the echoes of Olympus [. . .]."[25] In the evening, the firing began again, bringing "noise and the rattle of windows [like] a bombardment;" the ships on the Bosphorus, the palace, and mosque minarets were illuminated "like pale clusters of stars;" and "bands of Turkish 'music' paraded the streets; and a more barbarous noise no red Indians could ever perpetrate."[26] The first international telegrams were sent to and from Istanbul at the same time: the first international message, sent from Bulgaria and subsequently enshrined at the Istanbul Postal Museum, announced "The Allied Forces have entered Sevastopol" (Figure 2.1).

A few days later, the first international telegram sent from Istanbul announced the fall of Sevastopol to Ottoman embassies in Western Europe with this simple message: "Our telegraph begins under happy auspices."[27] Although these (no

[25] Lady [Emelia Bithynia] Hornby, *Constantinople during the Crimean War* (London: Richard Bentley 1863), 35.

[26] Ibid., 36.

[27] Uğur Akbulut, "Suriye'ye İlk Telgraf Hatlarının Çekilmesi," *History Studies: International Journal of History* 2 (2010), 3; and Roderic Davison, *Essays in Ottoman and Turkish History, 1774–1923: The Impact of the West* (Austin: University of Texas Press, 1990), 136. Davison sees in this brief note both innovation and tradition: "This victory telegram was the successor, in concise nineteenth-century style, to the traditional *zafername,* the often lengthy eulogy of a military success, and the *fetihname,* the sultan's victory announcement." The formal opening of the Edirne-Istanbul telegraph line took place on 15 September 1855, with a public demonstration lasting roughly twenty minutes. Cf.

doubt staged and ceremonious) messages mark the end of the Crimean War, they may also provide oddly fitting responses to the first Turkish telegraphic message on record, a pair of questions Sultan Abdülmecid posed as a test message in an 1847 experiment: "Has the French steamer arrived? And what is the news from Europe?"[28]

Continuing the festivities, Giacomo Panizza's 1855 opera, *L'assedio di Silistria* (or *Silistre Kuşatması*, as it was known in its translated Turkish form), was the first opera written specifically for Istanbul's opera house, the Naum Theatre in Pera (Beyoğlu).[29] Composed while the war was still underway, the opera in its premiere in the immediate aftermath of the Battle of Sevastopol nevertheless fulfilled a commemorative purpose. The opera thematically explores the heroism of garrison commander Musa Paşa, killed by shrapnel while performing prayers in Silistra, the site of an 1854 siege by Russian forces, an event that also inspired Namık Kemal's famous 1873 play, *Vatan yahut Silistre* (Homeland or Silistra).[30] Rounding out the official celebrations in February 1856, Sultan Abdülmecid attended the first balls thrown at the British and French embassies, where he was regaled with "God Save the Queen," Ottoman marches, and a forty-one-cannon salute by the British Navy, automated by electrical wire to fire at the moment when the Sultan crossed the embassy threshold.[31]

Less celebratory than some of these moments, but no less significant in demarcating the war's end, was a series of sonically rich events in 1856: the Hatt-ı Hümayun, or Reform Edict (*Islahat Fermanı*) in February; the Treaty of Paris and related negotiations in March and April; and in celebration of the Treaty's

also Nesimi Yazıcı, "Osmanlı Telgraf Fabrikası," *Türk Dünyası Araştırmaları*, (Feb. 1983), 69–81; and Asaf Tanrıkut, *Türkiye Posta ve Telgraf ve Telefon Tarihi ve Teşkilât ve Mevzuatı* (Ankara, Turkey: Efem Matbaacılık, 1984).

[28] Cyrus Hamlin, *Among the Turks* (New York: American Tract Society, 1877), 190.

[29] Adam Mestyan, "'A Garden with Mellow Fruits of Refinement': Music Theatres and Cultural Politics in Cairo and Istanbul, 1867–1892." PhD diss. (Central European University, Budapest, 2011), 343. For more on the Naum Theatre generally, see Emre Aracı, *Naum tiyatrosu: 19. yüzyıl İstanbul'unun İtalyan operası* (Istanbul: Yapı Kredi Yayınları, 2010). The death of Musa Paşa during prayers (or while preparing for them) became a rallying call for Ottoman soldiers and the cultural legacy of the Crimean War more generally; see Badem, *The Ottoman Crimean War*, 185–86.

[30] Figes highlights the circumstances of Musa Paşa's death as being infused with religious significance: "Every day, at morning prayers by the Stamboul Gate, the garrison commander Musa Pasha would call upon his soldiers to defend Silistria 'as becomes the descendants of the Prophet,' to which 'the men would reply with cries of "Praise Allah!"'" Without any clear citation, Figes then adds, "Their determination was given more religious force when Musa Pasha was later killed by a shell that landed directly on him while he was conducting evening prayers for divine intervention to save Silistria"; Orlando Figes, *The Crimean War: A History* (New York: Picador, 2010), 173.

[31] Badem, *The Ottoman Crimean War*, 334–35. Badem notes that the Şeyhülislam and the official historian Cevdet Paşa were invited but (after some confusion) opted not to attend. Badem underscores how unusual it was to have the Sultan attending *any* event at a foreign embassy, let alone a ball.

signing, Sultan Abdülmecid's celebratory banquet in July. Roderic Davison's account of "reform in the Ottoman Empire" opens with these events, noting especially the convoluted sentence of the Sultan's edict (which, in Davison's words, was "as complex as the question of reform itself"). The edict began:

> Wishing today to renew and enlarge yet more the new regulations instituted for the purpose of obtaining a state of affairs in conformity with the dignity of my empire and the position which it occupies among civilized nations, [. . .] I desire to increase well-being and prosperity, to obtain the happiness of all my subjects who, in my eyes, are all equal and are equally dear to me, and who are united among themselves by cordial bonds of patriotism, and to assure the means of making the prosperity of my empire grow from day to day.[32]

For the occasion, the edict, or rescript—which emphasized the equality of all Ottoman subjects (regardless of religion or nationality)—was then read aloud to its end, after which confusion ensued over who was to recite the *dua* invocation (the actual designee failed to attend).[33] A well-known reciter was recruited on the spot, but, as Davison recounts,

> his prayer contained no mention at all of reforms, of non-Muslims, or of equality. "O God," he beseeched, "have mercy on the people of Muhammad. O God, preserve the people of Muhammad." A chill fell on the assemblage. The minister of war whispered in the ear of a neighbor that "he felt like a man whose evening-long labors on a manuscript were ruined through careless upsetting of the inkpot."[34]

In so whispering, the minister himself metaphorically transduced the law from sound (reading the edict aloud) to writing (the spilled inkpot). Legal media consistently produce more of other media, even in metaphor. After the

[32] Roderic Davison, *Reform in the Ottoman Empire 1856–1876* (Princeton, NJ: Princeton University Press, 1963), 3.

[33] The Sultan's message of equality was prompted by Britain's and France's intercession on behalf of Christians in the Ottoman Empire. As Badem points out, these (largely unsuccessful) attempts to raise the social status of non-Muslims was one of the war's many conspicuous legacies in the region. See Badem, *The Ottoman Crimean War,* 405. Davison's account largely comes from the memoirs (*Tezakir*) of Cevdet Paşa, the official historian of the Ottoman Empire. Where my details go beyond Davison's (e.g., that a reciter had been appointed in advance but that he simply did not show up), I am drawing on Cevdet Pasa, *Tezâkir,* ed. Cavid Baysun, vol. 10 (Ankara, Turkey: Türk Tarih Kurumu Basımevi, 1953), 67–89.

[34] Davison, *Reform in the Ottoman Empire,* 3–4.

ceremony, attendees received printed copies of the edict and were sent on their way.

A week later, the "Concert of Europe" gathered in Paris to negotiate a peace treaty in the wake of the Crimean War—and the Ottoman Empire was finally invited (a formal stipulation of Article VII, in fact). Article IX offers one of the earliest formal responses to the Reform Edict:

> His Imperial Majesty the Sultan, having, in his constant solicitude for the welfare of his subjects, issued a firman which, while ameliorating their condition without distinction of religion or of race, records his generous intentions towards the Christian population of his Empire, and wishing to give a further proof of his sentiments in that respect, has resolved to communicate to the Contracting Parties [i.e., the European Powers] the said firman, emanating spontaneously from his sovereign will.
>
> The Contracting Parties recognize the high value of this communication. It is clearly understood that it cannot, in any case, give to the said Powers the right to interfere, either collectively or separately, in the relations of His Majesty the Sultan with his subjects, nor in the internal administration of his Empire.[35]

This passage has attracted considerable attention from historians, as it became precisely the aspect of the Paris Treaty most actively violated in the years to follow. But it also suggests certain ways of thinking about law medially: the Sultan issues a firman that simultaneously records (his intentions) and gives proof through its circulation (to the other "Concert" members). It is a document in the broadest sense: record, proof, legal evidence, and perhaps—for "the Powers"—also a so-called teachable moment (*docere*), one that trains the Ottoman state apparatus in post-Enlightenment democracy. The firman "emanat[es] spontaneously from his sovereign will"—it is literally an *edict*, a speaking forth that comes from inside the Sultan himself (proceeding on the assumption that his will is located in his person). In this light, the term "edict" is perhaps a better English translation of the original terms *ferman (firman)* and *hat* than the other common rendering, "rescript," which suggests a response rather than a spontaneous emission (itself a naive notion, given the overt pressures from other European states). Yet both terms suggest a medial quality: an edict is spoken (which this was), while a rescript is written (which this also was, emphasized with a term like

[35] J. C. Hurewitz, *Diplomacy in the Near and Middle East: A Documentary Record; 1535–1914*, vol. 1 (Princeton, NJ: Van Nostrand, 1956), 154.

hat, itself meaning "writing"). As with the Reform Edict, these legal statements articulate a particular media logic about themselves and other media with which they interface (that is, other laws, treaties, utterances of the Sultan).

The treaty was settled on 30 March 1856, with ratifications exchanged between the various Contracting Parties in Paris on 27 April. And so festivities could begin once more, reopening with a huge banquet to mark the Paris Treaty. Hornby was again in attendance and cast the event in ominous terms:

> A minute or two after the Sultan had retired we were startled by two frightful claps of thunder followed by a storm of wind and hail. The whole building seemed to shake, and in a moment the gas went out and we were in total darkness. The band dropped their instruments with a clash and fled. For some moments no one spoke, and then a thin, shrill voice was heard in French saying, "It wants but the handwriting on the wall and the words 'Mene, Tekel, Upharsin' to make of this a second feast of Belshazzar."[36]

Lady (Emelia) Hornby, the wife of our raconteur, adds several key details. The palace hall had a massive chandelier in the middle that "burns four hundred jets of gas" in addition to "about two hundred wax candles," which were soon to be extinguished:

> The Sultan's band was posted at one end of the hall, but, after playing one or two airs, the musicians grew frightened at the storm, which now crashed with great fury just over the palace, and ran away. These valiant men left a large door open in their flight, which producing a tremendous draught, half the candles were blown out.[37]

In Lady Hornby's memory, as recounted by her husband, the production of sound and of sight were coupled at this moment. His account highlights the not so subtle view held by the rest of the Concert of Europe that the Ottoman Empire was "the sick man of Europe"—an empire that, like Belshazzar's Babylon, was on the verge of collapse, as described in the Hebrew Bible. And yet the sensory modalities are upended: in Babylon, a silent hand produced the original "writing on the wall," while in Istanbul, the visual (including the written) was taken away, leaving a scene closer to *Macbeth*, with extinguished candles in a palace and vanished performers adding to the acousmatic "sound and fury" of the Bosphorus.[38]

[36] Lady Hornby in Davison, *Essays in Ottoman and Turkish History*, 4.

[37] Hornby, *Constantinople during the Crimean War*, 410.

[38] William Shakespeare, *Macbeth* (Surrey: Thomas Nelson, 1997), V.5.22–28.

The Reform Edict, as much as it heralded a new Ottoman future, was also haunted by its past; it was embedded in and emerged from a complex network of legal codes and utterances that would shape both what the 1856 Edict could articulate and also how it could do so. In 1839, the Gülhane (Rose Chamber) Edict, known in Turkish more commonly as the *Tanzimat* firman, was decreed as law. Like *ghazavat, tanzimat* is a plural noun that has both a general, dictionary meaning and a more specific meaning in the Islamicate world of the Crimean War. As scholar Ali Akyıldız has pointed out,

> *Tanzîmât*, the plural of the word *tanzîm*, having a literal meaning of "to put in order, to arrange, to reform," is used in writing to mean "the reforming and reorganizing of the civilian administration." It also describes the period in which this re-ordering was carried out.[39]

This reordering of things, building on the major (and sometimes violent) reforms implemented during Sultan Mahmud II's reign (1808–1839), has generally been interpreted as a gesture of westernization at odds with the prevailing Islamic leadership of the Ottoman Empire.[40] But some scholars see in it an act of Islamic renewal, perhaps driven by the rise of the Naqshbandi Sufi order—the same order that claimed the loyalties of Imam Shamil—among leading Ottoman circles.[41]

Regardless of its origins in "Western" or "Islamic" thought (as though such categories were ever stable and impervious to each other), the Gülhane Edict of 1839 opens with striking reference to Islam and proper governance through *shari'a*, foreshadowing the kinds of debates that arose in both the Ottoman Empire and the Caucasus Imamate in the decades to follow. Furthermore, in a recursive, media-like way—not unlike other documents discussed previously—the 1839 Edict theorizes its own legal impact and meaning:

> It is known to the all that since the first days of the Ottoman State, the lofty principles of the Qur'an and the rules of *shari'a* [*ahkâmı celîle-i kur'âniye ve kavânîn-i şer'iyyeye*] were always perfectly observed. Our mighty Sultanate reached the highest degree of strength and power, and all its subjects [the highest degree] of ease and prosperity. But in the last

[39] Ali Akyıldız, "Tanzimat," *Türkiye Diyanet Vakfı İslâm Ansiklopedisi*, 40 (2011), 1.

[40] For example, see Niyazi Berkes, *The Development of Secularism in Turkey* (Montreal: McGill University Press, 1964), 144ff.; and M. Şükrü Hanioğlu, *A Brief History of the Late Ottoman Empire* (Princeton, NJ: Princeton University Press, 2008), 73.

[41] Butrus Abu-Manneh, "The Islamic Roots of the Gülhane Rescript," *Die Welt des Islams* 34 (1994), 173–203.

> 150 years, because of a succession of difficulties and diverse causes, the sacred *shari'a* was not obeyed nor were the beneficent regulations followed; consequently, the former strength and prosperity have changed into weakness and poverty. It is evident that countries not governed by the laws of the *shari'a* cannot survive.[42]

The text not only affirms *shari'a* law but also posits a theory of law and society that the Ottoman government itself later challenges: that perfect observation of *shari'a* was necessary or desirable for the empire to survive. Whether by the Sultan's volition or under duress from the British and others, this articulation of a *shari'a*-centric model of governance is completely absent from the 1856 Reform Edict. Indeed, much of *tanzimat* can be seen as a paring back of the role of *shari'a* in Ottoman society, especially through limiting the activities of the *ulema*, the elite class of Muslim scholars tasked with interpreting *shari'a*, and by way of the establishment of alternative educational, judicial, and governmental institutions and the bolstering of non-Muslim communities.[43]

However, the role of *shari'a* here is complex: it figures not only in the outcomes of legal reforms (that is, whether or not a given reform adheres more or less closely to *shari'a*) but also in the legal utterance itself and its mediality—how it is articulated, and the kinds of techno-juridical operations it sets in motion. As with the Reform Edict, the performative context of that utterance is deeply relevant. Şükrü Hanioğlu opens his 2008 history of *Tanzimat* with the proclamation by the Foreign Minister on behalf of (and in front of) the Sultan and assembled dignitaries. Hanioğlu remarks that the decree was unprecedented in Ottoman history, its "singular importance" underlined by the Sultan's (proxy) oath within the ceremonial room in which the Prophet Muhammad's mantle was kept.[44] "The real novelty of the decree lay in its content," Hanioğlu writes, because it underwrote new laws ensuring life, property, prohibiting bribery, and regulating taxation and conscription applied to all subjects of the empire.[45] But even after avowing the importance of "content," Hanioğlu stresses the importance of

[42] This translation is based on my revisions of translations by E. A. Van Dyck in Hurewitz, *Diplomacy in the Near and Middle East,* 113–16, and the draft version at the *Electronic Middle East Sourcebook,* <http://sitemaker.umich.edu/emes/home>, accessed 15 Sept. 15, 2014), with reference to the Ottoman and original French translations in, *Tanzimat, I: Yüzüncü yıldönümü münasebetile,* ed. Hasan-Âli Yücel (Istanbul: Maarif Matbaası. 1940), 48ff. and a transliterated version in Şeref Gözübüyük and Suna Kili, *Türk Anayasa Metinleri: Tanzimattan Bugüne Kadar* (Ankara, Turkey: Ankara Üniversitesi Siyasal Bilgiler Fakültesi, 1957), 3–5.

[43] See Davison, *Reform in the Ottoman Empire,* and Berkes, *The Development of Secularism in Turkey.*

[44] Hanioğlu, *A Brief History of the Late Ottoman Empire,* 72.

[45] Ibid., 72.

form: "The declaratory value of the Rose Chamber Edict clearly outweighed its legal significance. [. . .] Its intended audience has been a matter of some debate. To an extent, the edict was directed at European ears," though it was "directed both inward and outward" to indicate "a serious commitment to reform" among Ottoman bureaucrats.[46] He sees the Islamic concepts as "cosmetic changes added as a sop to the ulema," while substantive change came in "the role of the Ottoman bureaucracy in drafting, codifying, and implementing the administrative reform."[47]

Hanioğlu's account highlights the Gülhane Edict's medial quality: it is both ceremony (gathering, oath) and utterance (proclamation, the text itself), directed at a certain "audience[s]" and "ears," and entailing a particular set of technical practices (drafting, codifying, implementing). Other historical narratives heighten this reading even further. Historian Niyazi Berkes, for example, notes some significant details: Sultan Abdülmecid "added in his own handwriting his statement of oath."[48] And yet, "a traditional practice was forgotten or ignored when the Charter [the Edict] was proclaimed; although it obviously concerned the Şeriat [*shari'a*] it was not accompanied by a *fetva* that would bring about its legitimization by the Şeyhul-Islâm," a gesture Berkes interprets as "the first formal breach between the 'temporal' and the 'religious'" that characterized *Tanzimat* secularism.[49] Once again, law comprises not merely content, but also certain kinds of actions and protocols—whether sonic, written, or both.

For historian Butrus Abu-Manneh, by contrast, the content of the Gülhane Edict *means* the opposite of Hanioğlu's and Berkes' reading (that the Edict really was intended as an Islamic revival). He grounds his interpretation in the other kinds of legal utterances (*hat*, or edict/rescript; *irade*, decree, but literally also "will" or "volition") that Sultan Abdülmecid made in his earliest months as sultan, as well as his audiences for such utterances.[50] These legal mediations were reciprocal: in late summer 1839, the Sultan summoned his highest-ranking officials (*Meclis-i Şūrā*) to meet and discuss the reinvigoration of *shari'a* in response to his *irade* a few weeks earlier. The gathering culminated in a petition being "drawn up and submitted to the Sultan which carried the seals of 38 dignitaries who apparently attended the meeting," which in turn precipitated another decree: "In the *irade* written at the top of the sheet of the petition, the Sultan expressed his satisfaction with the petition and ordered to have it proclaimed publicly as a *hatt-ı şerif*," or noble edict/rescript, which became the Gülhane

46 Ibid., 72–73.

47 Ibid., 73.

48 Berkes, *The Development of Secularism in Turkey*, 145.

49 Ibid., 147.

50 Abu-Manneh, "The Islamic Roots of the Gülhane Rescript," 190–91.

Edict just a few months later.[51] He notes that not only the Sultan but also the senior *ulema* and state functionaries took an oath to uphold the Gülhane Edict, with the result that the religious hierarchy was so involved in the process that "a *fetva* of the sheikh ül-Islam for [the Edict's] proclamation was as it seems not even issued."[52] While Berkes and Abu-Manneh disagree over *why* there was no *fetva* issued with the Gülhane Edict, they both understand it as (potentially) belonging to a media network of other legal utterances and—significantly—the performative proclamations of those utterances.

With this understanding of the 1839 Gülhane Edict in mind, the 1856 Reform Edict's medial qualities become more readily apparent. The somewhat convoluted sentence cited by Davison previously in this chapter ("Wishing today to renew . . .") is in fact not the first sentence of the decree—though perhaps it was the first sentence read aloud. The written text of the 1856 Edict is bookended by a personal comment to the Grand Vizier:

> Let it be done as herein set forth.
>
> To you, my Grand Vizier Mehmed Emin Ali Pasha, decorated with my imperial order of the *mecidiye* of the first class, and with the order of personal merit; may God grant to you greatness and increase your power [. . .].
>
> Such being my wishes and my commands, you, who are my grand vizier, will, according to custom, cause this imperial firman to be published in my capital and in all parts of my empire; and you will watch attentively and take all the necessary measures that all the orders which it contains be henceforth carried out with the most rigorous punctuality.[53]

The Grand Vizier is thus the addressee—the original audience—and intermediary for publication and dissemination of the edict. The same kind of reading of the body of the edict yields a similar emphasis on mediations: "privileges [. . .] shall be confirmed and maintained;" formation of an ad hoc commission "to discuss and submit" recommended reforms from different classes/religions; for building permits for religious structures, "the necessary authority must be *asked for*, through the *medium* of the patriarchs and heads of communities" (emphasis added); local public schools (in science, art and industry) may be established but with pedagogical methods and personnel decisions "under the control of a mixed council of public instruction," presumably issuing guidelines; any legal

[51] Ibid., 193.

[52] Ibid., 194.

[53] Hurewitz, *Diplomacy in the Near and Middle East*, 150 and 153.

designations differentiating between classes of citizens on the basis of "religion, language, or race, shall be forever effaced from administrative protocol;" and finally, "Penal, correctional, and commercial laws [. . .] shall be drawn up as soon as possible and formed into a code. Translations of them shall be published in all the languages current in the empire."[54]

These layered remediations went beyond "how to do things with words," as J. L. Austin notably put it—though these words did, of course, cause things to happen.[55] Lawmaking inaugurated a self-perpetuating set of utterances and uttered responses to those utterances. Bureaucracy fed bureaucracy: law issued forth (for example, as an edict) and in so doing called forth further issuances (confirmations of privileges, submissions of recommended reforms, etc.). Legal utterances thus became the audiovisual source material for (and also the enactment of) a cultural code—which had then to be remediated and circulated throughout the heteroglossal space of empire. Reversing the common-sense notion that power spreads outward, we might say that the utterance of an edict all but demanded that it should issue forth from the thresholds of governance: in this case, the Sublime Porte or Bâb-ı ʿÂlî, literally the exalted gate. As media theorist Bernhard Siegert has written, "D[d]oors are operators of symbolic, epistemic, and social processes that, with help form the difference between inside and outside, generate spheres of law, secrecy, and privacy and thereby articulate space in such a way that it becomes a carrier of cultural codes."[56] The speaking-forth of edicts like those of Tanzimat (1839) and Reform (1856) were the sites of "articulat[ing] space"—generating laws and transmitting cultural codes at one and the same time. Furthermore, these edicts became cultural codes in their own right, acting simultaneously as source, transmitter, and substance. The intertwining of legal uttering and the process it instigated across time and space constituted its jurisdiction—the intersection of law (*juris*) and utterance (*diction*).[57]

Law as Voice and Silence

What kind of medium is law? Through which sensory modalities is law articulated? These questions are central to—and generally unanswered by—jurisprudential writing, although attempts to address them frequently highlight

[54] Ibid.

[55] John Langshaw Austin, *How To Do Things with Words*, ed. J. O. Urmson and Marina Sbisà, 2nd ed. (Cambridge, MA: Harvard University Press, 1975).

[56] Bernhard Siegert, "Doors: On the Materiality of the Symbolic," *Grey Room* 47 (2012), 8.

[57] Cf. Justin Richland, "Jurisdiction: Grounding Law in Language," *Annual Review of Anthropology* 42 (2013), 209–26.

the audiovisual nature of legal production, particularly emphasizing the role of voice and silence in the absence of sight. For example, Ronald Dworkin's seminal work on jurisprudence, *Law's Empire* (1986), opens with precisely this question—not what law is, but rather how it is sounded and voiced, in terms that are only seemingly metaphorical. His preface begins,

> We live in and by the law. It makes us what we are: citizens and employees and doctors and spouses and people who own things. It is sword, shield, and menace: we insist on our wage or refuse to pay our rent, or are forced to forfeit penalties, or are closed up in jail, all in the name of what our abstract and ethereal sovereign, the law, has decreed. And we *argue* about what it has decreed, even when the books that are supposed to record its commands and directions are silent; we act then as if law had muttered its doom, too low to be heard distinctly. We are subjects of law's empire, liegemen to its methods and ideals, bound in spirit while we debate what we must therefore do.[58]

Dworkin then continues with this question: "How can the law command when the law books are silent or unclear or ambiguous?"—which he answers by arguing for legal reasoning as a form of interpretation. Yet Dworkin's question is more fundamental than he acknowledges. How *does* a law command? How can law books be silent or speak at all? Returning to the opening paragraph, we might ask, how does law utter (or mutter) anything, and what does it mean that it is (or is not) "heard distinctly"? And how does arguing—the word is emphasized by Dworkin—play into the question of decree?

While Dworkin's examples are largely concerned with the social and technical processes of law, he does point out several instances in which the question of voicing and silence—no doubt meant figuratively elsewhere—become central. One such instance is the case of the speed limit of 55 mph in California, a law that is grounded not in "what some ghostly figure had said or what was found on transcendental tablets in the sky" but "because a majority of that state's legislators said 'aye' or raised their hands."[59] In other words, law emerges not from an utterance from the sky, but from an utterance on the earth—indeed a rather literal, sounded one, at that. Going further, he notes that judges create active, new law with each case, as they "announce a rule or principle . . . that has never been officially declared before," offering " 'new' statements of law as improved reports of what the law, properly understood, already is."[60] Yet our more fundamental

[58] Ronald Dworkin, *Law's Empire* (Cambridge, MA: Harvard University Press, 1986), vii.

[59] Ibid., vii.

[60] Ibid., 6.

questions remain: how was law uttered, by whom and to whom? The heavenly voice of Law may be a convenient metaphor, but the need for law to be articulated and so brought into existence—and to do so as sound, with surprising frequency—is literal, and often urgent.

Dworkin's ruminations belong to a much longer narrative about law and voice, spanning from Socrates and the Book of Exodus, to Kant and Heidegger, as Mladen Dolar has traced.[61] Contemporary media theorists like Cornelia Vismann and Siegert similarly emphasize the question of audiovisual mediation in the act of law creation.[62] And although they both ultimately privilege visual and written practices over sonic ones, the sense that law is a kind of sounded utterance (or silence in the absence thereof) remains. Thus Vismann describes law's media operations in terms of silence about itself and its origins: "The law remains silent about its records. It works with them and creates itself from them. In other words, it operates in a mode of difference that separates it from the varying formats of files."[63] This algorithmic "silence" obscures precisely the moment and process of articulation, but that moment remains, whether embedded in files acting as media or as Dworkin's California speed-limit legislators. Legal scholars, for their part, have taken a slightly different approach to the uttering of law, emphasizing what James Boyd White has called "the rhetoric and poetics of law" or the "grounding" of law in the uttering of language, as Justin Richland writes in his article, "Jurisdiction"—which is to say, as mentioned previously, "juris-diction."[64] Marianne Constable has an even more sustained engagement with the question, with book-length discussions of "legal speech acts" and silence within and as law.[65]

All of these examples focus primarily on Anglo-American law, yet insofar as they point to a broader theory of law and its uttered mediations, they have strong relevance for Islamic law as well. Both legal traditions—Anglo-American and Islamic law—have deep histories of oral/common law practices with writing emerging as a later counterpart that has since become the dominant sensory modality of thinking through these traditions. For example, in *The Calligraphic State*, Brinkley Messick appears to argue, not unlike Vismann or Siegert, that law is indeed a media operation based in the cultural practice (or "cultural technique," as Siegert would have it) of writing.[66] (His account is of particular

[61] Dolar, *A Voice and Nothing More*, 97–102.

[62] Cornelia Vismann, *Files: Law and Media Technology* (Stanford, CA: Stanford University Press, 2008), and Siegert, "Doors," 13.

[63] Vismann, *Files*, 13.

[64] Richland, "Jurisdiction: Grounding Law in Language," 209–226.

[65] Marianne Constable, *Just Silences: The Limits and Possibilities of Modern Law* (Princeton, NJ: Princeton University Press, 2005), 2014.

[66] Messick, *The Calligraphic State*.

relevance, since he is recounting the history of Islamic law in Yemen within the broader narrative of nineteenth-century Ottoman reform and the collapse of that empire.) Yet even in attempting to place writing at the center of his narrative, he continually encounters the salient presence, both historically (a millennium ago or more) and in more contemporary settings (since the nineteenth century), of "audition" and oral forms of lawmaking, circulation, and transmission in ways that belie his larger claim of "textual domination" in Islamic law.

The idea of Islamic law remaining "silent" on an issue is widespread in contemporary scholarly discussions of Islamic law. Silence as a legal category in *shari'a* tends to manifest either on a diachronic-communal level, as an absence of legal precedent, or on a synchronic-personal level, when an individual refuses or fails to verbalize assent to some legal action. An interesting point of intersection in these two types can be seen in the legal condition of so-called tacit consensus:

> Generally speaking, tacit consensus occurs when one jurist issues an opinion and the remaining jurisconsults remain silent [. . .]. Their silence is taken thus as a vote of approval, when the term *ijmā' sukūtī* (consensus known by silence).[67]

In his study of Iranian law, Ashk Dahlén traces the depiction of the Qur'an as being "silent" in legal matters back to 'Ali ibn Abi Talib, the first follower of the Prophet Muhammad, who, according to tradition, called on believers, "Request this Qur'an to speak to you, but it will not speak to you unless I enlighten you about it."[68] Dahlén sketches an entire theory of "silent *shari'a*" and "speaking *shari'a*," as initially posited by scholar Abdolkarim Soroush, in which, strikingly, the silent version of law—the one found in the Qur'an, for example—is more perfect than the speaking one, which is "born out of human understanding, flowing and contradictory."[69] While Dahlén and Soroush are particularly interested in Shiite law, their usage of silence highlights how widespread and productive scholars of Islamic law have found the idea in conceptualizing the sonic and mediated natures of *shari'a*.[70] More broadly, the idea of silence touches in

[67] Sherman A. Jackson, *Islamic Law and the State: The Constitutional Jurisprudence of Shihāb Al-Dīn Al-Qarāfī* (Leiden, The Netherlands, New York, Cologne: Brill, 1996), xxxiv.

[68] Ashk Dahlén, *Islamic Law, Epistemology and Modernity: Legal Philosophy in Contemporary Iran* (London: Routledge, 2004), 293.

[69] Ibid.

[70] In one striking deployment of this trope of "silent" law, Intisar Rabb even describes law as potentially being "resoundingly silent" in cases in which interpretation was demanded because of "wily sources" (i.e., legal texts) that were "significantly vague" in their scope and potential applications; see Rabb, *Doubt in Islamic Law* (Cambridge: Cambridge University Press, 2014), 31.

particular on the foregoing discussion about *'adat* and *'urf*, forms of traditional law that come into existence when other legal sources are silent.

"Our Mother Tongue by Telegraph": Ottoman Media Archaeologies ca. 1855

Reactions to the Reform Edict of 1856 were complex, and often not positive, especially when they involved religious matters. Once again, sound was a site of contest, as in several instances involving Christians' use of bells. In Nablus, a mixed Christian/Muslim city, different groups interpreted the Crimean victory in divergent ways. Muslims saw it as a "Muslim victory," while Christians considered it an allied victory. To celebrate, Christians "raised French and British flags on their houses in Nablus and placed a new bell over the Protestant mission school"—measures viewed as provocations by local *ulema* (Muslim leaders), who warned their congregations at Friday prayers that "Muslims would soon be called to prayer by the English bell, unless they rose up to destroy the Christian churches, which, they said, would 'be a proper form of prayer to God.' "[71] Riots ensued, resulting in ransacking, arson, and even murder.[72] While Christians demanded to ring bells, the Ottoman authorities denied them, in the name of preserving public order. As Şükrü Hanioğlu suggests, the bell-ringing issue—along with countless others, which rendered "implementation of the Reform Edict of 1856 impossible"—was fairly widespread.[73] In 1859, a small group calling themselves the Society of Martyrs organized by a Naqshbandi Sheikh Ahmed and an army officer, Hüseyin Daim Paşa, was formed. It comprised officers and members of the *ulema* who wanted to assassinate Sultan Abdülmecid on account of "the little regard shown by the present government to the Holy law, the prescriptions of which according to them are trampled under foot."[74] The attempted assassination, known as the Kuleli Incident, failed and the group was arrested. When interrogated, the sheikh repeatedly gave the response, "My aim was not assassination (suikast) but to carry out the statutes of the sharia!"[75] These incidents suggest a

[71] Figes, *Crimea*, 429.

[72] The Friday sermons led immediately to a storming of the Protestant mission where the new bell was hung; the outbreak of violence was sparked by one Reverend Lyde, "a Protestant missionary and Fellow of Jesus College, Cambridge, who had accidentally shot a beggar attempting to steal his coat"; see Figes, *Crimea*, 429.

[73] Hanioğlu, *A Brief History of the Late Ottoman Empire*, 85.

[74] Florian Riedler, *Opposition and Legitimacy in the Ottoman Empire: Conspiracies and Political Cultures* (New York: Routledge, 2011), 15.

[75] Ibid., 17.

general consensus that, regardless of the Sultan's intents in the Gülhane Edict, his reforms had the effect of diminishing the place of *shari'a*.

Mid-twentieth-century histories like those of Roderic Davison and Nizami Berkes cast many of the central reforms of the *Tanzimat* years as a question of westernization pitted against *shari'a*, reform against tradition, young politicians against the *ulema*, and so on.[76] Such a narrative is compelling—and indeed, lies at the heart of my broader argument that political currents in the Caucasian Imamate and the Ottoman Empire were moving in different or even opposite directions with regards to *shari'a*. But given the medial qualities of these legal reforms I highlighted previously, I would like to suggest a subtler set of interactions between the state and *shari'a* in both cases. Certainly some of the aforementioned reforms like public (secular) schools and non-*shari'a* courts would have been an obvious challenge to the *ulema*-guided religious equivalents (*medrese* schools, *shari'a* courts) that had existed for centuries. Yet many of the reforms typically lumped into *Tanzimat*, especially those related with communications—the postal service (1840), the telegraph (1855), railways (1856), or even slightly older developments like the Ottoman state's *Tercüme Odası*, or Translation Bureau (1821, but more earnestly from 1833 onward)—seem to have produced much less severe reactions, while also producing more complex effects on law-as/in-media in the 1850s.[77]

The cultivation and emergence of Ottoman telegraphy offers a prime example of one such reform. As described previously, the beginnings of official use of the electric telegraph in the Ottoman Empire were closely bound to the Crimean War. But that was not an inevitable chronology. In 1839, just four years after Samuel Morse unveiled his first working model, a colleague of Morse's, Mellen Chamberlain, attempted to demonstrate telegraphy in Istanbul. After working in Istanbul with an American missionary, Cyrus Hamlin, he left for Vienna to replace some equipment, but died shortly afterward in a steamer accident on the Danube.[78] In 1847, Hamlin, along with two other Americans, was part of the demonstration of telegraphy for Sultan Abdülmecid, who was thrilled by the technology and immediately inquired about the possibility of installing a line from Istanbul to Edirne (Adrianople).[79] Hamlin's account suggests from the outset that the Sultan—then only twenty-four years old—had both great curiosity

[76] Davison, *Reform in the Ottoman Empire*; and Berkes, *The Development of Secularism in Turkey*.

[77] For an overview of many of these more technocratic reforms, cf. Carter Vaughn Findley, *Bureaucratic Reform in the Ottoman Empire: The Sublime Porte, 1789–1922* (Princeton, NJ: Princeton University Press, 1980).

[78] Hamlin, *Among the Turks*, 185–86; Davison, *Essays in Ottoman and Turkish History*, 133; Yakup Bektaş, "The Sultan's Messenger: Cultural Constructions of Ottoman Telegraphy, 1847–1880," *Technology and Culture* 41/4 (2000), 671.

[79] Hamlin, *Among the Turks*, 190ff.

and a fairly keen grasp of the apparatus: he wanted to understand the electric current used; he cheered at the demonstration ("He threw up both arms, saying, 'Mashallah! mashallah!' "); and he immediately grasped the kinds of relationships telegraphy created between alphabet and meaning.[80]

A few years later, the Allied Forces proposed laying a telegraph line to connect across Europe and underwater between Varna and Balaklava in the Crimea. The plan was executed in 1854–55, with later extensions to Istanbul and Edirne as well.[81] The connections to the Ottomans took several more months, but the response from Western Europe was striking. Clearly the telegraph maintained its ability to fire people's imaginations—Sultan Abdülmecid was not unique in this regard. One British writer, George Dodd, who wrote extensively about the Crimean War, observed the following after the completion of the underwater line:

> Thus were the means afforded for almost instantaneous communication between England and the Crimea: an achievement truly marvelous, although familiarity speedily brings down all such marvels to the level of everyday commonplace. It was curiously observed at the time that, "It would not now be difficult, by some little further novelty of invention, to cause the reverberation of the very cannon themselves, as it were, to be transmitted, in the shape of electric vibrations, through the 3000 miles of intervening wire, and heard, in still continuous vibrations, finally communicated to some acoustic apparatus in the British Houses of Parliament!"[82]

This is a blend of CNN-style embedded journalism and a strange piece of sound art: the speculative apparatus transmits cannon fire directly and continuously (unceasingly?) into the heart of Britain's political establishment. The telegraph thus took on a second life in the popular imagination during the Crimean War, fashioning an audile connection between distant war and metropolitan politics. Imagine, for instance, William Howard Russell's dispatches from the front, describing cannon fire and explosion, but now transduced and remixed in the unique acoustic space of Parliament: "the sullen roar of the heavy mortars which came booming upon the ear twice or thrice in every minute"; or "the steady, unceasing thunder of gun, and rifle, and musket"; or "the loudest thunder [resembling] some great convulsion of nature," all blasted into highly reverberant legislative chambers.[83] The emphasis on reverberation and vibration suggests

[80] Ibid., 188–190.

[81] Davison, *Essays in Ottoman and Turkish History*, 134–35.

[82] George Dodd, *Pictorial History of the Russian War 1854–5–6: With Maps, Plans, and Wood Engravings* (Edinburgh and London: W. & R. Chambers, 1856), 389.

[83] William Howard Russell, *The British Expedition to the Crimea* (London and New York: G. Routledge & Co., 1858), 216, 205, and 551.

not only electrical transduction but also the "vibratory materialism" of "sonic warfare," to borrow Steve Goodman's terminology.[84] For Russell, as for Dodd, the rumble of cannons is not the imperial triumphalism of can-do telegraphy, but rather brings home sonic dystopias of "fear and dread."[85]

Dodd summed up his position as follows: "The Black-Sea Submarine Electric-telegraph was, perhaps, the most wonderful gift ever made by the arts of peace to the purposes of war."[86] Presumably, he meant that telegraphy could hold politicians accountable, by transmitting up-to-date evidence from the front. His imagined model of sound itself vibrating along wires—an impossibility with the electric telegraph, which rendered all sound alphabetical, triangulated between sound, technology, and legislative space. In other words, although such telephonic transmission of sound was not yet possible, the use of the telegraph as a touchstone for imaginative technological speculation remained a hallmark of its spread in the Ottoman Empire, as seen from Sultan Abdülmecid's enthusiastic first encounter in 1847.

Responses to the Ottoman telegraph system were, unsurprisingly, varied. According to Cyrus Hamlin, part of the reason no telegraph was constructed between 1847 and 1854 was the resistance of mid-level government officials:

> The pashas united against it. They wanted no such tell-tale to report their doings every day, while in the distant interior. Six years later, the Crimean War made it a necessity; and the lines have become numerous, uniting Constantinople with all the world.[87]

Initially, most of those lines were built and staffed by French and British contractors, simultaneously training Ottoman engineers while extending the reach of the imperial communications network to India. That building process entailed a mixture of resistance, curiosity, and fear:

> Turkish engineers erecting lines in 1864 between Samsun and Amasya believed that the lines had to be put up on the poles wet, so they dipped the wire in mud puddles. It was reported that inhabitants of Edirne, where the telegraph office was located outside the city, did not want the wires coming into the city for fear that they would conduct lighting.[88]

[84] Steve Goodman, *Sonic Warfare: Sound, Affect, and the Ecology of Fear* (Cambridge, MA: MIT Press, 2009), xviii, 9.

[85] Ibid., xiv.

[86] Dodd, *Pictorial History of the Russian War*, 390.

[87] Hamlin, *Among the Turks*, 194. Roderic Davison notes Hamlin's lack of clear source here but concedes that the observation was probably correct; see his *Essays in Ottoman and Turkish History*, 139.

[88] Davison, *Essays in Ottoman and Turkish History*, 139.

Latin letter	Morse Code	Ottoman letter	Transliteration	Mustafa Efendi system	İzzet Efendi system
A	• —	ا	A	• —	• —
B	— • • •	ب	B	— • • •	— • • •
C	— • — •	پ	P	— • — •	— • • —
D	— • •	ت	T	— • •	—
E	•	ث	S	•	— • — •
F	• • — •	ج	C	• • — •	— • • • — — —
G	— — •	چ	Ç	— — •	— • • • — — —

Figure 2.2 Comparison of telegraphic codes, including International Morse Code and two competing Ottoman systems by Mustafa Efendi and İzzet Efendi. Excerpt, adapted from Nesimi Yazıcı, "Osmanlı Telgrafında Dil Konusu." *Ankara Üniversitesi İlahiyat Fakültesi Dergisi* 26 (1983): 751–64.

To protect the physical materials from vandalism and theft, a certain rhetorical strategy was developed and began circulating (it was probably started by British officials) that the new technology would "convey the messages of the Sultan," and so should be respected.[89] The rhetoric was apparently so successful that a local historian in Diyarbakır observed in 1860 that "all people watched the telegraph wires with great admiration, and prayed for Sultan Abdul Mejid most gratefully."[90] As with so many other empires, telegraphy allowed the Sultan and the central government to consolidate power, "enabling the central power in Constantinople to move the whole empire like a machine"—while also making it possible for inhabitants around the empire to contact the Sultan directly, whether as spies or as mobs.[91]

But the telegraph was not simply a tool of expansive/expanding imperial power—just as it was not simply a technology to be "transferred" and absorbed from "the West." Because of its linguistic diversity and the wartime origins of the telegraph there, the Ottoman Empire adopted multiple systems of Morse code (French, numerical ciphers), with Turkish added in May 1856. Two former employees of the state Translation Bureau, Mustafa Efendi and Vuliç Efendi, devised a Turkish Morse code (see Figure 2.2) that was to spread widely until it was replaced two decades later by a new system devised by İzzet Bey that was based on frequency of letter usage.[92] In drafting a short 1933 biography of Mustafa Efendi, called "Our Mother Tongue by Telegraph and Mustafa Efendi" (*Telgrafçılıkta Ana Dilimiz ve Mustafa Efendi*), Ahmed Baha suggests that this particular aspect of reform did not so much bring equality among Ottoman subjects as bring the empire into its own, relative to other empires: "While all other states

[89] Bektaş, "The Sultan's Messenger," 692.
[90] Ibid., 692.
[91] Ibid, 694–95.
[92] Nesimi Yazıcı, "Osmanlı Telgrafında Dil Konusu," *Ankara Üniversitesi İlahiyat Fakültesi Dergisi* 26 (1983), 760ff; and Davison, *Essays in Ottoman and Turkish History*, 151.

send telegrams in their own languages, the practice of sending French-language telegrams in the Ottoman state constitutes an oddity. Foreigners make fun of us."[93] Language reform (especially the formation of the Translation Bureau) thus enabled the full measure of telegraphic reform, allowing Ottoman subjects to send telegrams in Turkish rather than French. Technology, language, and performativity all percolated within a framework of a broader reordering of the state.

The Media Logics of Telegraphy and *Shari'a*?

By way of conclusion, I would like to look at one final Ottoman source on the telegraph. Religious reactions to telegraphy, like more general ones, appear to be mixed. One commentator from Damascus praised Sultan Abdülhamid for "grant[ing] a favor to all pilgrims by having a telegraph line installed between Damascus and Mecca." He continued, "The pilgrim may inform his family or they may inform him or inquire about him, in any place on the road having a telegraph branch. May Allah reward our master and sultan, the caliph, with the best of rewards. Amen!"[94] The sentiment here, although not expressed within the realms of the Ottoman Empire, is similar to one that opened this essay: as with the writer who compared Imam Shamil's divinely expedited passage to a telegram, this writer suggests that even among a more fundamentalist set, the telegraph was widely acknowledged as something miraculous.

An anonymous Ottoman cleric, probably from the late nineteenth century, picks up this thread in evaluating the miraculous developments of steamships, factories, and telegraphs.[95] This text is routinely marshaled by historians as vague evidence that such a religious reaction existed at all. Here I hope to use it to pull together some of the strands I have been drawing out: not as an authoritative interpretation of telegraphy, but as a plausible example of how a nineteenth-century thinker addressed issues of mediation and technological development from the perspective of *shari'a*.[96]

> Now, if you ask: "What is your opinion about the telegraph which the unbelievers have brought to light [in 1848–49 CE] and which can transmit

[93] Ahmed Baha, *Telegrafçılıkta Ana Dilimiz ve Mustafa Efendi* (Istanbul: Güneş Matbaası 1933), 14.

[94] Davison, *Essays in Ottoman and Turkish History*, 140. Originally in Jacob Landau, *The Hejaz Railway and the Muslim Pilgrimage* (Detroit: Wayne State University Press, 1971), 97.

[95] Translated in Rudolph Peters, "Religious Attitudes Toward Modernization in the Ottoman Empire: A Nineteenth Century Pious Text on Steamships, Factories and the Telegraph," *Die Welt des Islams* 26 (1986), 76–105. All page numbers that follow refer to this translation.

[96] Cf. Davison, *Essays in Ottoman and Turkish History*, and Bektaş, "Sultan's Messenger."

> messages from a far distance in a very short time to the extent that if one reflects about it, one is at a loss, because neither reason nor tradition can accept it?" [...] I answer with the help of God, Who is exalted: Certainly, but this also comes from God, Who is exalted. It belongs to the [category of] miracles, of which deceitful assistance is one of the varieties, which God has brought to light through sinners and unbelievers ... [I]t is there by God's creation and by His decree, by His knowledge and He has written it on the Tablet [i.e., *Umm al-Kitab*, the "Essence of the Book"]. Its qualification according to Divine Law is like that of correspondence, viz. like that of a letter. It can only furnish presumption and not certainty and it cannot serve as evidence for a legal verdict.[97]

The question, whether hypothetical or not, elicits a response that jumps almost immediately into the legal status of the telegraph: what would its qualification, or legitimacy, be in Islamic legal proceedings? Here, a series of mediations again emerge as the simplest explanation: the telegraph is like a letter in judicial settings. Furthermore, it was brought into existence by God's decree and (or because?) he has written it on the Tablet, or *Umm al-Kitab,*, the Essence of the Book where all worldly deeds and heavenly edicts are stored. Finally, it is a miracle but not necessarily a divinely sanctioned one. In his discussion of steamships, he responds in a similar sequence, which culminates in a comment on their legal status: "We have said according to Divine Law, because we are people of it, not people of secular law (*'ādah*)."[98] This distinction between *shari'a* (Islamic law) and *'ādah* (custom, or *'adat*) recalls the conflict in the Caucasus over traditional modes of lawmaking. *Tanzimat* opened up new forms of lawmaking, with the question of *shari'a* remaining central since the 1839 Gülhane Edict.

Yet the divine explanation prompts further considerations. Why, for example, did telegraphy not appear until the present? And what is the wisdom of its delayed appearance? The cleric responds at greater length to the latter, suggesting (among other things) that believers are being led astray not by the technology itself, but by the technology as a manifestation of greater power among the nonbelievers: "Therefore, if they see something like the aforesaid telegraph, they will flee away [from Islam]. Then they will accept neither treatment nor guidance. On the contrary, they are convinced that the unbelievers are people of merit and vision."[99] The cleric continues on, including the opinion that those who look at ships, steam engines, and telegraphs need a warning: "All of [God's] servants are incapable and can only do something through God's omnipotence. The use of tools, that have

[97] Peters, "Religious Attitudes toward Modernization in the Ottoman Empire," 93.

[98] Ibid., 90.

[99] Ibid., 94.

been created [by Him], belongs to Him and [also] the [mechanical] parts that have been created [by Him]."[100] Such tools, he continues, are essentially cheap imitations of godly creations, such as the human body, the heavens, and so on. On a related note, a believer should "be positive, as much as he can, with respect to everything he sees, hears, tastes and smells, that his creator and maker is God, and no one else."[101] Sensation, for this jurist, is both a site of contestation and also a baseline state enabling humans to judge the merits of technological mediations.

This response to the telegraph rehearses several key ideas: *shari'a* is not just a legal system but also a kind of measuring system, capable of taking on and evaluating any technological development (telegraph, steamship); although not identical to humans' use, God makes use of similar legal media/tools, like legal decrees and a tablet for recording; and finally, there are competing legal claims that can be made on the grounds of *shari'a*. On some level, *shari'a* begins to emerge not only as a way of evaluating new technologies; it also functions much like a kind of audio-visual technology. As Messick suggests (drawing on Marcel Mauss), *shari'a* can be understood as "a type of 'total' discourse," and one that emerges from a " 'discursive tradition that includes and relates itself to the founding texts.' "[102] But as mentioned previously, despite his avowed commitments to text, when Messick turns to his ethnographic sections, he begins with a chapter entitled "Audition"—the discourse must again be sonified. Along similar lines, Cornelia Vismann argues for conceptualizing (European) law as a kind of media network that emerges from unexpected sources.[103] In her case, she focuses on "files" as the archival impetus for law; but a similar argument could be made, both for *shari'a* and Western legal systems (with deep roots in orality), for a law emerging from sound. From this perspective, Islam shifts from being a "discursive tradition," as anthropologists like Talal Asad and Charles Hirschkind argue, rooted in sonic contestations about what constitutes divinely sanctioned behavior, to being something more akin to a "discourse network," Friedrich Kittler's notion of a writing-out system (*Aufschreibesystem*) that transforms such utterances into a media activity with its own rules for storing, processing, and transmitting them.[104]

These nineteenth-century examples of engaging with, against, and about *shari'a* highlight a back-and-forth play between Islamic sonic and visual cultures (which, of course, can never be fully isolated from each other). The "founding

[100] Ibid., 94

[101] Ibid., 99.

[102] Messick, *The Calligraphic State*, 3.

[103] Vismann, *Files*, 39–70.

[104] See Talal Asad, "The Idea of an Anthropology of Islam," *Qui Parle* 17/2 ([1986] 2009), 1–30; Charles Hirschkind, *The Ethical Soundscape: Cassette Sermons and Islamic Counterpublics* (New York: Columbia University Press, 2006); Friedrich Kittler, *Discourse Networks 1800/1900* (Stanford, CA: Stanford University Press, 1990).

texts" of Islam—the Qur'an, *hadith* oral traditions recounting the Prophet's life and deeds (*sunna*)—were sound before they were writing, and they continue to exist in deeply sonic ways, as Messick suggests in his account of Qur'anic recitation courses. The technological slippage happening during the Crimean War, as photography, telegraphy, and rail-based transit begin to take hold across a wide geography, created seams in the historical fabric. Thus the Caucasian Naqshbandis largely resisted the encroachment of technology and the Russian Empire through a deep devotion to *shari'a*. The Ottoman Empire, for its part, conceded the central role of *shari'a*, while pursuing other forms of order for the state.

I close with an image from a late-nineteenth-century Ottoman book of statistics on telegraph usage (Figure 2.3). The image—yet another remediation—depicts most of the communications and transportation technologies I have touched on here: trains, steamships, early telephones, and the telegraph. The

Figure 2.3 Techno-futurist telegraphic landscape, frontispiece from a collection of Ottoman telegraphy statistics for years AH 1299–1302 (1883–1886 CE). From Telgraf ve Posta Nezareti, "İstatistik: Telgraf kısmına ait" (Dersa'adet [Istanbul]: Matba'a-i Osmaniye, 1306 [1890]); in the public domain.

cityscape itself grows into these technologies. In a sense, it offers a view of these discourse networks, as urban architecture, religious space, and communications media collapse into a network of interconnected technologies and techniques. The image articulates a certain paradox about the relationship between *shari'a* and these media—they are always in flux and always diverse across time and place, and yet *shari'a* and media, law and audiovisual technologies, remain intimately interlinked. Long before the rise of networked culture, the Crimean War and other political upheavals of the era generated a rich conflux of legal-technical mediations, making it possible to revolutionize and reform these media cartographies.

3

Gunfire and London's Media Reality

Listening to Distance between Piano, Newspaper, and Theater

GAVIN WILLIAMS

> . . . [T]heirs was a paper empire: an empire built on a series of flimsy pretexts that were always becoming texts.
>
> —Thomas Richards (1993)

Acres of printed words issued forth from London's nineteenth-century printing presses, a thick seam that survives into our own time in ever greater accessibility. I begin with a fossil found along the coalface: a remnant of the popular music industry, a few unloved pages. Largely ignored since 1854, the year of its printing, the work was pressed into the British Museum Library's national archive, a copyright depository since 1814.[1] It was published on 6 November by sheet-music vendor and piano manufacturer Jewell & Letchford of Soho Square. An occasional work very much of its moment, this piano piece was the by-product of a newspaper story announcing Britain's first, victorious engagement in Crimea. Written for solo piano by one J. Mayer, a composer now virtually unknown, it bears the title "Grand Military Funeral March of the Battle of the Alma." The battle itself took place on 20 September, but news reached the metropolis only twenty days later, after a protracted journey over land and by sea.[2] Telegrams

[1] After 1814, British publishers were required to deliver a copy of all books for onward transmission to the British Museum. Following the Imperial Copyright Act of 1842, the Museum's reach was further extended. This new law required the "best issue of every book when it was first published" to be given to the Museum within one calendar month if published in London, within three months if published elsewhere in Britain, and within a year if published within the Empire's dominions. However, the farther from London publication took place, the more difficult enforcement of the law became. See Philip Rowland Harris, *A History of the British Museum Library, 1753–1973* (London: British Library, 1998), 148. See also Thomas Richards, *The Imperial Archive: Knowledge and the Fantasy of Empire* (London: Verso, 1993), 11–44.

[2] For the first installment of William Howard Russell's report on the Battle of Alma, see "From Our Special Correspondent," *The Times* (10 Oct. 1854), 7–8; an excerpt from Russell's report

had, of course, arrived much more quickly, but were not fully believed. In the wake of the Battle, further telegrams announced (falsely, as it would turn out) the capture of Sevastopol, prompting Britons to speculate that the Crimean War might already be coming to an end.[3]

Talk of war's end was smoothly replaced in the public mind by details of Alma's many casualties. In the wake of these reports, the "Funeral March" prepared to go to press.[4] But beyond its close relation to political events, more information is impossible to trace. We do not know where and when it was played in 1854. No sign of its reception survives, so far as I can tell: what chatter it may have generated may be forever lost. Like many other popular piano pieces of the period, it has enjoyed a peaceful entombment within Britain's national archive, remaining ignored and untouched since it was deposited there more than a century ago. This quick obscurity is no great tragedy, or even particularly surprising. Such loss is, after all, the rule for the kind of popular culture we have inherited from the nineteenth century, in which ripples both large and small tend to be short lived. Looking back now, though, we may be struck by the work's lively imbrication within Britain's popular, mediatized culture: evidence that seems strange, even uncanny, suggesting a fast-paced world now long dead. The environment within which the piece appeared offers clues as to its function and meaning. For one thing, the musicalization of a recent battle presented a quandary within what we would nowadays call public relations.[5] The decorative swirls and colored lithographs that usually adorned sheet music covers were banished: in their place came a stark, black-rimmed title page resembling a magnified death notice

appeared the previous day, "Arrival of the Wounded in the Bosphorus," *The Times* (9 Oct. 1854), 8; on the time taken for Crimean letters to reach England, see Elizabeth Grey, *The Noise of Drums and Trumpets: W. H. Russell Reports from the Crimea* (London: Longmans, 1971), 92–93.

[3] In early October, newspapers reported the fall of Sevastopol; see, for example, *Morning Post* (3 Oct. 1854), 4. However, three days later, they were obliged to admit that this was not in fact the case. For further discussion of this mistake, and its correction, see later in this chapter. *Morning Post* (3 Oct. 1854), 4; *Morning Chronicle* (3 Oct., 1854), 4–5; *The Times* (2 Oct. 1854), 6.

[4] The British Museum Library's entry stamp marks the "Funeral March" as received on 6 November 1854, but it must have been printed (and was perhaps being sold) earlier.

[5] As the "Funeral March" appeared on the sheet-music market, famous conductor Louis-Antoine Jullien presented a similarly topical "Alma Quadrille" at the Drury Lane Theatre. We can get a sense of the kind of snide comments that lighthearted commemorations of the battle could call forth from the review of a performance, published in *The Standard*: "The first performance took place last night, and there were multitudes of persons present to listen to the musical details of blood and slaughter, and applaud in proportion to the noise. [. . .] The bustle and motion of a fierce bodily struggle, mingled with peals of ordnance, and the shouts of victors, furnish M. Jullien with a caucus, which he fills up with a wonderful exuberance of detail; and there is every contrivable suggestiveness in the issue. The forces of the orchestra are of course multiplied for the occasion, and the power of sound, in its literal and material sense, can go no further." *The Standard* (4 Nov. 1854), 1.

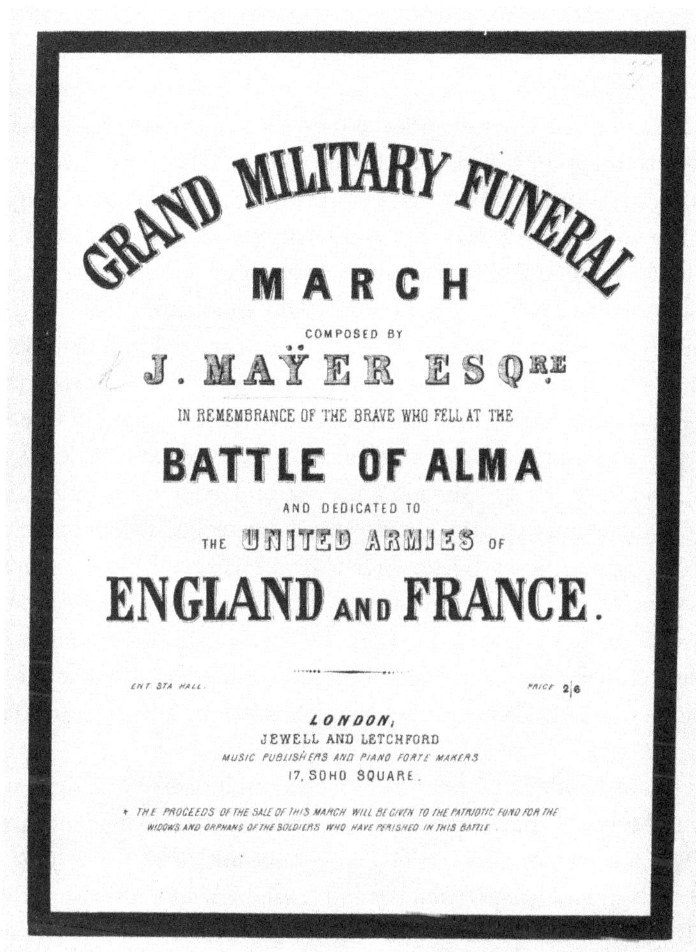

Figure 3.1 J. Mayer, "Grand Military Funeral March of the Battle of the Alma." Soho Square, London: Jewell & Letchford, 1854: cover image. © British Library Board h.723.n.(27.); reproduced with permission.

(see Figure 3.1).[6] Small print informed potential buyers that "the proceeds of the sale of this march will be given to the Patriotic Fund for the widows and

[6] Commemorative maps and illustrations were also published; see Ulrich Keller, *The Ultimate Spectacle: A Visual History of the Crimean War* (Australia: Gordon & Breach, 2001), 43. More widely distributed, and perhaps serving as a model for the March's typographical format, was the pamphlet containing the official list of the dead, wounded, and lost in battle, a publication that sold for a penny; see *List of Killed and Wounded at the Battle of the Alma on Wednesday, September 20, 1854, with the Official Sketch of the Battle Field, and Other Illustrations*, 5th ed. (London: Clarke & Beeton, 1854).

orphans of the soldiers who have perished in this battle"—perhaps an attempt to ward off anticipated complaints of opportunism and commercialism.

In some ways, this funeral march is an unremarkable example of the genre. It is couched in a stable minor mode and boasts a characteristically measured tempo; like so many funeral marches before and after, it is permeated by dotted rhythms; its harmony cleaves to the tonic, giving the usual sense of weight (see Figures 3.2a and 3.2b).[7] However, close details reveal a more unusual kind of musical mournfulness. The first bar initiates a repeated operation: a rising arpeggio is followed by a crashing minor chord whose resonance is enhanced by the sustaining pedal. This repeated booming gesture is glossed by a footnote: "The bass is a continual imitation of the sound of cannon."[8] These blasts punctuate every measure, although as the piece progresses they are varied in volume and register. Occasionally, the ominous arpeggio is inserted (mm. 6–7, 29–30) or removed (mm. 3–4, 25–26) to randomize the impact. In more lyrical sections (notably in the major-key Trio, not reproduced here, but also in mm. 5–8 and 14–21) the cannonade is quieter, higher in register, and hence—perhaps—farther away. Yet the booms threaten to intrude whenever the introductory arpeggios return (mm. 13–14, 17–18). The psychological mechanism behind these sonic cues is crude, the march becoming a macabre game and a virtual battlefield in which the listener tries to predict when the missiles might fall.

The survival of London's popular sheet music can broadly be attributed to the nineteenth century's archival impulse: the period's growing tendency to monitor, record, and memorialize itself.[9] Throughout this book we have seen how that archival impulse shaped sounds and memories across spaces of empire, selectively preserving, ignoring, and obscuring. By following paper trails emanating from London, we can home in on the ways such operations of memory played out at the heart of one particular empire. Sheet music, like much other

[7] For an in-depth discussion of the musical semiotics of the genre, see Lawrence Kramer, "Chopin at the Funeral: Episodes in the History of Modern Death," *Journal of the American Musicological Society* 54/1 (2001), 97–125.

[8] Elizabeth Morgan traces this piano gesture back to Franitsek Kotzwara's well-known battle piece, "The Battle of Prague," composed around 1788. See her "Combat at the Keys: Women and Battle Pieces for the Piano During the American Civil War," *Nineteenth-Century Music* 40/1 (2016), 7.

[9] Jürgen Osterhammel emphasizes the role of the archive in society's self-observation during the nineteenth century; see his *The Transformation of the World: A Global History of the Nineteenth Century* (Princeton, NJ: Princeton University Press, 2014), 1–44. By contrast, according to Thomas Richards's classic account, the archive should be primarily understood in terms of an imperial fantasy of control over distant parts of the world by means of data handling; see his *The Imperial Archive: Knowledge and the Fantasy of Empire* (London: Verso, 1993), 1–10. Both Osterhammel and Richards take inspiration from Michel Foucault's discussion of archival practices in *The Archaeology of Knowledge*, trans. A. M. Sheridan Smith (New York: Tavistock, 1986), 142–48.

Figure 3.2a J. Mayer, "Grand Military Funeral March of the Battle of the Alma." Soho Square, London: Jewell & Letchford, 1854, 1. © British Library Board h.723.n.(27.); reproduced with permission.

print ephemera, has usually been preserved as single copies of what were batch-produced items. In more recent decades, the wide circulation of this music has returned in the shape of digitization, a process that has made much printed music freely available online. More often, the priorities and budgets of libraries have meant that researchers encounter nothing but musical metadata, input

Figure 3.2b J. Mayer, "Grand Military Funeral March of the Battle of the Alma." Soho Square, London: Jewell & Letchford, 1854: 2. © British Library Board h.723.n.(27.); reproduced with permission.

by latter-day librarians and archivists, the labor sometimes outsourced to the so-called developing world. Such information has rendered music instantly locatable via salient details: its composer, title, publisher, place and date of publication, and so on. Encoded thus, vast swathes of popular music long since forgotten have returned to us, entering virtual spaces in which they begin to take on new contours.

In the context of the digital archive more broadly, musical metadata is a drop in the ocean: a fraction of nineteenth-century words available to us now. A useful point of comparison is with the period's newspapers, which represent a monumental achievement of early-twenty-first century digitization. In their virtual form, newspapers present us with huge, broken layers of information from the past.[10] Amid these countless millions of words, musicologists have to date been largely concerned with reviews of performances: literary, often formulaic accounts of concerts, operas and the like, although concert reviews are only the most obvious way in which the newspaper and music industries intersected.[11] Another angle is suggested by the "Funeral March." As we have seen, it was written in response to unfolding news; its dedication to the Patriotic Fund forestalled censure from the public and the press, although Jewell & Letchford probably also hoped for free publicity, since the Patriotic Fund was, for several months, the subject of a daily column in *The Times*.[12]

Digital searches I have attempted suggest—but do not prove—that this particular "Funeral March" was not among the many newspaper advertisements printed in 1854; nor does the company's name or the work's title appear in the long columns dedicated to contributors to the Patriotic Fund that year.[13] Yet the presence or absence of such data is ultimately less significant than the media environment within which it appeared. The "Funeral March" was a brief moment in a loop that began with the newspapers and ultimately aimed to return to them, and the sparks of recursion hint at larger patterns of the informational climate within 1850s news culture. On the one hand, we have an early phase in the history of the mass media: one that, according to Niklas Luhmann,

[10] However, the results of keyword searches on digitized documents may be far from comprehensive, with the accuracy rates of optical character recognition for historical newspapers varying widely between 71 and 98%. See Rose Holley, "How Good Can It Get? Analysing and Improving OCR Accuracy in Large Scale Historical Newspaper Digitisation Programs," *D-Lib Magazine* 15/4 (2009), <http://www.dlib.org/dlib/march09/holley/03holley.html>.

[11] There were countless newspaper ads for the sheet-music trade. For example, during winter 1854—the most intensive period of Crimean fighting—Jewell & Letchford's most publicized piece was "The Silvery Shower," a fantasia for piano that had little to do with the war in progress. An ad for the piece appeared in *The Times* throughout September and October 1854; see for example, *The Times* (27 Sept. 1854), 13.

[12] *The Times* was the newspaper with the greatest circulation at the outbreak of the war. During winter 1854, there were also regular updates on the Royal Patriotic Fund in the *Morning Herald, Morning Post,* and *Daily News*, as well as in the regional papers.

[13] At least one (perfunctory) advert appeared in a music journal; see "The Music Publisher's Circular," *The Musical World* 32/46 (London: Novello, 18 Nov. 1854), 765–66.

should be conceived as a socio-technical apparatus of cybernetic feedback.[14] On the other, the prominent role accorded piano music in London's media marks an point of divergence between the then and the now, opening up a space for the historical imagination. To put all this another way, the piano both was and was not a wartime medium; it came into physical and cultural proximity with visual- and text-based media such as newspapers, books, broadsides, maps, cartoons, and prints, sharing some of the burden of spreading news of the war, while seeking to conserve a degree of musical autonomy and apartness from the violence of everyday life.

These disparities between music, image, and text in 1850s media culture played out within domestic spaces. Sheet music and newspapers were often consumed in adjacent ground-floor rooms. Countless historians describe the solidly middle-class environments within which the nineteenth-century piano was to be found. The instruments were often luxurious and, in the 1850s, still largely handmade; the pianists were—or so we are often told—mostly women.[15] In nineteenth-century novels, women pianists all too often feature as social climbers and sexual self-promoters: mindless machinists who are simultaneously the objects of (hetero-) sexual consumption. The reality was, of course, rather different.[16] As Elizabeth Morgan has shown, focusing on the years of the American Civil War, piano music was composed and performed by women to a variety of ends. Female music making was of a piece with increased participation within the broader wartime economy, embodying patriotic commitment and sometimes enacting opposition to the war. In the same way, the "Funeral March" allows for a spectrum of female action and interpretation. Most straightforwardly, the title cues grief over the fallen men of the country, and perhaps also for more personal losses, or for one's own sake. Yet, the music might also have conjured up the battle, transforming its players into virtual bombardiers. And there is no reason to restrict this kind of vicarious participation, and the

[14] Niklas Luhmann, *The Reality of the Mass Media*, trans. Kathleen Cross (Stanford, CA: Stanford University Press, 2000), 1–9.

[15] While the social history of the piano in Britain has received a great deal of attention from scholars—the classic text here is Cyril Ehrlich's *The Piano: A History* (Oxford: Clarendon, 1990)—much less interest has accrued to the popular sheet-music industry the instrument supported. Some notable exceptions include Derek B. Scott, *The Singing Bourgeois: Songs of the Victorian Drawing Room and Parlour* (Milton Keynes, UK, and Philadelphia: Open University Press, 1989), 45–59; David Rowland, ed., *The Cambridge Companion to the Piano* (Cambridge: Cambridge University Press, 1998); Dave Russell, *Popular Music in England, 1840–1914* (Manchester, UK: Manchester University Press, 1998), 73–104; and Derek B. Scott, *Sounds of the Metropolis: The 19th-Century Popular Music Revolution in London, New York, Paris and Vienna* (Oxford: Oxford University Press, 2008), 15–37.

[16] See Ruth A. Solie, *Music in Other Words: Victorian Conversations* (Berkeley: University of California Press, 2004), 85–117.

performance of masculinity it entails, to women: such enactments were open to all middle-class performers and listeners at a remove from the battlefield.[17] In providing an opening for emotional participation, the "Funeral March" invites us to ponder afresh music's role within the 1850s news cycle: to ask why, in a news culture dominated by newspapers, musically simulated gunfire came to be a familiar virtual presence.

The archives surrender dozens more Alma-inspired compositions, including several that, although not engaging the battle topically, were published in its aftermath.[18] Almost all were written for piano.[19] They are mainly of the conventional "battle piece" variety, in that they commemorate a recent military achievement through a schematic narration of battleground events, and are thus connected to the tradition of eighteenth- and nineteenth-century battle symphonies. As Richard Will has pointed out, the battle symphony sought to draw listeners into formulaic retellings of an orderly, collective advance on the enemy. It called attention to precise rhythmic coordination among orchestral forces as a metaphor for lockstep discipline, leading to an inevitable crux: musical dissolution into chaos, often involving effects of gunshots and cannon fire.[20] This trajectory was also typical of nineteenth-century battle pieces for piano, their structure predicated on glorious victory (leaving mournful rumination to other genres).[21] Often strident, even triumphalist in tone, battle pieces were inextricably tied to news events—their bombast contained within more precarious cultural timetables. In this sense, battle pieces provide a musical analogue for what Mary Favret has described as the dislocating temporalities of wartime

[17] While I have not come across any accounts of contemporary piano music being exported, it is well known that print media in general flowed between Britain and Crimea. See, for example, Stefanie Markovits's discussion of John Dalbiac Luard's paining *A Welcome Arrival* (1857), which depicts the wall of an officer's mess covered in maps and prints; *The Crimean War in the British Imagination* (Cambridge: Cambridge University Press, 2009), 170–72.

[18] To list only some of them: W. D. Glyde and G. P. R. Pulman, "The Ballad of Alma" (London: Jewell & Letchford, 1854); Stephen Glover, "National Music Played at the Battle of Alma, the 'British Grenadiers,' 'Partant pour la Syrie,' 'La Marseillaise,' 'God Save the Queen,' and 'Rule Britannia'" (London: Charles Jeffreys, 1854); Charles William Smith, "The Alma Polka" (Liverpool, UK: John Smith, 1854); R. James and H. Austin, "The Battle of the Alma" (London: Jullien, 1854); William Wilson, "The Battle of the Alma Fantasia" (London: T. E. Purday, 1854); William Ireson, "The Alma March" (London: Campbell, Ransford, 1854); Charles Jefferys, "The Heroes of Alma" (London: Charles Jefferys, 1854).

[19] Quadrilles also sometimes included an ad libitum cornet part: for example, Philippe Musard, "Schamyl" (London: Campbell, Ransford, 1854). See also Scott, *The Singing Bourgeois*, 172.

[20] Richard Will, *The Characteristic Symphony in the Age of Haydn and Beethoven* (Cambridge: Cambridge University Press, 2002), 190–91.

[21] Ibid., 200–201.

at play within British culture more generally.[22] In reliving events not long past and usually far from home, battle pieces recruited visceral effects to inscribe a sense of war's distance.

Consider, for example, "Alma: A Battle Piece for the Piano Forte," written by Albert Lindahl, a prolific French composer of piano music. The piece was published by Jullien & Co. and released in London more or less simultaneously with the "Funeral March."[23] Triumphalism begins with the front cover (there would be no funereal restraint here: see Figure 3.3).[24] It shows British and French armies, historical foes until recently, advancing together up an improbably steep hill in their joint attack on the Russians. This maneuver became instantly legendary, perhaps because it was the most distinctive moment within newspaper coverage. The British played a supporting role in the advance: a fact reflected in Figure 3.3 by the French flag, prominent in the foreground.[25] Further off, plumes of gunpowder smoke rise; in the background well-drilled files of soldiers dot the valley floor.

On turning the page, this picture converts into musical storytelling. The piece begins quietly, as though from far away, with a heavily syncopated march-like tune, supported in the left hand by bounding leaps—a standard accompaniment figure, but one exaggerated here through reaching down into the instrument's lowest register. Further marking their unusualness, these cavernous notes (the low B-flat in Figure 3.4a) are not reinforced at the octave, or by any harmony notes, at least initially. Over the course of the piece, however, as the volume gradually increases and the troops figuratively advance, this deep register fills out. As with the "Funeral March," the score of "Alma" makes explicit that the low blasts should be "imitating cannons" (see Figure 3.4b). Chordal acciaccaturas send cyclical shockwaves through the instrument, also stressing—by way of slightly undermining—the synchronicity with the right-hand melody.[26] Following this iteration of the march theme, the volume suddenly drops and a surprising new melody begins, the Napoleonic hymn "Partant pour la Syrie," which initiates another long crescendo, this time culminating in the decisive encounter. Then follow yet more cannon, sounding alone but now interspersed with urgent,

[22] Mary Favret, *War at a Distance: Romanticism and the Making of Modern Wartime* (Princeton, NJ: Princeton University Press, 2012), 49–52.

[23] The score received the British Library's entry stamp on 9 November 1854, three days after the Mayer's "Funeral March;" Albert Lindahl, "Alma: A Battle Piece for Piano" (London: Jullien, 1854).

[24] On the visual representation of the Battle of Alma in prints, see Keller, *The Ultimate Spectacle*, 41–70.

[25] This maneuver was led by Algerian Zoauves; see Orlando Figes, *The Crimean War: A History* (New York: Picador, 2010), 209.

[26] This was a typical strategy from representing military lockstep in the battle symphony; Will, *The Characteristic Symphony*, 193.

Figure 3.3 Albert Lindahl, "Alma: A Battle Piece for Pianoforte." London: Jullien, 1854: cover image. © British Library Board h.1350.(1.); reproduced with permission.

darting scales.[27] The battle's conclusion is announced by horn signals and the victorious outcome affirmed by a medley of French and British national airs. To close, the march theme returns, but peters out in steady undulations in the low register as a final reminder of the blasting guns.

Beyond the piano works discussed so far, there were abundant, cannon-rich evocations of the Battle of Alma published and performed around this time. The

[27] Fast scales were conventional at moments of crisis in early-nineteenth-century melodrama; see Mary Ann Smart, *Mimomania: Music and Gesture in Nineteenth-Century Opera* (Berkeley: University of California Press, 2004), 64–65.

Figure 3.4a Albert Lindahl, "Alma: A Battle Piece for Pianoforte." London: Jullien, 1854: 1, excerpt. © British Library Board h.1350.(1.); reproduced with permission.

Figure 3.4b Albert Lindahl, "Alma: A Battle Piece for Pianoforte." London: Jullien, 1854: 2, excerpt. © British Library Board h.1350.(1.); reproduced with permission.

ubiquity of cannon fire is not in itself surprising: it was a conventional signifier of the battlefield in late-eighteenth- and early-nineteenth-century orchestral music and became a topos of the sublime.[28] What is more interesting here is the shift from orchestra to piano: a transposition of musical medium, involving a culturally specific complicity between noisy signifiers and news culture.

In the broadest sense, musicalized cannon fire called into being a national imagined community in a manner familiar in British cultural life since the eighteenth century (albeit one newly open to identification with French neighbors).[29] I have already noted the crude mechanism by which pianistic thuds fostered a sense of wartime patriotism, engendering emotional participation through performance of military masculinity. However, more fundamental interactions between musical actions and political events appear to have played out across the wider public sphere. In the wake of the Battle of Alma, the piano cannonade seems to have acquired a double signification—one poised between distinct modes of communication implied by "music" and the "news." Bringing sounds

[28] Will, *The Characteristic Symphony*, 197–98.

[29] Benedict Anderson, *The Imagined Community: Reflections on the Origin and Spread of Nationalism* (London: Verso, 2006), 48. As Linda Colley points out, Britain's patriotic imagined community was (and continues to be) inextricably bound to particular wars; see her *Britons: Forging the Nation, 1707–1837* (New Haven, CT: Yale University Press, 2005), 5.

of war into the home, domesticating them both physically and mentally as musical entertainment, was one way of understanding them in 1854: as a staged repetition of noises originally conveyed by newspaper reports that filtered through the metropolis. In response to wartime news, pianos were imaginatively retooled to provide a sonic analog to the verbal signifiers of noise being channeled into the home by textual media, musical mimicry enhancing and enabling the battlefield realities communicated by printed news.

This interpretation takes its cue from cultural theories that stress the ways in which mimetic operations bubble beneath the surface of objective systems of representation. For Homi Bhabha, as for many other critics, mimicry provides the productive difference that endows the original with its prior status, thereby facilitating repetition and identity.[30] Mimicry fixates on a conspicuous detail of the original, establishing a relation to the represented object that conserves a power to challenge its identity. In a journalistic context such as the Battle of Alma, sounds supplied a part-for-whole relation to reality as it was reported and imagined: sounds *of* the battlefield contained within them the potential to disrupt the representational order introduced by the wartime newspaper press, giving rise to an ironic discourse that revealed journalistic reportage to be without secure foundation.

Within the wartime news ecology, the reverse also held true: musical representation of gunfire helped determine the delicate sense of reality constructed by the news. Not only did pianos and pianists channel wartime news, but, more fundamentally, news was also shaped by them. This notion goes against a critical tradition that has prevailed since the 1980s: theorists more used to focusing on visual rather than sonic media have repeatedly shown that news media have been, and continue to be, important for generating shared "views" of the world: consensual and coercive pictures of reality.[31] In the case of wartime media, critical theory has tended to undermine the distinction between reality and its mediation. In Paul Virilio's celebrated claim, "the history of battle is the history of radically changing fields of perception." By "changing fields" he meant constant innovation in the visual technologies that have been used to render war visible, whether on the battlefield or elsewhere.[32]

Drawing inspiration from Virilio's ideas, and adapting their twentieth-century emphases for the purposes of the mid-nineteenth, Ulrich Keller has argued that

[30] Homi Bhabha, *The Location of Culture* (London: Routledge, 1994), 121–22. See also Michael Taussig, *Mimesis and Alterity: A Particular History of the Senses* (London: Routledge, 1993), xiii–xiv.

[31] See Jean Baudrillard, *Simulacra and Simulation*, trans. Sheila Faria Glaser (Ann Arbor: University of Michigan Press, 1994), 59–60.

[32] Paul Virilio, *War and Cinema: The Logistics of Perception*, trans. Patrick Camiller (London, Verso: [1984] 1989), 7.

the Crimean War was coextensive with its simulacrum in prints, cartoons, maps, and paintings: that the campaign took place not only on the battlefield, but also in the imagination of its spectators. He says, "armed conflicts are shot through with *signs*, and the processes of signification are shot through with *conflict*; war is, among other things, an aesthetic enterprise, and art, among other things, a site of battle."[33] Yet Keller's insistence that visual media continued the war by other means—in the imaginations and on the bodies of distant spectators—risks undermining radical differences between different wartime experiences. What is more, his position translates awkwardly if we are examining the war's sounds, which suggest a different priority. Rather than assert the identity between fact and fiction, we need to inquire after the codes that made them legible in the first place, as obviously fictional indices of battlefield sound, on the one hand, and as symbolically real in the context of an ongoing war, on the other.[34] In 1850s London, guns and their sonic signifiers (verbal and/or musical) functioned as fuzzy objects that were useful for making sense of news; they could alternately blur and sharpen culturally sanctioned distinctions between reality and its mimed repetitions. Musical, literary, and (as we will see) theatrical noises were not the same as the sounds of gunfire in Crimea, even while their vibrating materiality partook of a fascination with their deadly origin.

The reality-bestowing power of noise could be felt, and heard in the imagination, in news stories such as those that reported the Battle of Alma. My opening paragraphs have already mentioned these stories, but to get at their sonic dimension we need to immerse ourselves in once sensational though now obscure details. As noted earlier, news of Britain's victory at Alma was preceded by false reports of the capture of Sevastopol. The *Morning Post* announced the city's fall on 3 October in a leader column on page 4: a column that, according to the usual format, followed the advertisements, domestic/commercial/shipping news, the weather forecast, and the theatrical listings.[35] The article reported 18,000 Russian casualties and the taking of 22,000 prisoners (reports in the coming days revealed both figures as vastly inflated). But the number of British and other casualties remained unknown. Tentatively—given the uncertainty over the scale of national grief to come—the *Morning Post* urged the country to look

[33] Keller, *The Ultimate Spectacle*, xiv, original emphasis. Later on Keller reformulates a similar position: "The middle class addiction to visual sensation was the motor which charged authentic reportage, in spite of itself, with volatile surplus values." Ibid., 38.

[34] In his study of cinema sound, James Lastra has suggested along similar lines that an investigation into sense-making (and in some cases truth-telling) is more pressing than denouncing mediatic representation as inherently falsifying. See his *Sound and the American Cinema: Perception, Representation, Modernity* (New York: Columbia University Press, 2000), 15.

[35] *Morning Post* (3 Oct. 1854), 4.

forward to peace and restored prosperity. It cautiously suggested that business as usual would soon be resumed throughout the Empire, while in Sevastopol, there ought to be a brisk clean-up operation, before British soldiers quit the region:

> When we have done with the débris of the fortress, have cleared off our men, and disposed of our prisoners, we take it that we shall leave the shadow of peace to smile over the spot where now the quick strife of war follows the clang and *dash* of our soldiers.[36]

However, four days after the *Morning Post* had reported the capture of the town, that same newspaper was obliged to retract, or rather change location: victory had been achieved on 20 September, but at Alma rather than Sevastopol. As the *Post* explained, the error was due to a telegram:

> The truth is, that the inventor [of the story] has signalised himself in a manner which he only failed to make famous by the omission of his name, and that all the world may fairly be included in the list of believers in an achievement which the character and *dash* of the assailants made but too probable. The amended account of what did happen at Sevastopol is less plain than was the account of what, as it is proved, did not happen.[37]

After this tortuous apology followed a series of renewed speculations. If Sevastopol had not yet been taken, then surely it was about to be, or perhaps it was being taken as the paper went to press. Such uncertainty stimulated journalistic invention: "[a]fter all," the *Post*'s editor admitted, "in our present imperfect information, it is all imagination."[38]

Not until a few days later, when *The Times* published a blow-by-blow account from "special correspondent" William Howard Russell, were the events of 20 September established—fixed in the condition endlessly reported by historical

[36] Ibid.

[37] Ibid. In this quotation, as in the last, the mythical "dash" of British soldiers departs from the word's traditional association with ranking officers and the aristocracy. We can track this changing meaning through the writings of Joachim Hayward Stocqueler, a contemporary journalist and entrepreneur, who suggested that dash was particular to Western soldiers: "For their system of military discipline, the Russians have to thank their Emperors. It makes soldiers, but fails to make them think and act like men. The soldier who is ignorant of the existence of everything outside his own company or division, can never have heroic courage, that self-confidence, that 'dash,' which distinguishes the soldiers of the West in the present struggle." Stocqueler, *The British Soldier: An Anecdotal History* (London: Orr, 1856), 281.

[38] *Morning Post* (3 Oct. 1854), 4.

accounts even today.[39] It is worth rereading Russell for the audiovisual coordinates that enliven his report. As he recounted from within the British camp, the day began before dawn with the rousing of the army. The reveille did not sound: "They were marshalled silently; no bugles or drums broke the stillness, but the hum of a thousand voices rose loudly from the ranks, and the watchfires lighted up the lines of our camp as though it were a great town."[40] Tens of thousands of British and French troops marched along the Crimean coast, shadowed at sea by huge warships, until they arrived at the delta of the River Alma. Across this river, and high above them on a steep mountainside, was the Russian front line.

To create vividness in reporting on unfamiliar terrain, Russell compared the Alma delta with Richmond Hill, a site more familiar to Londoners.[41] He invited readers to picture the enemy stationed on top of the mound and facing the Thames, and—adjusting for scale—imagine the river "shrunk to the size of a Hampshire rivulet."[42] By placing these well-known (and commonly represented) environs before his readers' imaginations, Russell primed their senses for the battlefield actions he was about to narrate. Allied armies advanced across the flat land to the north. The generals rode in front, loudly cheered by the infantry. At 1:20 p.m. the first shells were launched from French ships; the Russians responded with heavy fire. Less than half an hour later, as the British began to descend into the valley, French and Turkish troops were already scaling the heights, making a surprise attack over a ridge—this was the famous attack depicted in Figure 3.3. But the climax of the battle came later in the afternoon, when the Guards crossed the river and began to storm the mountain:

> Their line was almost as regular as though they were in Hyde Park. Suddenly a tornado of round and grape rushed through from the terrible battery, and a roar of musketry from behind thinned their front ranks by dozens. It was evident that we were just able to contend against the Russians, favoured as they were by a great position. At this very time an immense mass of Russian infantry were seen moving down towards the battery. They halted. It was the crisis of the day.

[39] See, for example, Andrew Lambert, *The Crimean War: British Grand Strategy against Russia, 1853–6* (London: Ashgate, 2013), 150–54.

[40] "Our special correspondent," *The Times* (10 Oct. 1854), 7.

[41] While views of the war proliferated in maps and prints, Londoners would have to wait until December for the first Crimean panorama. The earliest was Robert Burford's Panorama in Leicester Square, where an impressive Battle of Alma opened in time for Christmas 1854. See Keller, *The Ultimate Spectacle*, 60–61.

[42] "Our special correspondent," *The Times* (10 Oct. 1854), 7.

Coming face to face with the enemy, Britain's Lord Raglan was forced to make a momentous decision. He chose to advance, swiftly bringing the battle to a conclusion and saving countless lives (or so Russell reported, in a rare moment of approbation for the army's commander-in-chief). When the British dead and wounded at Alma had been counted, their number was, Russell reported, just below 3,000.

With that imprecise figure, Russell's chronicle comes to a halt. His account presents readers with an overwhelmingly detailed narrative in which a huge cast of officers and (typically anonymous) soldiers—also horses, guns, and bullets—sporadically appear and then disappear amid the tumult. The boom of cannon, the whizz of round shot, the bursting of shells, with the noises associated with grape, Minié musketry, canister and case shot: all these are common within the unfolding events, sometimes emerging as the active subject of sentences. In the long passage cited a moment ago, a tornado of munitions "rushed" from the Russian battery, while the roar of musketry "thinned" the British line. These sound effects are intent on distracting us from human actors firing weapons. Our attention is diverted instead toward the impression the scene has on the columnist-observer. A journalistic reality effect is achieved through bearing witness to a hostile and unreadable multiplicity of bullets.

Russell rarely if ever mentions a high-ranking officer doing anything so unworthy as firing a gun. The class association of guns is conspicuous, leading us to wonder whether his focus on bullets and their noises stands in for the impossible-to-represent collective acts and experiences of the common soldiery.[43] After all, this representational dilemma was fundamental to elite war narratives in which, at least initially, heroic generals were said to have inspired troops to victory.[44] It was a political exclusion, of course: one premised on the notion of a country led into war, as elsewhere, by the upper classes. What's more, it was an ideology that became untenable as the war went on, as the ordinary, gentle but manly soldier emerged as the major player in Crimean War stories.[45] It was partly through the writings of Russell, and other pioneer war correspondents such as Thomas Chenery, that these alternative stories became

[43] Yuval Noah Harari detects a paradigm shift in the representation of the common soldier during the late eighteenth century, as "Western culture began for the first time to solicit and listen attentively to the authentic voices of the common soldiers themselves." This new receptivity to soldiers' accounts coincided with the rise of what Harari's calls the "revelatory" interpretation of war, according to which the experience of battle granted soldiers access to higher spiritual and moral truths—truths barred to those who were not present. Harari, *The Ultimate Experience: Battlefield Revelations and the Making of Modern War Culture* (London: Palgrave, 2008), 190.

[44] Keller, *The Ultimate Spectacle*, 1–40.

[45] See Holly Furneaux, *Men of Feeling: Emotion, Touch and Masculinity in the Crimean War* (Oxford: Oxford University Press, 2016), 1–21.

well known. But for now, ordinary heroes would have to wait. Around the time of the Battle of Alma, low-born actions tended to be hidden within the crowded auditory channel. Much more conspicuous, and incongruous, was the bravery of the journalist and the kinds of looking and listening he performed in the act of bearing witness.

The figure of the war journalist arose as both target and cipher on London's travesty stages in late 1854. As a wartime medium, theater proved highly responsive to developments in Crimea, and so it was perhaps inevitable that the (relatively) novel presence of noncombatant battlefield correspondents would come in for satire and parody. Yet these on-stage journalists had strange, often humorously confusing implications for theatrical economies of representation. Particularly interesting for the purposes of this chapter is the synergy between journalists on stage and theatrical sound effects, especially when it came to representing battlefield noise. Sooner or later all of the city's theaters offered a Crimean spectacle; all of these shows were, to judge from the stage instructions, saturated with imitations of cannon fire.[46] Among the first to appear were shows at the Adelphi, Strand, Victoria, and Britannia Saloon, which were prompt to respond to Britain's declaration of war in March. In parallel with the music industry, theatrical productivity dropped over the summer, to be revived by news of Alma later in the year. The fresh tranche of plays made in response to Alma ranged from sentimental melodramas, such as the Surrey's *Bond of Love*, to the Marylebone's farcical *Sebastopol from Our Own Correspondent*, a play that led a trend for satirizing the presence of journalist noncombatants on the battlefield.[47]

The Battle of Alma, at Astley's Royal Ampitheatre, followed suit in placing a journalist in a prominent role. The theater boasted the definitive Crimean show.[48] Astley's preeminence lay in its size—it could accommodate 2,500 spectators—and its specialization in military-equestrian shows: it was here that

[46] See for example, "The War in Turkey, or The Struggle for Liberty" (Apr. 1854, Britannia Saloon, British Library MS 52946 KK); R. B. Borough, "The Overland Journey to Constantinople" (Apr. 1854, Adelphi Theatre, British Library MS 52947 C); "Bartelemy" [Barthélémy Deville], "Le bombardement d'Odessa" (June 1854, Soho Theatre, British Library MS 52947 V); E. Stirling, "Sebastopol from Our Own Correspondent" (Oct. 1854, Marylebone Theatre, British Library MS 52949 Y); J. P. Simpson, "Schamyl, the Warrior Prophet" (Nov. 1854, Princess's Theatre, British Library MS 52950 D); F. F. Cooper, "The Soldier's Wife" (Nov. 1854, Strand Theatre, British Library MS 52950 W).

[47] The title of the Surrey's play was changed from *Bond of Love* to *Brothers in Arms* to *The Battle of Alma*; see William G. Knight, *A Major London "Minor': The Surrey Theatre, 1805–1865* (London: Blot, 1997), 253.

[48] Jacky Bratton, "Theatre of War: The Crimea on the London Stage," *Performance and Politics in Popular Drama,* ed. David Bradby, Louis James and Bernard Sharratt (Cambridge: Cambridge University Press, 1980), 119–38.

the Battle of Waterloo had been more or less consistently commemorated in theatrical simulation during the first half of the nineteenth century.[49] *The Battle of Alma* bid to outdo this precedent, its hyperbolic billboard promising accuracy of "costume, scenery, properties, decorations, and mechanical effects," as well as reproduction of the battle's "complicated 'materiel.' "[50] There was to be a specially enlarged 700-foot stage to accommodate unseen numbers of troops and horses. Military auxiliaries included dozens of actively enlisted soldiers from the 1st Royal Fusiliers, while the band of the Coldstream Guards supplied an authentic musical soundtrack.[51]

Our trickster journalist makes his appearance near the outset of the show, which opens with a scene staged for real only months ago, as the Guards play the popular song "The Girl I Left behind Me" and British troops embark a steamer at Southampton docks. Aiming at the heartstrings, an emotional farewell between a private and his mother, father, wife, and young daughter (whose only line is "Goodbye Daddy") ensues; and this scene is complemented by a lighthearted episode featuring Biddy Flanagan, a comic Irish woman, who recalls the sixteen husbands she has lost to the wars.[52] Soliloquizing to the side of the stage, the journalist—yet to reveal himself as such—interjects, "Touching Scene! Must make a note of it. At this moment the signal is given to embark when an interesting Irish Female um-um-um." Stage instructions record that, as his speech tapers off in ums, the journalist "writes in his pocketbook." The purpose of his scribbling, as yet undisclosed, is revealed shortly afterward when the British commander asks the "person in Civil Garb" who he is:

> My name is Montague Quillet Esquire by courtesy. My profession is literature. In fact, I'm a man of letters, a humble follower of Johnson, Hume, Pope and so forth, called by patriotism and the personal necessities of the hour, which are not worth mentioning. I have accepted the appointment of our own special correspondent at the seat of war, and I am anxious to be permitted to embark with the brave Army, share its toils, and record its glories. [. . .] My Lord I'm the humble but I hope the intelligent and faithful representative of the Encyclopedia of War and

[49] Helen Stoddart, *Rings of Desire: Circus History and Representation* (Manchester, UK: Manchester University Press, 2000), 39.

[50] The billboard was quoted by the review of the show in *Morning Post* (24 Oct. 1854), 4.

[51] Ibid.

[52] This script for the play, like all the others mentioned previously, survives in a handwritten version that was submitted for censorship. *The Battle of Alma* was received by the censor on October 21 1854; a license for the play to be performed was granted the same day. British Library, Add MS 52950 H

> the Illustrated Blood and Murder Penny Herald, happy to book you for a week's subscription.[53]

As Quillet and the soldiers make their way from Southampton to Constantinople, the scene cuts to the palace of Prince Alexander Sergeyevich Menshikov, the Finnish-Russian commander much maligned in the British press of the time. Like newspaper readers, Astley's audiences were presented with a Menshikov reckless, delusional, and eager for glory. He repeatedly shuns the appeals of his wife, who begs him not to resist the combined might of Britain and France, so committing a crime against humanity. However, Menshikov's greatest offense, at least in the play, is in sending false dispatches of surprise victories to the Tsar, who bestows honors on him and his men. Because of Menshikov, the Russians are plagued by a constant flow of misinformation—a view of the enemy perhaps calculated to reassure British audiences concerning their own imperfect knowledge of happenings abroad.[54] Through the constant referencing of news, both in Britain (scene 1) and Russia (scene 2), Menshikov comes to be formally contrasted with Quillet, thus setting up the usual tension between the villain and the comic, which later becomes a moral contest over the abuse of information.

A close interplay between fiction, recent history, and the news continues throughout the play: next stop in the recap of "real" events is Gallipoli, where the British passengers disembark at the Ottoman camp. What happens here can illustrate the complex nature of music and sound within this particular economy of theatrical representation. After some awkward mingling between British and French soldiers, a member of the Guards launches into a song in tribute to the alliance—the first component in a song-and-dance routine completed by a troupe of so-called Circassian girls. While further details of their "Grand Dance" are not supplied by the play's manuscript, its narrative framing—by an onstage audience of ogling British soldiers—makes clear its status as exotic entertainment. As the Circassians begin their dance, Quillet picks up his notebook and exclaims, "Here's a Scene for my new Spectacle!"—exhorting Astley's audiences to imagine the music and dance they are about to witness (and, by extension, *The Battle of Alma* as a whole) as a fictional byproduct of his "real" activities as a news correspondent.

My summary has so far stressed the dynamic absorption of contemporary news culture into the theater. As might already be obvious, what was absorbed at

[53] Ibid.

[54] One scene featured him premeditating false telegrams: "We shall crumple them up in a week. While Gorschikoff [based on Prince Mikhail Dmitrievich Gorchakov, Russian commander-in-chief in Moldovia and Wallachia] dates his dispatches from Calcutta to announce the conquest of the East I shall date mine from the Bank Parlour in the City of London."

Astley's was not so much information already conveyed by newspapers, but the epistemological dilemma brought about by such knowledge. The doubt attending happenings abroad was an urgent problem in staging the play: at least two changes were made to the title and several substantial rewrites undertaken as uncertain news gradually became clear.[55] Within the play, as already mentioned, the untrustworthiness of media tends to be displaced onto the enemy and *their* regressive political system, while the British press—an embodiment of democratic if not liberal values—is seen to save England from a similar fate. These broader cultural aspirations emerge forcefully in the final act, which converts Quillet from a figure of fun into the play's unlikely hero. His moment of glory comes just before the final battle tableau. Taken prisoner by Menshikov, he protests by refusing to eat, declaring sympathy with starving Russian soldiers:

> I would crave my liberty—As for the delicacies which your bounty has prepared for us I feel that it would ill become "our own correspondent" of the Illustrated Blood and Murder Penny Herald to eat of the fat of the Land while the brave Russian Army is condemned exclusively to the lean.

Thus *The Battle of Alma*—along with some other contemporary plays in London's theaters that put journalists in starring roles—exploited the novelty of the battlefield correspondent toward political and comic ends. Yet I want to suggest that Quillet can also serve as a reminder that what audiences were witnessing—the clamor of the fighting—had a reality outside the theater, and that it had been brought to Astley's via the newspapers.

In this half-light of the mediated public sphere, noisy special effects were particularly useful. On the one hand, such sounds encouraged audiences to imagine the battle being placed before them. As historian Jacky Bratton has argued, Astley's cultural role was to create "an image of the event which became its reality in the popular imagination."[56] On the other, these sounds pointed beyond themselves, and we can see that the spectacle provided Londoners with something other than an exciting simulacrum. Through the eddying of journalism and theatrical effect, sounds were constantly suturing fiction to known events—and making them "realer" through contradistinction with their travesty on stage. Another case of this strange mimetic process followed from Quillet's moral triumph over Menshikov. A cannon booms and several gunshots are fired offstage: Menshikov cries, "Ha! The Battle has begun!" A Cossack officer then announces that French

[55] Bratton, "Theatre of War," 128–29.

[56] Ibid., 130.

troops have already climbed the ridge near the Russians and "their ships of War bombard our position"—an echo of Russell's report of what had only recently taken place on Alma's heights. The officer's yell (like Menshikov's knowing cry) smacks of theatrical contrivance, to be sure. Yet it also cues an inversion of perspective, encouraging spectators to imagine the advancing British army through enemy ears and eyes.

Further echoes of newspaper coverage can be detected in the unusually specific instructions for special effects in the closing battlefield tableau:

> The heights are seen crowded with Russian Artillery and Infantry—Ladies—and the Carriage of Prince Menshikov are also seen. Russian Riflemen descend from the heights—Cavalry ditto—and scour the stage and Arena—but cautiously as if watching the coming enemy—after a time the music changes to "the British Grenadiers" and then a French March—Cannon is heard—the Russian troops retire rapidly—the French and English troops enter the arena—the Rifles and Chasseurs leading it open order—And firing upon the retreating Russians—then the Cavalry—the Line—the Artillery—then more of the line—After a few manoeuvres they are formed and the Battle begins—The whole advances in line—firing as they go—the Russians come down again—A Grand Struggle between Cossacks and Cavalry—the Cossacks are driven back. As the English and French advances—the wounded are brought to the rear—the women and sailors attend them—band up their wounds and carry them off—Attack dies—A Caravan is dismounted—in fact all the incidents of war must be observed—Finally the British Commander appears to consult with the French General and one Grand Charge is ordered up the heights which are carried and cleared and the English and French colours hoisted amidst loud cheers and God save the Queen.[57]

In this last scene, the complicated "materiel" promised on the show's billboard was evidently on display. The *Morning Chronicle* reported that the "piece concludes with the storming of the heights of Alma, which, after a due

[57] J. H. [Joachim Hayward] Stocqueler, ""The Battle of the Alma: Grand Military Spectacle in Two Acts," British Library, Manuscripts Collection (52950 H). A full catalogue of theatrical manuscripts stored by the British Library and received for approval by the Lord Chamberlain between 1852 and 1863 is available online: <https://www.royalholloway.ac.uk/dramaandtheatre/research/researchprojects/lordchamberlainsplays/thelordchamberlainsplays,1852-1863.aspx>, accessed 25 Apr. 2015.

consumption of gunpowder and a terrific loss in killed and wounded, are gallantly taken at the point of bayonet."[58] Along similar lines, though lavishing more attention on the kinds of ordnance on offer, the *Morning Post* wrote that in this last scene the firing "becomes formidable, and, amid the roar of musketry and the exploding of mines, shells are thrown from one side to the other."[59] We can round out this soundscape by adding noises mentioned in the script: the shouts of soldiers, the clatter of hooves, the singing of the national anthem by chorus and perhaps audience, and of course the musical contributions of the Coldstream band.

We can only guess what contributions the band made, or what role their music might have played in relation to other sounds in the theater. Looking back with twenty-first-century ears, we might try to feel our way into the past by imagining the Coldstream band as a film soundtrack. In such an audiovisual scenario, music embraces other sounds within the fictional world while occupying a space just outside it. Sometimes labeled non-diegetic music, it functions as a binding agent, drawing into itself the disparate sounds occasioned by dialogue, props, and other noises. The Coldstream band might conceivably have fulfilled this binding function. But there are many other possibilities here. As I have noted, "real" bandsmen brought military music into the theater, embodying the movement of sound from Crimea into the show. And this embodiment in turn enabled—or at least allowed for the possibility of—an imaginative projection back there. The Coldstream's music, like noises produced by elaborately described bullets, propelled the show 3,000 miles east, weeks into the past, to a remote Crimean elsewhere.

Within the crucible of Astley's Theatre, the disparate sounds examined in this chapter came into contact. *The Battle of the Alma* made audible the battlefield reported by newspapers, and in the process was itself transformed into a complex mediatic node (if not a news medium in its own right). Drums, bugles, gunshots, bands, and marches—all conventional military signifiers—signaled toward a distant reality as they were joined with plots that playfully turned on media emblems: false telegrams; the earnest pronouncements of a journalist; the shout of a Cossack. There are both similarities and important differences between the kind of reality effect created here, in the theater, and the representation of gunshots in piano music and in journalism. In the theater, sound stimulated the imagination of news in a general sense; on the page, and in the home, sound functioned more as a phatic index, as particular

[58] *The Morning Chronicle* (24 Oct. 1854), 3.

[59] *The Morning Post* (24 Oct. 1854), 4.

noises impossibly attempted (as though urged on by a journalistic moral imperative) to mark the fact that "this" happened, there and then. Yet comparisons of this kind are, in the end, less important than the networked relationships between them: the links between newspaper, theater, and printed music—between eye, ear, and finger—which elevated gunshots to a pervasive cultural theme.

What can such a network, the evidence of such sensorial networking, tell us? Can it say anything about a broader condition of wartime as it was felt? Sound as an abstract quality or generic domain of experience means little here, or is endlessly fractured by the different audiences and the different kinds of listening engendered by a newspaper article, a piano piece, or a military-equestrian melodrama. Yet, across these coarsely linked sites of representation, if in dissimilar ways, particular sounds may have served a vital role. This may be due to the ambiguity inherent in the resonant materiality of signifiers of sounds, an ambiguity that can be productive. Whether sonic signifiers are verbal (perhaps silently read and imagined), musical, or even sounds themselves, such ambiguity can allow us to break with the story, to crumble distinctions between here/there and then/now, even to create a bridge between spaces and times. There were no doubt multiple motives in opening sonic portals of this kind during the Crimean War, not all of them aimed at promoting empathy and cosmopolitanism. Indeed, the fetishistic nature of bullets and their sounds in Russell's journalism—as in *The Battle of Alma* and in the piano pieces I have discussed—may lead us to suspect that gunfire held less than ethical fascinations for mid-century British Victorians. Through its insistence across and between diverse sites, these sounds seem to grasp at the untouchable, to attempt to take command over the dangerous power of the war's munitions.

And so to say that sound mimicked the battlefield can be only part of the story. Mimicry fluidly converted into mimesis: into what Michael Taussig once described as the power to bring distant things close by way of their replicas. For Taussig, mimesis is "the faculty to copy, imitate, make models, explore difference, yield into and become other"—it constantly reminds us of the entanglement of subjects and objects, a prior state of mixture that is awkward for rational systems of knowledge to accept, or even acknowledge.[60] So it is that post-Enlightenment subjects constantly disavow their own mimesis, projecting its allure onto primitive peoples and *their* savage ways of thinking.

> Whether taking its flight through the air, or encountering more solid obstacles, a round shot of course must be always obeying strict, natural

[60] Taussig, *Mimesis and Alterity*, xiii.

> laws, and must work out the intricate reckoning enjoined by conflict of power with absolute, servile exactness; but between the "composition" of "forces" maintained in our physical world and the fixed resolve of a mind made up under warring motives there is always analogy, with even sometimes strange resemblance; and to untutored hearers a formula set down in algebra would convey less idea of the path of a hindered, though not vanquished cannonball than would the simple speech of a savage who, after tracing its course (as only savages can), has called it a demon let loose. For not only does it seem to be armed with a mighty will, but somehow to govern its action with ever-ready intelligence, and even to have a "policy." The demon is cruel and firm; not blindly, not stupidly obstinate.[61]

This was how Alexander Kinglake described the experience of coming under fire in his magisterial, eight-volume *The Invasion of the Crimea* (1863–87), which remains the most important English-language chronicle of the war. The "strange resemblance" he perceived between the phenomenology and the physics of a flying cannonball sheds light on the sounds I have been unearthing in this chapter. For while sound does not feature much in this passage, the unintelligible yet perfectly comprehensible speech of "a savage" signals a bizarre, now alien, representational strategy for missiles and the damage they do. More than a period detail of British imperial consciousness, Kinglake's primitive voices betoken an awareness of the fate of bodies in industrial warfare.

A point of comparison may be with the deranged speech of British, German, and other soldiers returning from the First World War: a pathological utterance understood through medical discourses of traumatic neurosis and male hysteria.[62] Such categories and definitions were incipient at the time of the Crimean War in the shape of industrial diseases such "railway spine," but had yet to be transferred to battlegrounds of Alma, Inkerman, and Sevastopol. Historical and political conditions were not conducive to mapping industry onto war, and were to remain so for more than a century: campaigns for the medical recognition of trauma caused by industrial warfare continued intermittently until well after the Second World War. Not until 1980, in the wake of the Vietnam War, was post-traumatic stress disorder (PTSD) accepted as a clinical concern by medical

[61] Alexander Kinglake, *The Invasion of the Crimea: Its Origin, and an Account of Its Progress Down to the Death of Lord Raglan*, vol. 8 ([new ed.], London: Blackwood and Sons, 1888), 151.

[62] Daniel Lerner, *Hysterical Men: War, Psychiatry, and the Politics of Trauma in Germany, 1890–1930* (Ithaca, NY: Cornell University Press, 2003), 61–62.

institutions such as the American Psychiatric Association. The modern notion of trauma is glossed by historian Ruth Leys:

> . . . owing to the emotions of terror and surprise caused by certain events, the mind is split or disassociated: it is unable to register the wound to the psyche because the ordinary mechanisms of awareness and cognition are destroyed. [. . .] The experience of the trauma, fixed or frozen in time, refuses to be represented *as* past, but is perpetually re-experienced in a painful, dissociated, traumatic present.[63]

As Leys observes, PTSD, both as an illness and as a wider representational strategy, has by now become pervasive within many cultures across the globe. Not confined to a set of bodily symptoms, PTSD is "fundamentally a disorder of memory," one that deeply effects our narratives and experiences of wartime today.[64]

Back in the 1850s, trauma may thus seem to us conspicuous by its absence. Of course, it is not hard to find people who were horribly traumatized by the war; there are even those who exhibited classic symptoms of trauma, such as hypervigilance and loss of affect.[65] But in an age before "trauma" there were different representational strategies to accommodate these people: other means by which to select, inscribe, and recall battlefield experience. I have already mentioned the nineteenth-century British political context that initially favored representation of heroic deeds of the upper classes and only later in the war titled toward non-elite experience. Holly Furneaux has demonstrated the crucial role of gender in shaping these Crimean War narratives, which recast military masculinity in stories that emphasized acts of kindness and self-sacrifice.[66] Yet another way in which pre-traumatic experience of the battlefield could be narrated, as we have seen, was through sound: through a networked web of medially differentiated symbols that both recognized and disavowed battlefield realities.

A broad conclusion for this chapter, and media network it has unearthed, is that sounds carried implications for wartime memory, giving structure to contemporary battlefield experience through its dialectical relationship with distant sensations and archives. While these "experiences" are unrecoverable, then as

[63] Ruth Leys, *Trauma: A Genealogy* (Chicago: University of Chicago Press, 2000), 2.

[64] Ibid.

[65] James J. Reid, *Crisis of the Ottoman Empire: Prelude to Collapse 1839–1878* (Stuttgart: Franz Steiner, 2000), 406–15, 431–35.

[66] Furneaux, *Men of Feeling*, 1–21.

now, we can observe the ways in which Britons in the 1850s dealt with this representational crisis as a material practice. In other words, wartime experience and its narration was transformed through printed music and other memorial matter, which inscribed and archived events of their moment. Copiously if not compulsively churned out, such printed matter attests to a formidable archival impulse in British culture of the mid-nineteenth century—one not specific to, but nevertheless spurred on by the Crimean War. Yet amid the mountainous buildup of paper, we can also observe the workings of wartime memory on a much smaller scale, sometimes in minute peculiarities of documents.

I am conscious that this mode of uncovering larger habits of mind in obscure details will not appeal to everyone who reads this chapter. I have already given plenty of examples in this vein. Yet consider one last piece of music—this time a song with piano accompaniment—published as the Crimean War entered its final stages in early 1856. By this point, victory for the allies seemed all but guaranteed, and already we can detect the campaign beginning to slip into the historical past. In the world of sheet music, as elsewhere, the dominant mood began to shift (somewhat ahead of unfolding events) from celebration to reminiscence: countless commemorative songs were published whose lyrics referred to episodes of the war in the past tense.[67] Among the many was "Whistling Dick," a song that recorded in a mock colloquial idiom the characteristic cry of the battlefield:

> We thought it sport, as from each port
> The shells flew pretty quick,
> 'My eyes,' cried Bill, 'Look out my boys
> For here comes Whistling Dick!'

Released in May, before the war was over, "Whistling Dick" both set to music and committed to paper the sailor's exclamation to warn of incoming round

[67] Henry Farmer and Edward Farmer, "The Battle of the Alma" (London: Leoni Lee, 1855); Stephen Glover and Henry Abrahams, "A Voice Was Heard in England" (London: D'Almaine, 1855); Frederick R Shrivall and H. Montagu, "Alma, 'Tis Sweet for Our Country to Die," a musical number from *A Trip to the Crimea: A New Musical Entertainment* (London, Addison Hollier & Lucas, 1857). There are several much later songs that commemorate the Crimean War—for example, James Smyth, "The Warrior's Return from the Crimea" (London: Lafleur & Sons. 1874)—as well as some printed later that purported to date from the time: Matthew Henry Weetman, "England's Bygone Days: Song Composed at the Siege of Sebastopol by British Soldiers" (London: B. Williams); Charles Osborne, "On Duty: A Tale of the Crimea" (London: R. Maynard, 1893); Samuel Liddle, sung by Mr. Plunkett Green, "The Kerry Recruit (An Irish Recruiting Song of the Time of the Crimean War)" (London: Stainer & Bell, 1938).

shot. The cry itself was illustrated on the front cover (see Figure 3.5a) and further explained by a note at the top of the first page (see Figure 3.5b):

> These shells have done our works and guns much damage; but the Sailors, who are principally treated to these agreeable missiles, have got quite accustomed to them. "Bill" cries one fellow to another "look out, here comes 'Whistling Dick!' " Vide Russell's correspondence from the Crimea.

Figure 3.5a George Ricardo and J. E. Carpenter, "Whistling Dick (Crimean Song)." London: Campbell, Ransford, 1856: cover image. © British Library Board h.1764.(41.); reproduced with permission.

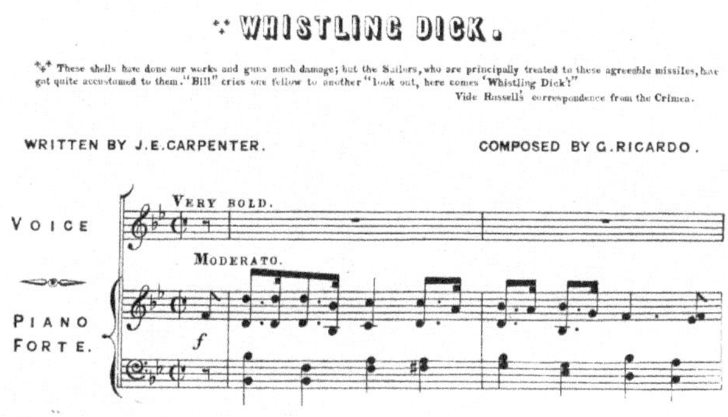

Figure 3.5b J. E. Carpenter and George Ricardo, "Whistling Dick (Crimean Song)." London: Campbell, Ransford, 1856: 1; upper half of the page. © British Library Board h.1764.(41.); reproduced with permission.

This detail reveals that, as the patina of history was being applied—before the war had been won—the link between Crimea and London, via newspaper report and piano transcription, came to be memorialized in a bow to Russell. Before long "Whistling Dick," like so much other sheet music, was itself to be pressed into Britain's national archive. It discloses a society organizing its memories in creating a monument to the savage, who heard demons let loose in the flight of cannon.

4

Overhearing Indigenous Silence

Crimean Tatars during the Crimean War

MARIA SONEVYTSKY

The Crimean War (1853–1856) was the first global conflict marked by the rapid dissemination of representational content beyond the battlefield. New media technologies such as photography and telegraphy allowed information to circulate with unprecedented speed, transmitting visual and narrative information to populations far removed from the theater of war. Yet, amidst the din of wartime communication (much of which has been preserved in imperial archives), there is a notable gap in representation, a figurative silence that marks the experience of the Crimean Tatars, the indigenous population of Crimea who fled the war-ravaged region in unprecedented numbers between 1854 and 1863.[1] While acknowledging the silence of the Crimean Tatars in imperial accounts, we know that Crimean Tatars, like all peoples, were not silent among themselves. Rather, through centuries of mass emigration, exile, and return, Crimean Tatars voiced their stories—through sounds of devotion, poetry set to melody, and courtly instrumental traditions—utilizing sound to create an archive of indigenous memory that challenges the "charged silence" of Crimean Tatars in histories of the Crimean War.[2] This chapter takes on the possibility of recuperating Crimean Tatar narratives of wartime loss through musical sounds that are indirectly evoked, implied, or referred to in historical and contemporary accounts. In part, this attempt at recuperation is also an extrapolation: it is an exercise

[1] It is estimated that 200,000 people—approximately two-thirds of the total Crimean Tatar population—fled as a consequence of the Crimean War. Official Russian documents from the mid-1860s recorded "784 deserted villages and 457 abandoned mosques"; see Orlando Figes, *The Crimean War: A History* (New York: Metropolitan Books, 2010), 424.

[2] Ana María Ochoa Gautier, "Silence," *Keywords in Sound*, ed. David Novak and Matt Sakakeeny (Durham, NC, and London: Duke University Press, 2015).

in overhearing as over-hearing, that is, in imaginatively listening through and beyond the noise of clashing empires for evidence of indigenous subjects who were more than just the silent victims of war.[3]

Listening as over-hearing offers a challenge to the "repeated histor[ies] of silencing through denial, negation, and abuse" that mark indigenous encounters with colonial powers globally. Ana María Ochoa Gautier writes that "music often permits a process of unsilencing that involves political, acoustic, aesthetic, sensorial, and bodily explorations . . . [but that] highlights and mobilizes rather than resolves the tensions and contradictions between the personal, the juridical, and other political aspects of histories of redress and recognition."[4] Delving into the scant evidence that exists of Crimean Tatar musical practices around the time of the Crimean War does indeed highlight the tensions between the personal (as enunciated in émigré song texts that complicate the individual's relationship to place) and the political (as embodied in the collective, indigenous memory constructs of the late and post-Soviet era, in which territorialized identity is conceived of as innate).[5] Yet charting musical practice through the "blackest periods" of indigenous Crimean Tatar musical history also reveals a parallel story of resilience that extends beyond the nineteenth century. For contemporary Crimean Tatar activists who face what they consider to be a "cultural genocide" under the current Russian occupation of Crimea, it is instructive and emboldening to discover that nineteenth-century Tsarist policies to "expel and exterminate" Crimean Tatars during and after the Crimean War failed in part because of

[3] Rory Finnin's reading of how the "captive Turks" of Crimea figured in the Turkish (and Pan-Turkic) literary imagination of the nineteenth and twentieth centuries—reading against the dominant narrative of Turkish silence and inaction with regard to the abuse of Crimean Tatars by Russian Imperial and Soviet powers—shares in the spirit of attempting recuperative history with regard to this embattled indigenous minority. See Rory Finnin, "Captive Turks: Crimean Tatars in Pan-Turkist Literature," *Middle Eastern Studies* 50/2 (2014), 291–308.

[4] Ochoa, "Silence," 187–88. For a case study of how the politics of "silencing" manifest in the vocal repertoires of Rongelapese victims of nuclear experimentation in the Marshall Islands, see Jessica A. Schwartz, "A 'Voice to Sing': Rongelapese Musical Activism and the Production of Nuclear Knowledge," *Music & Politics* 6/1 (2012), published online at <http://dx.doi.org/10.3998/mp.9460447.0006.101>.

[5] Pursuing both primary and secondary sources for this project has been a challenge for logistical reasons. After considerable effort, I have found no sources from the period of the Crimean War written from the Crimean perspective; if such materials exist, they may be in Turkish archives. The secondary journalistic and scholarly materials that I refer to come from the Gasprinsky Library in Simferopol, Crimea. I am indebted to Nadjie Yagya and Zarema Islyamova of the Gasprinsky Library, who assisted me during a brief research trip to Simferopol in June 2015, and to Milara Settarova for her assistance in translating texts from Crimean Tatar to Russian.

the indigenous memory stoked through song and sonic practice.[6] Thus the imaginative listening that I propose necessitates a degree of creativity in its approach; this chapter will stretch across time periods, positioning nineteenth-century examples of perceived indigenous silence along a continuum next to contemporary examples in which Crimean Tatar responses to trauma and erasure are asserted vociferously through musical sound.

Listening attentively for the symbolic resonance of music as a vehicle for "indigenous memory" undoubtedly projects modern ways of listening onto the seemingly blank spaces of history; here it is used as a recuperative strategy in the absence of existing primary sources.[7] I refer to "indigenous memory" as a set of memory practices that emphasize an intimate connection to place and foster a territorialized sense of collective identity.[8] The term asserts the potential of memory as a "ritual of power," which also may work as a "weapon of the weak."[9] "Indigenous" [*korennoi*] is understood as a key term of twenty-first-century politics, utilized by Crimean Tatars as part of their battle to secure rights as the indigenous people of Crimea in accordance with the protections designated in the *Declaration on the Rights of Indigenous Peoples* adopted by the General Assembly of the United Nations in 2007.[10] As such, "indigenous memory" is distinctly

[6] Dzhemil Karikov, a prominent Simferopol-based musician and revivalist of music from the Khan's period, told me that the nineteenth century was a "black period" in Crimean Tatar music because of the constant war, emigration, and discriminatory Tsarist policies against the local population. He contrasted this period with the "golden age" of Crimean Tatar courtly instrumental music that existed before Catherine's annexation of the peninsula in 1783, and which he attempts to reconstruct in contemporary Crimea (personal interview, 14 June 2015). "Expel and exterminate" are the terms used by Enver Ozenbashli in "Otrazhenie tragicheskoi sud'by krymskikh tatar v pereselencheskikh pesniakh XIX veka," *Golos Kryma* 2004.

[7] Since the 1990s, indigenous Crimean Tatar historians have attempted to reconstruct a narrative of this war and other conflicts from the perspective of the indigenous population by examining Russian imperial archives, which they did not have access to during the Soviet era.

[8] In the successor states of the Soviet Union, including contemporary Ukraine and Russia, the designation of "indigenous" status is posed against official "minority" status; while the latter term is legally protected, the former remains contested. For a history of these terms, see Natalya Belitser, "'Indigenous Status' for the Crimean Tatars in Ukraine: A History of a Political Debate," *"Fuzzy Statehood" and European Integration in Central and Eastern Europe*, ed. Judy Batt (University of Birmingham: Economic and Social Research Council, 2002).

[9] The queer theorist J. Jack Halberstam has argued, "Memory is itself a disciplinary mechanism that Foucault calls 'a ritual of power'; it selects for what is important (the histories of triumph), it reads a continuous narrative into one full of ruptures and contradictions, and it sets precedents for other 'memorialization.'" Judith Halberstam, *The Queer Art of Failure* (Durham, NC, and London: Duke University Press, 2011), 15. For the analysis of peasant resistance in which the notion of "weapons of the weak" was introduced, see James C. Scott, *Weapons of the Weak: Everyday Forms of Peasant Resistance* (New Haven, CT: Yale University Press, 1985).

[10] The *Declaration* has been reaffirmed a number of times since its original adoption in 2007. The full text of the *Declaration* can be viewed and downloaded through the website of the United Nations:

postcolonial and post-Soviet, shot through with the late-twentieth-century discourse on indigenous rights situated within the Western framework of human rights.[11] Furthermore, "indigenous memory" is a form of cultural memory defined by "imagination and interconnection."[12] It functions as a resource that can assert claims to places, and it may harden into political agendas, mobilizing "counter-memories" to challenge hegemonic discourses.[13] Greta Uehling's study of memory practices among Crimean Tatar returnees to Crimea in the 1990s supports such an analysis: she explains that, especially for generations born in exile, "their return is beyond the memory they use to explain it."[14] Thus indigenous memory is emergent but also strategic; it provides a way to *do things* with memory.

Writing on the "past and future of Crimean Tatar music" in 1925, Arkady Konchevsky said that "songs are the living history of a people. This is especially true for Tavrida" (the Russian Imperial name for Crimea).[15] Songs, as a "sensuous modality" through which counter-memories may be encoded, reimagined, and circulated, function as a conduit for indigenous memory by fusing and also eliding historical events with sentiment.[16] A song's semiotic slipperiness affords the potential of reimagining the historical through the affective dimensions of musical sound, while the documentary evidence of song texts also introduce a narrative. If "songs are a historical record," as a Crimean Tatar singer told me

<http://daccess-dds-ny.un.org/doc/UNDOC/GEN/N11/467/34/PDF/N1146734.pdf?OpenElement>, accessed 15 Aug. 2015.

[11] In this sense, my use of indigenous memory differs from an understanding of "indigenous knowledge" as the unimpeachable legacy of indigenous communities and their traditional ways of life. Twenty-first-century scholarship has critiqued such notions of indigenous knowledge for supporting an understanding of contemporary indigenous communities as "ecologically noble savages." See Ter Ellingson, *The Myth of the Noble Savage* (Berkeley and Los Angeles: University of California Press, 2001); Elizabeth A. Povinelli, *The Cunning of Recognition: Indigenous Alterities and the Making of Australian Multiculturalism* (Durham, NC, and London: Duke University Press, 2002).

[12] Diana Taylor, *The Archive and the Repertoire: Performing Cultural Memory in the Americas* (Durham, NC, and London: Duke University Press, 2003), 82.

[13] Michel Foucault, *Language, Counter-Memory, Practice: Selected Essays and Interviews*, ed. Donald F. Bouchard, trans. Donald F. Bouchard and Sherry Simon (Ithaca, NY: Cornell University Press, 1977).

[14] This notion of "indigenous memory" has been deployed strategically as historical justification for the Crimean Tatars' right to return from exile: following their mass deportation in 1944, they were granted the right to return in 1987. Subsequently, hundreds of thousands of Crimean Tatars moved from Central Asia to reclaim territories in Crimea in the last years of the Soviet empire. See Greta Lynn Uehling, *Beyond Memory: The Crimean Tatars' Deportation and Return, Anthropology, History, and the Critical Imagination* (New York: Palgrave Macmillan, 2004), 17.

[15] A. K. Konchevskii, "Proshloe i nastoiashchee v pesniakh Kryma," *Obshchestvenno-nauchnyi zhurnal* 1 (1925), 31.

[16] Paul Stoller, *Sensuous Scholarship* (Philadelphia: University of Pennsylvania Press, 1997), 47.

during an interview in 2009 (unconsciously echoing Konchevsky), and if, as the anthropologist Michel-Rolph Trouillot proposed, "each historical narrative renews a claim to truth," then what kind of claim to truth can musical sounds provide?[17] Can musical sound help us reimagine the disenfranchised indigenous victims of grand political maneuverings and war as agents in the production of history?

Overhearing the Crimean Tatars During the Crimean War: A British Account

> The Tatars are a rapidly diminishing race; and failing numbers is accompanied with declining moral energy. This melancholy fact is referable to their position as a conquered people, spoiled of territorial wealth, social and political importance and exposed to the harassing peculation of subaltern agents of government. It is painful to reflect, that the present war must be an additional disaster to them.
>
> —Reverend Thomas Milner, 1855[18]

The present war was indeed an additional disaster in a long string of disastrous events. At the time of the Crimean War, notions of a cohesive "Crimean Tatar" populus were nascent. The ethnonym "Crimean Tatar" was considered derogatory by ethnographers, whereas the term *Kırımlı* was probably more frequently used to refer to "Crimeans."[19] More likely, Crimean Tatars of the mid-nineteenth century would have self-identified as one of three separate sub-ethnic groups: the northern steppe dwellers (*Nogai*), the mountaineers (*Tats*), and the southern coast traders (*Yalıboyu*, whose position at the end of the Silk Road allowed for a melting pot of diverse ethnic groups, including a substantial population of Genoese and Greek merchants).[20] These three geographically distinct groups were differentiated by livelihood, spoken dialect, and expressive practices. In the mid-thirteenth century, during the Mongol invasions by the armies of Batu Khan, these three groups were brought together for the first time under one regime, united as indigenous Crimeans.[21] As the Golden Horde's grip over

[17] Michel-Rolph Trouillot, *Silencing the Past: Power and the Production of History* (Boston: Beacon Press, 1995), 5; personal interview with Rustem Memetov in Simferopol (11 May 2009).

[18] Cited in Mara Kozelsky, "Casualties of Conflict: Crimean Tatars during the Crimean War," *Slavic Review* 67/4 (2008), 888.

[19] E. Ozenbashli, *Krymtsi: Zbirnyk prats' z istoriji, etnohrafiji, fol'kloru ta movy* [Crimeans: Collected works in history, ethnography, folklore and language], 2 ed. (Simferopol, Crimea: DOLIA, 2006), 3.

[20] Hakan Kırımlı, *National Movements and National Identity among the Crimean Tatars (1905–1916)*, Ottoman Empire and Its Heritage 7 (Leiden, The Netherlands: Brill Academic, 1996).

[21] However, these subgroups were so distinct that, until the deportation of the Crimean Tatars in 1944, intermarriage between the three groups rarely occurred.

Eastern Europe began to disintegrate in the early 1400s, the first Crimean Khan, Haci Giray, established an independent Crimean Khanate in the early 1440s.[22] This Crimean Khanate lasted until Catherine the Great's annexation of Crimea into the Russian Empire in 1783. The dissolution of the Crimean Khanate forced the Giray dynasty and many elites to flee to the Ottoman Empire. Orlando Figes estimates that "by 1800, nearly one-third of the Crimean Tatar population, about 100,000 people, had emigrated to the Ottoman Empire with another 10,000 leaving in the wake of the Russo-Turkish war of 1806–12."[23] By the time of the Crimean War, the great majority of Crimean Tatars remaining in Crimea were peasants, poor and illiterate.[24]

The first widely memorialized mass migration encompasses a vast historical period stretching from Catherine's 1783 annexation through the period following the Crimean War.[25] Greta Uehling explains that this swath of time "is retrospectively refigured as the 'first exile,' linking it symbolically to the exile they experienced later, under the Soviets."[26] As Crimean Tatars fled to the relative safety of the Ottoman Empire, they often traveled westward along the coast of the Black Sea, leaving a trail of settlements along the coasts of modern-day Romania and Bulgaria, where a politically active Crimean Tatar diaspora still exists today. It has been estimated that as many as 400,000 Crimean Tatars left their homeland for the Ottoman Empire in the eighteenth and nineteenth centuries.[27]

During the Crimean War, many Russian officials suspected and accused Crimean Tatars of treason, though Imperial policies with regard to the Tatars neither protected nor officially persecuted the population.[28] Often too impoverished to leave, many Crimean Tatars were exploited by both Russian and Allied forces for provisions, livestock, and transportation. When Russian surveillance of the Tatars began to pick up momentum in September 1854, some Tatars began to flee. Emigration escalated throughout the war years and peaked

[22] The exact relationship of Haci Giray to the Golden Horde, and the contract that the Crimean Khanate entered into with the Ottoman Empire beginning in 1475, are contested historiographical problems. Alan Fisher, *The Crimean Tatars* (Stanford, CA: Hoover Institution Press, 1978), 3–5.

[23] Figes, *The Crimean War*, 21.

[24] Kozelsky, "Casualties of Conflict."

[25] Mustafa Jemilev, *A History of the Crimean Tatar National Liberation Movement: A Sociopolitical Perspective,* ed. Edye Muslymova, trans. Sanoma Lee Kellogg and Inna Pidlusk (Simferopol, Crimea: Odjak, 2005), 51.

[26] Uehling, *Beyond Memory*, 35.

[27] Brian Glyn Williams, *The Crimean Tatars: The Diaspora Experience and the Forging of a Nation.* Brill's Inner Asian Library (Leiden, The Netherlands, and Boston: Brill Academic, 2001).

[28] Wartime allegations that the Crimean Tatars were traitors and suggestions to remove them from the peninsula were to anticipate Stalin's deportation of the entire population in 1944, during which they were branded "enemies of the Soviet people." Kozelsky, "Casualties of Conflict."

between 1860 and 1863. Aside from a small number of petitions written by Crimean Tatars resisting arrest, nearly all Russian-language materials that depict Crimean Tatars were produced by Russian authorities involved in disciplining the local population.[29] As such, "the subjective, faulty, disorganized surveillance stored in Crimean archives reveals less about the 'mood of the Tatars' than about the mood of those people gathering information and making policy."[30] This mood was overwhelmingly—but not uniformly—punitive.

British representations of the Crimean Tatar wartime experience include those of the English writer Alexander William Kinglake, whose description of sonic clues is particularly interesting. The first of his eight-volume series titled *Invasion of the Crimea*, published in London between 1863 and 1887, begins with this following description of the local population:

> Along the course of the little rivers which seamed the ground, there were villages and narrow belts of tilled land, with gardens, and fruitful vineyards; but, for the most part, the Chersonese was a wilderness of steppe or of mountain range much clothed toward the west with tall stiff grasses, and the stems of a fragrant herb like southernwood. The bulk of the people were of Tartar descent, but they were no longer in the days when nations trembled at the coming of the Golden Horde; and though they were of Moslem faith, their religion has lost its warlike fire. Blessed with a dispensation from military service, and far away from the accustomed battle-fields of Europe and Asia, they lived in quiet, knowing little of war, except what tradition could faintly carry down from old times in low monotonous chants. In their husbandry they were more governed by the habits of their ancestors than by the nature of the land which had once fed the people of Athens, for they neglected tillage, and clung to pastoral life. Watching flocks and herds, they used to remain on the knolls very still for long hours together, and when they moved, they strode over the hills in their slow-flowing robes with something of the forlorn majesty of peasants descended from warriors. They wished for no change, and they excused their content in their simple way by saying that for three generations their race had lived happily under the Czars.[31]

The "low monotonous chants" that Kinglake hears may refer to Islamic religious practices (such as the *dua*, or supplications) of the predominantly Sunni Muslim

[29] Ibid.

[30] Ibid., 872.

[31] Alexander William Kinglake, *The Invasion of the Crimea: Its Origin, and an Account of Its Progress down to the Death of Lord Raglan*, 8 vols. (London: Blackwood and Sons, 1888), vol. 1, 25–26.

Crimean Tatars; in this passage, he underscores the perceived link between Islam and "the warlike fire" inherent in the religion, a trope of orientalizing nineteenth-century accounts of Islam. At the end of this description ("for three generations . . ."), Kinglake may be referring to the loyalty oath taken by Crimean Tatar *beys* to the Czar in 1854, during which they emphasized their peaceful existence under the Russian crown since 1783. Local Crimean Tatar populations did provide assistance to the British and other Allied forces in some forms, however.

More often, British soldiers encountered Crimean Tatars *in absentia*, as homes left abandoned became sites of exotic discovery for imperial forces. Kinglake depicts such a moment toward the end of the first volume:

> Pure ignorance of the invaded country gave charm to every discovery tending to throw light upon the character and pursuits of the inhabitants; and if our soldiery had found in the villages high altars set up for human sacrifices, they would scarcely have been more surprised than they were when, prying into the mysteries of this obscure Crym Tartary, they came upon traces of modern refinement and cultivated taste. In some of the houses at Kentugan there were pianos; and in one of them a music-book, lying open and spread upon the frame, seemed to show that the owner had been hurried in her flight.[32]

Kinglake's description of the shock at the presence of upper-class commodities—a piano and music book—in the abandoned homes of "this obscure Crym Tartary" highlights the presence of European-inflected musical traditions among elite segments of Crimean Tatar society. It seems likely that such musical traditions would have been imported as part of the general trend among Russian Imperial elites (including Crimean Tatar ones) in the post-Catherinian period to appropriate the habits of Western European nobility in dress, language, and musical practice. It is intriguing to imagine elite Western traditions in parallel with the sounds of Islamic worship; this notion constitutes a step toward recovering the indigenous contributions to the wartime soundscape.

Kinglake offers evidence for both a religious tradition of song and an elite tradition of European music, providing one mode of overhearing the Crimean Tatars of the Crimean War. This act of listening contrasts with the twentieth-century song texts collected by ethnographers, both imperial and Soviet, texts that are considered in this chapter's next section. These collections feature many *destan*—epic ballads and laments—that make concrete reference

[32] Ibid., 423.

to events of the seventeenth, eighteenth, and nineteenth centuries.[33] One such "émigré song," a *destan*, was believed by some scholars to refer specifically to the Crimean War.

"Speaking, I Weep": Émigré Songs

Multiple historians and ethnographers have identified "émigré songs"—those songs whose texts describe the act of leaving one's home in Crimea—as the genre within Crimean Tatar musical folklore that functions particularly well as a historical document. The prominent ethnographer Enver Ozenbashli wrote that "émigré songs allow us to understand the reasons for the mass emigrations of Crimean Tatars: above all it was the cruel colonial politics of tsarism, directed at the extermination and expulsion of Crimean Tatars from Crimea, and also some of the traitorous activities of local hereditary nobility and clergy."[34] Writing in 1925, the Soviet Russian ethnographer Arkady Konchevsky noted that émigré songs are the "cry of the people" ["плач народа"]. For that reason, he explains, a common refrain in émigré songs is, "speaking, I weep!"[35] As the singer speaks and weeps, the song consciously calls attention to its sonic materiality, to its own noisiness. It resists silencing by repeatedly invoking the "I" that speaks and weeps, suggesting both agency (the sheer will to perform an utterance) and affect (the utterance that provokes tears).

In 1910, a slender volume titled "The Songs of the Crimean Turks" was published by the ethnographer Olesnitskii, who, while vacationing in Alupka (which he refers to as one of the "charming corners of the Russian Riviera"), decided to travel through Crimea to record the songs of its local population.[36] He used the material to publish his first book, which came out in Moscow.[37]

[33] See Fevzi Aliev, *Antologiia krymskoi narodnoi muzyki* [Anthology of Crimean Folk Music] (Ak'mesdzhit [Simferopol]: Krymuchpedgiz, 2001); Y. Sherfedinov, *Zvuchyt' Kaytarma* [the Qaytarma sounds] ed. L. N. Lebedinskij (Tashkent, Uzbekistan: Izdatel'stvo literatury i iskusstva imeni Gafura Guliama, 1979).

[34] Ozenbashli, "Otrazhenie tragicheskoi sud'by krymskikh tatar v pereselencheskikh pesniakh XIX veka." See also Brian Glyn Williams, "*Hijra* and forced migration from nineteenth-century Russia to the Ottoman Empire," *Cahiers du Monde russe* 41/1 (2000), 79–108, for a description of the religious-emigrant songs (*muhacir destan*) and their value as evidence in recuperating Crimean Tatar experiences of the nineteenth century.

[35] In Crimean Tatar, the phrase is "Ай-тыр-да аглерим." Translated into Russian, Konchevsky and others render it as "говоря, я плачу!"

[36] Olesnitskii writes about the problem of ethnonyms of the people he calls "Crimean Turks": the term "Tatars" is "unscientific and prone to ostracism" (ix).

[37] Aleksei Olesnitskii, *Pesni krymskikh turok* [Songs of the Crimean Turks] (Moscow: Typografia N-v' Gattsuka, 1910), VI.

Initially, he intended to travel the entire peninsula, but after numerous hostile encounters with local police and "un-neighborly" local communities, he decided to confine his study to fifteen villages located on the Southern Black Sea coast. (In his introduction, he expresses frustration over not being able to access the "heart of Tavrida.") Among the songs he recorded and transcribed, he identifies one as dating from the time of Catherine's annexation, with the rest coming from the mid-nineteenth century, from the years leading up to what he called the "Sevastopol War," or from the 1870s, when the Tsarist policy of general conscription began. He details a number of émigré songs that lament the loss of Crimea with common lines such as "I am leaving my grapevines behind," but identifies only one as coming from the period of the Crimean War.

This text, which Olesnitskii called the "Destan of Emigration from Crimea" and which he believed to have been written in 1855–56, came under scrutiny by prominent Soviet Turkologist Alexander Samoylovich (1880–1938), who challenged many aspects of Olesnitskii's analysis. Enumerating the erroneous phoneticization of dialect, the incorrect words that confounded the meter of the poetic text (which he identified as a folk-Turkic syllabic meter known as the *murebba*), and the inconsistencies between the literary Crimean Tatar of parts of the text and the vernacular language used in other passages, Samoylovich concludes that the "Destan" "became folklorized by the time of Olesnitskii's recording," though he suspects that it originated from the "pen of an educated man." (Olesnitskii noted that that particular *Destan* was sung in many of the villages he visited.)

Further, Samoylovich concluded that the text was not "about separation from the motherland as Olesnitskii suggests, but about the separation of sons who are being drafted as soldiers, as thoughts about eviction cause suffering not because the separation from the motherland causes pain, but because the implementation of these intentions meets many obstacles."[38] He scrutinizes the text for historical clues, and prepares a "plan of the destan" that summarizes the six main sections of the text, as follows:

I. By the order of the Russian Tsar, army conscription in Crimea is announced. As a true Muslim, the poet explains this misfortune as a predestination coming from God's plan and seeks help in the first place from God, the prophets and saints (I–V).

II. The Poet, referring to the exile of Muhammad from Mecca to Medina, convinces his readers and listeners—Muslims, that their religious obligation

[38] A. N. Samoylovich, *Izbrannye trudy o kryme* [Collected works on Crimea] (Simferopol, Crimea: DOLIA, 2000), 62.

in the expectation of Mohammed's protection and absolution at the Last Judgment (*sunnet*) is to follow the lesson of the Prophet and commit to emigration (*khidzhret*). Tatars begin to campaign for the eviction and send a petition [about it] to the Russian Tsar, which ends up being futile (VI–XIV).

III. The military duty is imposed [on the Crimean Tatars]: [it includes] the building of barracks (*kyshla*) in Bakhchisaray, the recruitment of soldiers without exemptions; some Tatars voluntarily give up their sons; the people moan, dress in mourning clothes; the [Tatar] soldiers are brought to Simferopol for training; soldiers say farewell to their parents; they have no time for *namaz* prayers (XXXVI); there's fear of Christian evangelization (XX)—(XV–XXXVI).

IV. Hope for help from Turkey and the sultan; petitioners are sent to Turkey asking for permission to relocate; a firm decision to emigrate [is made] (XXXVII–XL).

V. The hopelessness of the situation: Russians are on one side, but on the other—the sea; the farewells of children to parents, of parents to children; the accusation of the nobility's (*murza*) sympathies to the Russian order and the statement, that the majority of ordinary Tatars agree to conscription, as though it were a test, sent down from above.

VI. Conclusion. Appeal to pray to God and defend the Muslim faith.[39]

Even though Samoylovich concludes that dating the text to 1855–56 "was a mistake," situating it instead to the 1870s, it nonetheless articulates an indigenous perspective on the tragic consequences of the Crimean War in Crimea. Furthermore, that such songs were gathered in the early twentieth century from among the rural populations that had not emigrated from Crimea suggests a history of circulation between the diaspora of Crimean Tatars in the Ottoman Empire and Crimea proper. These émigré songs from the mid-nineteenth century assert an indigenous voice sonically, poetically, and metaphorically, offering another method of overhearing that was already heard by twentieth-century ethnographers. Contemporary scholars, meanwhile, transmute émigré songs from all periods of Crimean Tatar exile into the substance of indigenous memory.

[39] Ibid., 62–63. Translated from the Russian by Olga Voronina and the author. We have preserved the original document's punctuation and formatting.

"The Memories We Learned in Songs": From Crimean War to Putin's Annexation

Historian Mara Kozelsky describes the Crimean War as one that "involve[d] much more than European power plays, advances in military technology, or strategic battle maneuvers: the war dramatically changed the Crimean landscape and has left a legacy that local populations on the peninsula are still struggling with today."[40] This legacy is felt deeply in the ongoing dispute over Crimea. Putin's twenty-first century act of annexation (or, in his Messianic rhetoric, the "restoration" of Crimea to Russia), much like the Stalinist twentieth-century deportation of the Crimean Tatars (1944), the ravaging of the peninsula by war in the nineteenth century (1853–56), and the eighteenth-century Catherinian conquest of Crimea (1783), brought about the same result: an underrepresentation of Crimean Tatar perspectives on the events that occurred because of the depopulation of Crimea by its indigenous population.[41] In every instance, narratives of war and occupation amplified the positions of the powerful, while muting those of the peripheral. Since their return to Crimea in the late 1980s, the Crimean Tatar community has been battling against, among other things, silence.

In 2008, Abduraman Egiz was leader of the local youth group Bizim Qirim [Our Crimea] in Simferopol, and spoke to me about his relationship to Crimean Tatar songs of exile. As a spokesman for the activist subset of the young generation of Crimean Tatar repatriates, Egiz was emphatic about the vital role that music has played in instilling his sense of belonging to and ownership of Crimea, a land that he saw for the first time as a child.

> As children, we didn't understand what fatherland is, as we do now, but we knew that it was ours and we must return. . . . We didn't know what Crimea was, but we knew it was our land. We couldn't explain how we found ourselves in Uzbekistan, and why we were born there, but the "dream of Crimea"—there's no other way to explain it—also lived in us

[40] Kozelsky, "Casualties of Conflict," 891.

[41] The 2013 conflict in Crimea gave rise to what Alexei Yurchak calls "a curious new political technology—a military occupation that is staged as a non-occupation" orchestrated through the presence of the "little green men" who became a ubiquitous presence on the peninsula leading up to the sham referendum that declared Crimea's independence from the Ukrainian state and resulted in the *de facto* absorption into Putin's Russian Federation. "Little Green Men: Russia, Ukraine and Post-Soviet Sovereignty," http://anthropoliteia.net/2014/03/31/little-green-men-russia-ukraine-and-post-soviet-sovereignty/, accessed 15 Aug. 2015.

> and we always wanted to return to Crimea. When I came to Crimea it was as a child but I understood—and I saw in trees, and in nature, that it was ours, and I was searching in nature for something that was ours—like the memories we learned in songs—and I understood that this is ours, we have returned, it was our real fatherland and we are going to live here. So even as children we loved Crimea, even before we saw it, we loved it. And when we arrived we held as an axiom that this is ours. And we returned, and this was our axiom, and we didn't ask if there were other variants. Other variants did not exist. And this is the foundation upon which the new generation was built.[42]

Egiz went on to describe the first time he felt the winds of the Black Sea on his face, and how it strongly connected him to his experience singing songs with his family, triggering what he calls "the memories we learned in songs." Born in exile, Egiz inherited a sentimental relationship to a place that neither he nor his parents had physically seen, smelled, or touched.[43] Yet, upon seeing that place for the first time, his received sensory memories of Crimea heightened and deepened his feeling of ownership over the land. Songs forge links between intimate subjectivities and territorialized identities that reach back, beyond the memories of one individual, one family, or one experience.[44] The reinforcement

[42] Egiz preferred to conduct this interview (and most encounters with me) in the Ukrainian language. In part, this was a gesture of generosity directed at me, since my Ukrainian is far more sophisticated than my Russian. But it was also a savvy political gesture that Egiz consciously used with my audience in mind. In Crimea, Russian is by far the most dominant language. As an aspiring politician sensitive to the politics of language choice, Egiz believes that Crimean Tatars should avoid speaking in Russian, which he regards as the language of their oppressors, as much as possible. He aimed to show "respect for their home country"—Ukraine—by privileging Ukrainian in contexts where Crimean Tatar cannot be spoken. With other fluent Crimean Tatar speakers, Egiz and all of the members of Bizim Qirim spoke solely in Crimean Tatar. (Personal interview, 21 Nov. 2008.)

[43] His parents were also born in exile, in the Samarkand region of Uzbekistan. The children or grandchildren of Crimean Tatar deportees share much in common with children born to parents outside of any home country, though specific reasons for emigration can vary widely. Scholars of diaspora have written extensively about the universal phenomenon of displacement in its particular manifestations, tackling subjects such as trauma, loss, denial, nostalgia, and the "myth of homeland." For a geographically and theoretically diverse sample of such discussions, see Svetlana Boym, *The Future of Nostalgia* (New York: Basic Books, 2001); James Clifford, "Diasporas," *Cultural Anthropology* 9/3 (1994); Ingrid T. Monson, *The African Diaspora: A Musical Perspective* (London: Routledge, 2003); ed. Kristen E. Schuze, Martin Stokes, and Colm Campbell, *Nationalism, Minorities and Diasporas: Identities and Rights in the Middle East* (London and New York: I. B. Tauris, 1996); William Safran, "Diasporas in Modern Societies: Myths of Homeland and Return," *Diaspora: A Journal of Transnational Studies* 1/1 (1991); Mark Slobin, "Music in Diaspora: The View from Euro-America," 3/3 (Winter 1994).

[44] The intergenerational transmission of "[counter]memories learned through song"—especially salient when trauma operates as a form of cultural inheritance—has been richly documented by

of indigenous memory, one that links personal sentiment to place, is done potently through repeated performances of songs that evolve in text, melody, and meaning.

In summer 2014 and 2015, I met with the Crimean Tatar leaders of *CrimeaSOS* at offices in Kyiv and L'viv. Spurred by the lack of reliable reporting coming from Russian, Ukrainian, European and American media during the Russian incursion into Crimea, a group of three young Crimean Tatar journalists formed *CrimeaSOS* to produce reliable news to challenge the disinformation and propaganda they perceived in media representations of Crimean unrest. In summer 2014, *CrimeaSOS* became a partner of the UN Agency for Refugees (UNHCR), and their activities today are largely focused on assisting Internal Displaced Persons (IDPs) from Crimea as well as fostering understanding between new communities of Crimean refugees in various locations throughout Ukraine. In large part, these campaigns to promote dialogue and understanding center on the exchange of expressive practices, including musical collaborations between Ukrainian and Crimean Tatar musicians. Both leaders of *CrimeaSOS* in Kyiv and L'viv asserted the potency and possibility of musical exchange to strengthen the relationship of Ukrainians with Crimeans, and also to maintain the solidarity of Crimean Tatars in exile.

Since the 2014 annexation of Crimea, Crimean Tatar musical performances in Ukraine have operated as performances of resistance and solidarity with the Ukrainian state and the Maidan's ensuing "Revolution of Dignity." With the proliferation of recorded Crimean Tatar music in a wide variety of folkloric and popular music genres (including jazz, reggae, hip-hop, punk) since the mid-2000s, diverse repertoires are being heard in new ways, now redefined as "protest" music. This use suggests one last modality for overhearing that is worth considering: the kind of listening involved in hearing beyond the immediate cues of a song for its broader social resonance. Far from providing a monolithic notion of "Crimean Tatarness," contemporary Crimean Tatar musical practices in Ukraine articulate multiple ways that Crimean Tatars imagine their affective

ethnomusicologists. See, for example, Adelaida Reyes, *Songs of the Caged, Songs of the Free: Music and the Vietnamese Refugee Experience* (Philadelphia: Temple University Press, 1999); and Kay Kaufman Shelemay, *Let Jasmine Rain Down: Song and Remembrance among Syrian Jews*, Chicago Studies in Ethnomusicology (Chicago: University of Chicago Press, 1998). Joshua Pilzer has asserted the need for critical musical studies to engage with "survivors' music," arguing that "the musical means by which survivors stay alive and come to terms with their experiences are overlooked" when the long-term manifestations of trauma are not considered; see Joshua D. Pilzer, "The Study of Survivors' Music," *The Oxford Handbook of Applied Ethnomusicology*, ed. Svanibor Pettan and Jeff Todd Titon (Oxford and New York: Oxford University Press, 2015), 484. Crimean Tatar songs of emigration and exile, passed down through families and communities, may also be considered as a form of intergenerational "survivors' music."

ties to self, nation, and place. These musical practices open a space for Crimean Tatar "claims to truth" to emerge and mount a challenge to official histories. Such songs work as icons of indigenous memory, claims on territorialized identity, and memorials of wartime loss. In the wake of these contemporary overhearings, expressive practices from the era of the Crimean War may also suggest that amidst the noise of empires at war, indigenous Crimean Tatar voices were straining to be heard.

PART II

VOICE AT THE BORDER

5

Orienting the Martial

Polish Legion Songs on the Map

ANDREA F. BOHLMAN

Jeszcze Polska nie zginęła, Kiedy my żyjemy. Co nam obca przemoc wzięła Szablą odbierzemy.	Poland has not yet perished, So long as we still live. What the foreign force has taken from us, We shall retrieve by sword.
Marsz, marsz, Dąbrowski, Z ziemi włoskiej do Polski. Za twoim przewodem Złączym się z narodem.	March, march, Dąbrowski, From Italian soil to Poland. Under your command Let us rejoin the nation.

—"Dąbrowski March," Józef Wybicki (1797)

The chorus of the most well known of Polish "military" songs—the Dąbrowski March—concludes with a resounding imperative: "Let us rejoin the nation!" Penned by the poet-patriot Józef Wybicki (1747–1822) while he was fighting with Polish Legions in the 1797 Napoleonic campaigns in Italy, the popular song galvanized two audience demographics in the nineteenth century: Poles home and abroad. Throughout the century, Poland was partitioned among the Russian, Prussian, and Austro-Hungarian Empires. The populations of these lands became imperial subjects, although some fled suppression by living in exile. A primary project for the former elite ruling class became maintaining—and even nurturing—the idea of Poland in the wake of its deletion from the map. Across European front lines and scenes of revolution, Polish generals and soldiers fought for their nation's independence. One audience—the combat-inclined—would have heard the refrain as a summons: the popular song is a call to arms, directed at individuals who might join the ranks of Polish soldiers

Figure 5.1 Kajetan Saryusz-Wolski, postcard with the "Dąbrowski Mazurka." National Library of Poland, ca. 1906, DŻS XII 8b/p.52/1; in the public domain.

fighting abroad.[1] As a simple tune, the march allows listeners to imagine that this is functional music, accompanying the legion's charge in the field and organizing the step-by-step advance of an army (see Figure 5.1). Its status as military symbol was eventually inscribed in the law, when the forceful melody officially became the Polish national anthem in 1926.[2]

[1] The music remains unattributed, but has been fairly stable throughout adaptations of the text to contemporary political tensions; see Dioniza Wawrzykowska-Wierciochowa, *Mazurek Dąbrowskiego: Dzieje polskiego hymnu narodowego* (Warsaw: Wydawnictwo Ministerstwa Obrony Narodowej, 1974).

[2] For a history of the song as national anthem see Maja Trochimczyk, "Sacred/Secular Constructs of National Identity: A Convoluted History of Polish Anthems," *After Chopin: Essays in Polish Music,*

During the time of partition, the meaning and affordances of Wybicki's legion song expanded beyond martial overtones to rally nationalist sentiment more generally. Anyone, anywhere, joining in song could (at least potentially) hear the legions' military might in the firm rhythmic tactus and stepwise rise that is the backbone of the chorus. This rhythm—unmistakably that of a mazurka—makes the song doubly "Polish" by encoding it in the dance topic that gained cosmopolitan status over the course of the nineteenth century.[3] As Barbara Milewski has compellingly shown, the mazurka's presumptive "folk" status ideally positioned the dance for transformation beyond the village dance hall: within the plays performed on Warsaw's stages, in sheet-music arrangements for parlor gatherings, and, of course, in art-music miniatures by the likes of Józef Elsner and his student Frédéric Chopin.[4] Some mazurka settings were dances; others, like the Dąbrowski March, merely danceable. The legion song stretches Milewski's study of the robust national music tradition to include military networks at the intersection of cultural and political nationalism.[5] The anonymous musical setting of the text renders war audible at a distance, amplifying the text's message as an instrument of military recruitment. The song thus functions as a diffuse weapon for developing collective national identity in the absence of nationhood. Singing and circulating the song facilitates the survival of the divided, dispersed, and disempowered nation, deleted from the map.[6]

There is much at stake in rehearing this popular song, which both commands its audiences to build a nation and promotes itself as a fundamental cultural symbol of that very nation. For the purposes of this essay, I emphasize the territorializing aspirations that depend on what might be considered, at base, a primitive musical technology. A hearing of this song that is invested in music's

ed. Maja Trochimczyk, Polish Music History Series 6 (Los Angeles: Polish Music Center, 2000), 263–94, in particular 288–90. Trochimczyk compares patriotic sentiment in the mazurka-march to the other contenders for official anthem status.

[3] It was over the course of the period of partition that the mazurka as an art-music dance became the "default musical-national genre" within Poland, in part through the performing hands and publishing network of Frédéric Chopin (Barbara Milewski, "Mazurkas and National Imaginings," PhD dissertation, Princeton University, 2002), 3.

[4] Barbara Milewski, "Chopin's Mazurkas and the Myth of the Folk," *Nineteenth-Century Music* 23/2 (1999), 113–35. Milewski cites 1,500 mazurkas—by 300 composers—printed in Warsaw in the first half of the nineteenth century, a statistic culled from Wojciech Tomaszewski, *Bibliografia warszawskich druków muzycznych 1801–1850* (Warsaw: Biblioteka Narodowa, 1992), quoted on p. 130.

[5] Andrzej Walicki, *Philosophy and Romantic Nationalism: The Case of Poland* (Oxford: Clarendon, 1982).

[6] I listen in on the mazurka from within the Polish nationalist context, complementing Jeffrey Kallberg's study of the foreign encounter with Poland through the nationalist genre. See Jeffrey Kallberg, "Hearing Poland: Chopin and Nationalism," *Nineteenth-Century Piano Music*, ed. R. Larry Todd (New York: Schirmer, 1990), 223–57.

potential for galvanizing nationalism might, for example, take more seriously the limits of minor military powers—the kind of limits pointed out by Jean-Jacques Rousseau in his influential "Considerations on the Government of Poland (1771–72)." In this work of political theory, the philosopher made explicit the weakness of the Polish army in any comparison with Russia. In response, he suggested that Polish citizens take the protection provided by the spiritual seriously, instructing, "The virtue of [Poland's] citizens, their patriotic zeal, the particular way in which national institutions may be able to form their souls, this is the only rampart which will always stand ready to defend her, and which no army will ever be able to breach."[7] Rousseau counseled a turn toward the embodied, a turn inward. In doing so, he proscribed the human spirit—within every individual—a greater capacity to preserve the nation than any institution. The Dąbrowksi March performs a similarly atopic mapping: the fury of Polish martyrdom stirs in this tune of fervent patriotism.

As malleable potential, these sounds lay in wait. Thirty years later, they would be partially activated through their citation in the rousing lectures by the poet and national bard Adam Mickiewicz. (Mickiewicz will play a central role later in this chapter in triangulating the sounds of Crimea from a Polish perspective.) Addressing Poles living in Paris, his *The Books of the Polish People and of Polish Pilgrimage* (1832) positioned the occupation of Polish lands as a fundamentally *religious* sacrifice for Europe. Recollecting Rousseau, the poet and orator—a prophetic voice celebrated as a "seer" (*wieszcz*)—rallied the public around the song. He took it as a foundational truth: "These words mean that within themselves people have the essence of Polish nationality and that they are able to extend the existence of their homeland irrespective of the political circumstances—and that they may seek to restore it."[8] Poland's unstable geopolitics were to continue through the First World War; during the Crimean War, émigré communities in London and Paris advocated for the parlay of the "Polish Question" into military strategy. Their rhetoric recalled the international solidarity against Britain during the American Revolutionary War, when engineer and military commander Tadeusz Kościuszko (1746–1817) sailed across the Atlantic to consult on American defensive strategy.

After the insurrection of 1830 (the "November Uprising") was quashed by the Russian Imperial Army, the allure of a revolutionary battle for independence provided a focus for anti-Russian sentiment in the 1830s and 1840s in both England and France. Communities of Poles abroad were concentrated in Paris and London, where their failed revolutions drew the attention and critique

[7] Jean-Jacques Rousseau, *The Government of Poland*, trans. Willmore Kendall (Indianapolis, IN: Hackett, 1985), 11.

[8] Adam Mickiewicz, *Dzieła: Wydanie narodowe*, vol. 10 (Warsaw: Czytelnik, 1948), 267–68.

of various political agitators seeking to alter power relations in Europe.[9] Adam Czartoryski (1770–1861), the leader of the insurrection, was, like many Polish noblemen and members of the intelligentsia, forced into exile as a result of his participation in the military action. From France, he cultivated relationships with French and British politicians while organizing a standby liberal government, known as Hôtel Lambert (after its seat on the Île Saint-Louis). As Sławomir Kalembka has written, "tensions between empires, and, even more so, the announcement of [the Crimean War], raised hope for independence amongst émigrés."[10] Lord Palmerston's early agitation for Western European alliance argued that an independent Poland would serve as a protective buffer against Russia—an idea that gave hope to émigrés, who thought the tactics might spur the diplomatic negotiation of independence. However, the rhetorical strategy of Polish prophylaxis was to fade as the theater on the Black Sea increased in importance.[11]

And so, the Polish involvement in the Crimean War was restricted to the participation of military legions. Any attention to the sounds of the war from the perspective of Polish history thus necessarily falls upon the legion song: as galvanizing force, tool for military action, and mode of telling battlefield stories. Only 1,500 men fought in the Polish legion organized to fight against Russia in Crimea; meanwhile, an unknown number were conscripted to fight against the allies by the Tsar.[12] Soldiers assembled from across Europe to band together. As legions moved, so did their songs, bearing witness to Kevin Robbins's provocative assertion that any history of popular music ought to be attuned to place and displacement. He writes, "Popular musical forms have always been migrating forms."[13] I begin with Wybicki's anthem in this sketch of the martial in Polish music in and around the nineteenth century—an attempt to put Polish popular music on the map by concentrating on a song that itself maps history and nationalist imaginations and also moves among genres.

Philosopher Eduardo Mendieta has issued the helpful reminder that the geography of war extends beyond the battlefield: "The art of waging war was always about technologies of controlling territory, of surveying spaces, traversing

[9] Karl Marx weighed in on the importance of understanding Polish demands an 1848 speech. See Marx, "Communism, Revolution, and a Free Poland," *Political Writings*, by Karl Marx, vol. 1 (New York: Penguin, 1973), 102–4.

[10] Sławomir Kalembka, *Wielka emigracja 1832–1863* (Warsaw: Wiedza Powszechna, 1971), 384.

[11] Orlando Figes, *The Crimean War: A History* (New York: Picador, 2010), 78–85.

[12] Figes, *The Crimean War*, 333.

[13] Kevin Robbins, "Migrating Music and Good-Enough Cosmopolitanism: Encounter with Robin Denselow and Charlie Gillett," *Migrating Music*, ed. Jason Toynbee and Byron Dueck (London and New York: Routledge, 2011), 150.

topographies, and circumnavigating the world on the surfaces of the sea."[14] He cautions against scholarly fixation on battlefields as the primary sites of war, urging us instead to consider the journeys traveled, resources depleted, and people displaced by the efforts of warfare. Following Mendieta, we might understand the travels and travails of Polish legions during partition as a "resource," even as the mobile machinery of warfare. In 1855, a company of Polish exiles visited their fellow countrymen who had fought in Crimea at their camp in Burgas on the Ottoman coast of the Black Sea in 1855. (This legion location will become important for reimagining soldiers' music making in what follows.) Notably, one visiting officer remarked as he stood beneath a red and white banner at the camp, "We are in our own place, on Polish land."[15] Legion songs enable such close associations with place—here a reclamation of belonging—to be performed through time and space. Many instances of singing at the camps or in battle—for example, in the encounter in Burgas—are lost because of the ephemeral and routine nature of musical performance in the legion context. Hear again the distance from Italy to Poland inscribed in the steps of Wybicki's refrain. The anthem's trepidation emerges out of the displacement its author experienced after fighting in the 1794 uprising against Prussia, after which he settled in France to escape political persecution.

With these moments of political instability and geographic dispersion in mind—many more could be cited—it becomes possible to hear the peculiar inflection in the final line of the Dąbrowski March's refrain as the core of this song.[16] It may be no exaggeration to say that the song—as melody, text, and sentiment—became the basis of patriotic popular music in Poland through the nineteenth century. During a visit to Carlsbad in 1835, Chopin notated a harmonization of *this* chorus in a Stammbuch; numerous accounts of the composer improvising on the tune at the piano record the significance of the distinctive mazurka rhythm for performer and émigré audience alike.[17] The melody resonates through both the absence of nationhood and the presence of Polish legions fighting in the 1848 revolutions (spread across at least seven locations), in the

[14] Eduardo Mendieta, "War the School of Space: The Space of War and the War for Space," *Ethics, Place, and Environment* 9 (2006), 208.

[15] Władysław Mickiewicz, *Żywot Adama Mickiewicza podług zebranych przez siebie materiałów oraz z własnych wspomnień*, 2nd ed. (Poznań, Poland: Poznańskie Towarzystwo Przyjaciół Nauk, 1929–31), lxxxiii. They also feel at home because a Polish nurse has traveled with them.

[16] The song had many lives in the nineteenth century, including as the Pan-Slavic anthem ("Hej, Slovane") in a slower contrafact.

[17] M/1690, National Chopin Institute, Warsaw, Poland. For a discussion of the piece in Chopin's improvisational practice and Polish audiences see Halina Goldberg, "'Remembering That Tale of Grief': The Prophetic Voice in Chopin's Music," *The Age of Chopin: Interdisciplinary Inquiries*, ed. Halina Goldberg (Bloomington: Indiana University Press, 2004), 63–70.

Crimean War, in the American Civil War, and in the 1863 uprising across the Russian partition.[18] Its immediate popularity and wide circulation from 1800 onward set the tone for the mythological importance of the Polish soldier as a *cultural* hero.[19] As M. B. Biskupski observes, "This anthem specifically associates national liberation with legions formed abroad, and equates patriotism with military volunteerism."[20]

Polish legions fought in other empires' wars with the longer goal of reclaiming independence for their own nation. Legion formation—which included recruitment and travel to sites of training—constructively configured networks among the Polish intelligentsia and planted Poles alongside former enemies and allies in battle. That is, Poles understood the legions as "complex, multiple objects," however "micro" in the scale of the anticipated battle; their constitution was an assemblage of the nation.[21] The Dąbrowski March lines this transformation out: the soldiers begin away from home, in the reality of a European war, and stride toward the fantasy of a reclaimed homeland at the heart of the continent. The movement connects the two audiences I imagined for the popular song at the start of this chapter: combatants and "those living through but not in a war," to borrow Mary A. Favret's apt description of those at home.[22] Reconstituting the absent nation-state by claiming martial agency and through collective imagination, the march tells the story of the century to come, mapping heterogenous possible responses to Poland's geographic precariousness. As historian Larry Wolff glosses the song, drawing attention to its characteristic "not-yet":

> The nineteenth century, which has always been considered the crucial century for the development of nationalism in Europe, could thus be

[18] Janusz Sikorski, ed., *Zarys dziejów wojskowości polskiej do roku 1864*, vol. 2 (Warsaw: Wydawnictwo Ministerstwa Obrony Narodowej, 1966). On the 1948 revolutions see particularly 469–76.

[19] I reference here Maria Janion's influential scholarship on romanticism, in particular her assertion that the Polish tradition is founded in the transmutation of uprisings and military themes into literature, which situates material loss as incidental to moral triumph. See Janion, *Płacz Generała* (Warsaw: Wydawnictwo Sic!, 1998).

[20] M. B. Biskupski, "The Militarization of the Discourse of Polish Politics and the Legion Movement of the First World War," *Armies in Exile*, ed. David Stefancic (New York and Boulder: Columbia University Press and East European Monographs, 2005), 72.

[21] I lean on the Deleuzean notion of assemblage as developed for music in Georgina Born, "On Musical Mediation: Ontology, Technology and Creativity," *Twentieth-Century Music* 2/1 (2005), 13. Manuel DeLanda has most rigorously extended this notion in his *A New Philosophy of Society* (London: Continuum, 2006). See in particular his discussion of heterogeneity and the micro, pp. 4–5.

[22] Mary A. Favret, *War at a Distance: Romanticism and the Making of Modern Wartime* (Princeton, NJ: Princeton University Press, 2010), 9.

> bracketed by the intersecting sentiments that Poland was not yet dead, but Polish national identity was not yet established.[23]

The refrain's emphatic desire to reassemble Poland captures an unease—an attempt to bridge a gap—that is laced through the text and in the disjuncture between the soldiers' bipedal march and the song's mazurka in three. With our ears attuned to the politics of geography, the confidence on the surface of Wybicki's anthem may be heard as compensatory forward propulsion, almost drawing attention to the myth of national homogeneity by reminding us of Poland's multiethnic parts.[24] At the close of the eighteenth century, this imagined nation is not the culmination of nineteenth-century vernacular cultures, *pace* Benedict Anderson, but a response to territorial loss and military subjugation.

Legion Songs and Mobilities

Despite its now exceptional status as the national anthem, the Dąbrowski March is only one example of the popular genre of legion songs (*pieśni legionowe*). They are playful and upbeat, built of rhythms that signal movement.[25] This strand of popular music, preserved most often through citations and traces in concert and stage music, slips through the net of traditional studies of Polish music in the nineteenth century, which, as Halina Goldberg has perceptively noted, tend to focus on urban musics and Polish dance topics as measures of patriotic import.[26] Like music written for the battlefield, the songs have also suffered from scholarly stigma against so-called functional music entrenched with ideology. Even in first-person narratives, they are buried amid the more common anecdotes that mention drumming and singing; occasionally a wartime account refers to tunes by name rather than merely alluding to music's presence. For example, I have scoured legion soldiers' reminiscences for accounts of the Dąbrowski March or the "March of the Zouaves" (discussed later) in vain. Yet these canonic

[23] Larry Wolff, "Revising Eastern Europe: Memory and the Nation," *The Journal of Modern History* 78/1 (Mar. 2006), 93.

[24] An iconic study of the orchestration of Polish nationalism through the discourse of the public sphere is Keely Stauter-Halsted, *The Nation in the Village: The Genesis of Peasant National Identity in Austrian Poland, 1848–1914* (Ithaca, NY: Cornell University Press, 2001). For an overview of the theme of "multiethnic Poland" through modern European history see Timothy Snyder, *The Reconstruction of Nations: Poland, Ukraine, Lithuania, Belarus, 1569–1999* (New Haven, CT: Yale University Press, 2003).

[25] Andrew Haringer, "Hunt, Military, and Pastoral Topics," *The Oxford Handbook of Topic Theory*, ed. Danuta Mirka (Oxford and New York: Oxford University Press, 2014), 194–213.

[26] Goldberg, "'Remembering the Tale of Grief,'" 63.

Figure 5.2 Ludwik Gliński, "Marsch des Generals Uminski." National Library of Poland, 1831, BN Mus. II. 17. 834; in the public domain.

legion songs appear prominently, for example, in Edward Elgar's *Polonia* (1915), among many other quotations and arrangements in art music.

Anthologized, circulated on broadsides, and continually reprinted by the nascent newspaper industry, legion songs spurred on Polish soldiers fighting for foreign armies during the time of partition. These reprints—see one example in Figure 5.1—spoke to those both at home as patriotic tunes performed through urban salon culture and, equally, on the streets wherever they were sung. Composers of this incidental music capitalized on an expanding pantheon of military heroes. Through the complex media of battle commemoration, the legion song is evoked with depictions of drums and trumpets in idealized battlefields. Following the 1830 uprising, for example, Ludwik Gliński compiled a collection of picture postcards and marches to honor its leaders.[27] The specifics of these dedications—each sketch places the generals nobly on their stallions—is undermined by a picturesque landscape that ostensibly hosts their efforts (see Figure 5.2). The insurrectionary Jan Nepomucen Umiński (1778–1851), who escaped from a Prussian prison to join in the efforts, is depicted as commanding

[27] Published individually, these items are held at the National Library in Warsaw (BN Mus. II.17.834).

a legion that includes drummers. Portraits of fierce leaders continued crucially to shape musical media as new technologies elaborated on the songbook format that had originally circulated these texts and, sometimes, their tunes. By the late nineteenth century, legion-song sheet music was to be flanked with photographs of legion heroes. By the time the military leader and future chief of state Józef Piłsudski (1867–1935) converted the legions fighting for Austria-Hungary in the First World War into the backbone of the Polish Army in 1918, sound recording technology enabled the crude preservation of his booming voice in a rendition of a legion song.

Songs gain their historiographic potential from their position at the interface of art and popular music as well as word and text in the nineteenth century. As Ana María Ochoa Gautier suggests, "song [. . .] brings together the heightened orality of poetry with the aurality of music." Brief, portable, and translatable, they are intimate and material, "particularly structured to pull the strings of affect."[28] The musicologist and music critic Tadeusz Kaczyński captured this meaning's relevance for Polish national songs in nineteenth-century Poland in conversation with Katarzyna Korczak. He echoed Mickiewicz and Rousseau's interior emphasis:

> [You need to] find yourself and understand that which is most frequently forgotten of human nature. Everyone is born in a particular place; you should know who your mother, father, [and] grandfather are. You should remember a couple of dates with history and know a few songs. And the more these historical dates and songs you know, the richer you are, the deeper your sense of national identity.[29]

Speaking in the 1980s, Kaczyński connected the didactic historical work of the songs in their original circulation with his own curatorial work as the director of an ensemble of actors and singers devoted to this repertory. The Polish language flags the importance of song by using two words: *pieśń* and *piosenka*, the latter effectively a "little song." Within a specified genre, *pieśń* and *piosenka* indicate scope and weight. For example, legion songs (*pieśni legionowe*) tell histories of battles and carry burdens of blood loss in their balladic forms, while the lively refrains of war songs (*piosenki wojenne*) more charmingly portray soldiers homesick and in love.

[28] Ana María Ochoa Gautier, *Aurality: Listening and Knowledge in Nineteenth-Century Colombia* (Durham, NC: Duke University Press, 2014), 80.

[29] Katarzyna Korczak, "Pieśń ujdzie cało," Joanna Kaczyńska, personal collection, Warsaw, Poland.

Only rarely do songs move between these two categories of distinction, crossing boundaries on the basis of who is singing them and why; yet this dissolution of genre is what takes place in the marches I discuss in this essay. It is as though the hybrid legion songs that emerge out of the Crimean context demanded of soldiers that they imagine new identities—even that they think of themselves as not Polish. In these marches there are moments of lyrical smoothness: moments that leave behind the jagged, step-organizing propulsion of the genre, giving way to hymnic steadiness in low registers. In other words, these songs invite their singers and listeners to connect the heightened dramatic action of war with the heightened emotion of its stakes. They are "little" songs to stake out military power and "big" songs for the heart.

The failed uprisings in Poland have been a major focus of military histories and cultural commemoration because they home in on the defense of the captured fatherland. Similarly, some legion songs have long been understood as aids for displaced communities to reimagine this same lost Poland—as the cultural artifacts that participate in building the martyrdom that dominated Polish nineteenth-century nationalism. The trauma of the captured nation under partition has triggered compensatory historical narratives tasked with recuperating this geographical loss through cultural reflection upon the self as victim—a strategy employed by politicians in the 1840s through the present.[30]

By contrast, the project of the nineteenth-century Polish legion in exile captures the spirit of the recurring movements for Polish independence; and (as we will see in the section that follows) there are also legion songs that embody this exile insurgency. This alternative history of the emergent nation avoids the entrenched compensatory narratives by foregrounding diplomacy and collaboration rather than the failure of the uprisings against Russia. Renegade and independently organized legions represented Polish engagement in the European political arena: they were a constantly available metonym for political action, prosthetics for nationalist aspirations that were continually being displaced and deferred. Crucially, for Poland, defining itself as "European" meant continually reorienting itself upon the map, forever distinguishing itself from "less enlightened" eastern populations. As the first partition of Poland began in 1770, Polish nobility organized the transfer of soldiers to Turkey in order to fight alongside Ottoman forces against Russia. Two legions fought during the Napoleonic Wars and émigrés in France fought with the Liberals in Portugal (1828–34). A Polish legion formed in Hungary with the 1848 revolution, and when the insurrection in Poland failed in May of that year, Adam Mickiewicz formed a legion in Rome,

[30] Kristin Leigh Kopp, *Germany's Wild East: Constructing Poland as Colonial Space* (Ann Arbor: University of Michigan Press, 2012), 204.

gathering together Slavic deserters alongside Polish soldiers to fight for the liberation of Italy. "No Pole should live any other way besides in a uniform," he wrote of his mid-century organizing efforts.[31] Some who had fled to the United States after fighting in the January Uprising fought in legions for the Union and Confederacy in the American Civil War.[32]

Poland, Crimean Battlefields, and the "March of the Zouaves"

Polish Legions fought in Crimea, but we know little about the success or sacrifice of their military actions. Nevertheless, they have long had a symbolic role in the telling of the war's multiethnic (and religious) battles. The context of the Crimean War clarifies the stakes of legion fighting as a means of articulating an independent Poland at the crossroads of Europe and a barbaric East. Many Polish émigrés volunteered, joining the French, Sardinian, and English forces against Russia.[33] Michał Czajkowski, known as Sadyk Pasha after his conversion to Islam, formed a legion of Polish émigrés as the "Sultan's Cossacks" to fight in Crimea.[34] The Sultan's Cossacks were a potent cultural symbol by which to reconfigure Polishness through the eastward gaze engendered by the Ottoman stakes in the war in Crimea.[35] After battle, the Ottoman government recognized the legions' sacrifice and welcomed their resettlement in Ottoman lands, extending the politics of allegiance spurred by the war.[36] As a nation self-branded as

[31] W. Mickiewicz, *Żywot*, vol. 4 (Poznań, Poland: Drukarnia Dziennika Poznańskiego, 1895), cvii.

[32] It is worth noting that members of Polish legions fighting on behalf of the Confederacy during the American Civil War specifically held up the Zouaves, whether in the United States or in Crimea, as models for tenacious soldiers. See Sigmund H. Uminski, "Poles and the Confederacy," *Polish American Studies* 22/2 (1965), 102.

[33] Sikorski, *Zarys dziejów*, 476–77.

[34] On Sadyk Pasha and the Cossacks see Candan Badem, "Sadyk Pasha in the Light of the Ottoman Archives," *The Crimean War 1853–1956: Colonial Skirmish or Rehearsal for World War? Empires, Nations, and Individuals*, ed. Jerzy W. Borejsza (Warsaw: Wydawnictwo Neriton, 2011), 91–110. His battle memoirs have also been published as Michał Czajkowski, *Moje wspomnienia o wojnie w roku 1864*, ed. Józef Fijałek (Warsaw: Wydawnictwo Ministerstwa Obrony Narodowej, 1962).

[35] I refer here not only to the war that is our focus in this volume, but also of the 2014 annexation of Crimea by Russia and the Yalta Conference. The contemporary (2016) discourse on Ukraine centers on the power and expansion of the European Union, which exerts economic (Poland maintains its own currency) and cultural (in particular with respect to questions of identity politics) pressures upon individual citizens as well as policy makers. In 1945, the "Polish question" was at the center of discussion at the Yalta Conference. As in the 1850s and 2013–14, Polish diplomats were absent, and the nation's interests went unrepresented.

[36] Neil Kent, *Crimea: A History* (London: Hurst, 2016), 106.

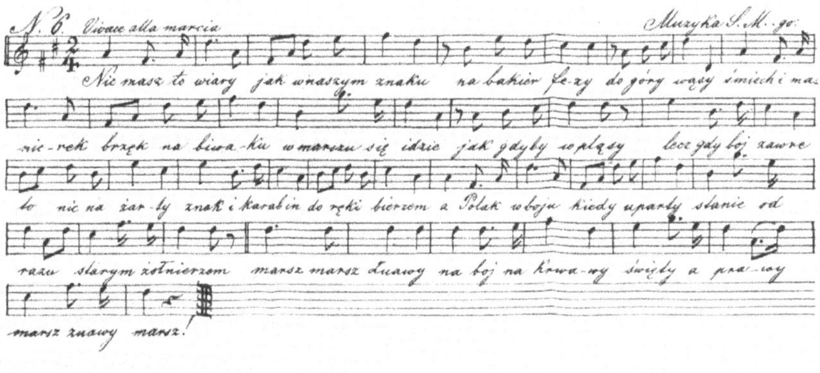

Figure 5.3 Włodzimierz Wolski, "March of the Zouaves." In *Śpiewy Powstańcze*, 2nd ed. (Paris: Drukarnia Polska, 1863). Jagiellonian Library, Musicalia 5686; in the public domain.

East Central European, Slavic, and Catholic, Poland easily served a pawn in the alluring and reoccurring division of the globe into two halves. Locating Poland at the margins of both of these domains—"inventing Eastern Europe"—has been crucial to configuring empires past and present.[37]

If the Sultan's Cossacks represent the Polish presence in Crimea, it is harder to locate evidence of the war's mediations within Poland. In Warsaw, Russian censors seem to have kept wartime news out of print, thus providing a point of resistance amid the increasing mediatization of the war by means of letters, newspapers, photography, and telegraphy (whose transmissions were elsewhere voluminous across Europe and the Ottoman Empire). Instead, the clearest mediation of the fighting by Poles in Crimea comes in the "The March of the Zouaves" (see Appendix 5.1 for the complete text). This snapshot of one legion, published in pamphlets in Warsaw, Paris, and Brussels, was written by the patriotic writer, poet, and activist Włodzimierz Wolski for a collection of songs during the January Uprising in 1863, in which he fought.[38] The edition presents a folded loose-leaf insert with the melody (see Figure 5.3) alongside, suggesting the tune's greater popularity by the time of the second edition of the French printing.[39] It contains a spurious attribution to the opera composer Stanisław

[37] Larry Wolff, *Inventing Eastern Europe: The Map of Civilization on the Mind of the Enlightenment* (Stanford, CA: Stanford University Press, 1994). A good example of the self-perpetuation of this borderland rhetoric contemporary with the Crimean War is in Juliusz Rawicz Przyjemski, *Sketches of the Polish Mind* (London: William and Frederick G. Cash, 1857), vii–ix.

[38] Only the Parisian version survives from the early printing (Włodzimierz Wolski, *Śpiewy powstańcze*, 2nd ed. [Paris: Drukarnia Polska, 1863]).

[39] This edition has formed the basis of revival concerts of these songs: for example, by Tadeusz Kaczyński's late-twentieth-century ensemble, the Traugutt Philharmonic, which took a hero of the January Uprising as its namesake.

Moniuszko, for whom Wolski had written libretti. The inclusion of a musical engraving helps it stand out from other songs in Wolski's compendium.

The collection might be considered a playlist for battle, beginning with preparatory verses—including a prayer that choreographs a moment of spiritual devotion and reflection. Wolski presents a roll call of those who are there, describing the combat prowess of each unit, beginning with Warsaw's poor. Five marches rally particular constituencies and build toward the "March of the Uprising's Leaders" (Marsz Powstańców). Each march makes sure to clothe the fighting forces in Warsaw in official military attire and conjures a sensorium of the battlefield. (Although the lineup is to some extent inclusive, ranging from the poor to sharpshooters to the rebellion leaders, it is decidedly an all-male assortment of civilian forces.) The bound edition of the collection alludes to the anonymously composed tunes to which they are to be sung. There is a long tradition of battle cries being used to organize and energize attack, and these songs, too, set units in motion. However, these songs are neither prompts nor accompaniment; they are to be embodied as battle.[40] The introductory song (*przedśpiew*) makes the connection between military movement and singing explicitly. The poet begins, "In vain do the sober platitudes ramble on/[Telling us] that when action speaks, song falls silent."[41] The opposition between art and concrete deeds is assumed and repeated, the poet suggests. But it should not be so. Wolski makes bold claims in verse: "The army needs music for battle." The collection is that music, explicitly framed as instrumental to the fight.

The "March of the Zouaves" celebrates the "Zouaves of Death," a legion of fighters led by Frenchman François Rochebrunne, who had served in the Zouaves' light infantry regiment in the Crimean War. From their mustaches to their uniforms to their feisty spirit, Wolski's portrait of the foreign legion—which was initially made up of North African soldiers—exoticizes their difference in a manner that would have suggested to readers and listeners a kind of "authentic" military might. None other than Roger Fenton, the notorious war photographer in Crimea, fixated on the markers of difference in his own portrait series of the Zouaves, even posing himself in their uniform.[42] The specific enduring power of the Zouaves of Death in Varsovian culture lies in their inclusivity and mimicry: two values that had resided at the heart of the Sarmatian movement among

[40] For an extended discussion of music that inspires combat see Jonathan Pieslak, *Sound Targets: American Soldiers and the War* (Bloomington: Indiana University Press, 2009), 46–49. Pieslak also analyzes repertories as inextricable from mediation, in particular the personalization and mobility afforded by mobile music devices from radio to the iPod.

[41] Wolski, *Śpiewy Powstańcze*, 3. Thanks to Tul'si Bhambry for assistance with translation.

[42] In a 1855 photograph by Marcus Sparling, No. 21a in the Roger Fenton Crimean War photograph collection, Library of Congress Prints and Photographs Division Washington, DC, USA. The collection contains numerous portraits of Zouaves.

Polish nobility (*szlachta*). From the sixteenth through the eighteenth centuries, Sarmatism turned to myth to assert that *szlachta* descended from ancient Sarmatian tribes.[43] A defining aspect of Sarmatism was its fashions: the nobility adopted supposedly Eastern customs and dress, particularly hats and mustaches (matching those of Wolski's Zouaves).

At mid-century, overtly orientalist dress continued as a strain of Polish nationalism, both out of necessity for travelers and as a means of performing Turkophilia.[44] On board a steamboat heading to Turkey, the poet Adam Mickiewicz was struck by the rubber mustaches Polish soldiers affixed to their lips as they headed to Crimea.[45] Singing the march in Warsaw in 1863 thus extended a specifically Polish self-orientalizing custom that was visual more than it was sonic.[46] It also connected those fighting an obviously losing battle against Russia in the January Uprising with the self-sacrificing spirit of the Zouaves, who proclaimed their desire to fight to the death. The identification with numerous cultural "others" merged with message of inclusivity that those driving the uprising used to recruit fighters. One of the widest circulating publications by the Polish rebels explicitly outlined "the Catholic, the Orthodox, the Uniate, the Protestant, the Armenian, the Jew, the Muslim—all are Poles despite religious differences."[47]

In "March of the Zouaves," Wolski illustrates the power of naming, shouting, and the sounds of war, placing them in juxtaposition with the more restful and soothing landscapes of nature. He leads with the most striking characteristic of the notoriously bloodthirsty unit (who were forced to pledge that they would either win or perish): their orientalist costuming. The musical setting, by contrast, ploughs a furrow already well established by earlier legion songs, more than it mirrors Wolski's rich sound world of immersed struggle and complex social negotiation. Indeed, the pedantic march mutes the harsh sounds of the verses. Meanwhile, the hook of the Dąbrowski March echoes through the words "March, march Zouaves." Throughout the settings of the collection jagged arpeggiations loop through march rhythms, and the boundaries between these

[43] Maria Bogucka, *The Lost World of the "Sarmatians": Custom as the Regulator of Polish Social Life in Early Modern Times* (Warsaw: Polish Academy of Sciences, 1996).

[44] On travel and costume see Adam Geczy, *Fashion and Orientalism: Dress, Textiles and Culture from the 17th to the 21st Century* (London and New York: Bloomsbury, 2013), 96–100.

[45] W. Mickiewicz, *Żywot*, cxiv.

[46] For a discussion of postcolonial theory in the Polish context see Stanley Bill, "Seeking the Authentic: Polish Culture and the Nature of Postcolonial Theory," *Nonsite* 12 (2014), <http://nonsite.org/article/seeking-the-authentic-polish-culture-and-the-nature-of-postcolonial-theory>, accessed 16 Dec. 2015.

[47] Quoted in Brian Porter-Szűcs, *Faith and Fatherland: Catholicism, Modernity, and Poland* (Oxford: Oxford University Press, 2011), 216.

miniatures begin to blur. In the musical material, the remarkable generic stability of patriotic marches—even those like the "March of the Zouaves," which glorifies death in orientalist terms—is conspicuous.

Toward Crimea under Mickiewicz's Command

Like the "March of the Zouaves," which captures a backward glance at Crimea from Poland, the final portion of this essay traces a route from the exile community toward the Black Sea. This is an attempt to draw more concrete connections across the marches that have framed the Polish orientations toward Crimea, and to thereby locate the legion songs not only in surviving evidence but also in historical practice. The key traveler for this orientation of Crimean sounds through the idea of Poland is Adam Mickiewicz (1798–1855), whose life neatly fills the historical gap between the Dąbrowski March and Wolski's retrospective valorization. At the moment of his lectures in Paris, mentioned previously, Mickiewicz earned respect for his visionary performance of poetic verse in terrific improvisations that impressed audiences more than did his poetry.[48] None other than George Sand understood his charisma in musical terms: "Not since the prophets of Zion has a voice risen with such force to sing on a topic so expansive as the collapse of a nation."[49] Through his formation of legions—first in Italy and then in Turkey with ambitions for Crimea, he planned to push beyond the "sung inspiration" behind which most writers sheltered themselves.[50] He died before he could realize his vision. Let us begin by seeing and hearing the peninsula through the writer's own eyes and ears, from his first encounter in 1827 until his passing in Istanbul, where he walked among the war's wounded in November 1855.

The bard conjured up Crimea—as both a location and as an idea—for the Polish romantics. Gazing upward, his cloak blending with the shadows of a rocky summit, Mickiewicz appears the paradigmatic bard in Walenty Wańkowicz's iconic painting of 1827–28 (see Figure 5.4). Cheeks flushed, his hand clutching

[48] Wiktor Weintraub, "The Problem of Improvisation in Romantic Literature," *Comparative Literature* 16/2 (1964), 119–37. On the relative favor of poetic improvisation over musical see Dana Gooley, "Saving Improvisation: Hummel and the Free Fantasia in the Early Nineteenth Century," *The Oxford Handbook of Critical Improvisation Studies*, vol. 2, ed. George E. Lewis and Benjamin Piekut (*Oxford Handbooks Online*, http://www.oxfordhandbooks.com.libproxy.lib.unc.edu/view/10.1093/oxfordhb/9780199892921.001.0001/oxfordhb-9780199892921-e-006, accessed 3 Jan. 2016).

[49] Quoted in Stanisław P. Koczorowski, *Adam Mickiewicz et la pensée française, 1830–1835* (Paris: Gebethner et Wolff, 1929), 42.

[50] Maria Janion, "Mickiewicz's Jewish Legion," *Hero, Conspiracy, and Death: The Jewish Lectures*. Trans. Alex Shannon (Frankfurt am Main: Peter Lang, 2014), 179.

Figure 5.4 Walenty Wańkowicz, lithograph of Adam Mickiewicz. National Library of Poland,1828, G.6227/III, held at National Museum in Warsaw; in the public domain.

his heart through a flamboyant scarf, he gazes as though to channel the mystic inspiration that fires the improvisations at the heart of his epic poetry. Mickiewicz projects solid might and artistic sensibility. The portrait depicts the writer in exile after fighting for Polish independence as a student, perched upon Ayu-Dag in Crimea. His writings elicit the early-nineteenth-century fantasy of Crimea when he described his arrival there in rapture. "I saw the Crimea!" he writes a friend, "I saw the orient in miniature."[51]

[51] Quoted in Roman Koropeckyj, *Adam Mickiewicz: The Life of a Romantic* (Ithaca, NY: Cornell University Press, 2008), 68.

During his first visit to Crimea Mickiewicz composed a series of love poems: the *Crimean Sonnets* (1826). These related his adventures in central Asia and told of his passionate love affairs. He dramatized impressions of his journey through recurring characters who represented Eastern perspectives (the Tatar noble, or Mirza) and Western encounter (the Pilgrim). Across the work, sound and speech play a particular role in inflecting cultural closeness and mimicry. The noble speaks foreign words; the Pilgrim hears Tatars at remove:

From out the mosques the pious wend their way;
  Muezzin voices tremble through the night;
Within the sky the pallid King of Light
  Wraps silvered ermine round him while he may,
And Heaven's harem greets its star array.
  One lone white cloud rests in the azure height—
A veiled court lady in some sorrow's plight—
  Whom cruel love and day have cast away.[52]

Mickiewicz's orientalism was strongly influenced by the relationships he had cultivated with Russian writers immediately preceding his journey to Crimea during an extended romp with nobility in Odessa.[53] "If the Sonnets are received well," he wrote to his mentor, "I intend to compose something more extensive in the oriental style; if, on the other hand, those minarets, namazes, izans, and other such barbarian sounds do not find favor in the classicists' delicate ear, if . . . then I'll say [. . .] I'm chagrined, but I'll keep on writing."[54]

Before the siege of Sevastopol, Crimea marked the furthest border of the imagined Poland "to be." Mickiewicz's *Sonnets* emerge in a nationalist discourse obsessed with geography. Upon the dissolution of the Polish-Lithuanian Commonwealth, the nobility and intelligentsia initially developed a vision for a new Poland that focused more on lost territory than lost people; they dwelt on lost power more than on imperial presence, with the boundaries of the former kingdom presumed as the "natural" borders of Poland.[55] The emphasis on places rather than peoples is particularly marked in the writings of poet-bards whose "songs" adumbrate diverse regions, such as Ukraine, Podolia, and Lithuania, as

[52] Adam Mickiewicz, *Sonnets from the Crimea*, trans. Edna Worthley Underwood (San Francisco: Paul Elder, 1917), 11.

[53] Roman Koropeckyj, *Adam Mickiewicz: The Life of a Romantic* (Ithaca, NY: Cornell University Press, 2008), 63–67.

[54] Roman Koropeckyj, "Orientalism in Adam Mickiewicz's *Crimean Sonnets*," *The Slavic and East European Journal* 45/4 (2001), 662.

[55] Serhiy Bilenky, *Romantic Nationalism in Eastern Europe: Russian, Polish, and Ukrainian Political Imaginations* (Stanford, CA: Stanford University Press, 2012), 27.

one ideological home. Based in Paris, the poet and geographer Wincenty Pol surveyed one such circumscribed nation from sea to sea in "Janusz's Song" (1835):

> I was in Lithuania and in the Crown Poland,
> I was in this and that side,
> I was here and there;
> From the [Carpathian] Mountains to the Baltic Sea coast,
> From Lithuania as far as to [the Dnieper rapids]
> I know the entire Poland.[56]

In his study of nationalism, Serhiy Bilenky has emphasized, contrary to later nineteenth-century thought and its legacy, that free Poland was not predicated on unified nationality or the question of whether all those living in Poland were Polish—or whether all Poles lived in Poland. What is more, Polish political life was organized by those military generals and members of the intelligentsia who had been displaced by exile to Paris and London and who were acutely aware of their own place while able only to imagine life within the partitions. For Mickiewicz, the Black Sea was a site of multiethnic cosmopolitanism that did not create friction with his sense of Polishness. During his stay in Odessa, for example, he attended Carl Maria von Weber's *Der Freischütz* with Russians. They saw French and Italian performers in action. The opera caught Mickiewicz's ear: he asked a friend to send for the score to the Hunter's Chorus, an anecdote that lends credibility to my hunch that the poet was listening carefully as he traveled.[57]

As I have already suggested, songs take on a peculiar agency in my materials, and it is with song that I stretch my listening to history. Songs accommodate text and music throughout diverse contexts of nineteenth-century Polish nationalism. They serve as a medium (both real and imaginary) for the Herderian voices of the folk and for the sophisticated metric improvisation of bards like Mickiewicz. The nobility prized the ability of poets to sing history into existence, to stimulate politics with recitation. One Polish nobleman in England mourned his exile in a memoir. To do so, he felt compelled to explain the instrumentality of song: "To the heart of a Pole, the coarsest rhyme that pays homage to the heroes of his country, recited by a witness and partaker of the scenes recorded, written with their swords, and dedicated to Poland, will be welcomed as an Homerian strain."[58] The obvious profusion of romantic metaphor in such comments can lead to their dismissal, but I would suggest that we not let hyperbole prevent

[56] Adapted from Bilenky, *Romantic Nationalism in Eastern Europe*, 19.

[57] Leon Gomolicki, *Dziennik pobytu Adama Mickiewicza w Rosji, 1824–1829* (Warsaw: Książka i Wiedza, 1949), 75. In a sense, the larger scope of my work on Crimea emerges as a study of artists' wartime hearing, perhaps as an exilic sensory reorientation.

[58] Przyjemski, *Sketches of the Polish Mind*, 29.

us from taking the text of patriotic song seriously. This exilic vision invites us to rehear patriotic songs as a technology that may elude the order of conscious action and predictable, mechanical effect that Wolski exhorted in the 1860s.

Sentenced to death after leading a failed military uprising, one of Mickiewicz's heroes pushes song's power beyond the body. Before the legionnaire-protagonist Konrad Wallenrod commits suicide, he bids farewell with the hope: "Where I cannot roam, my song will take wing."[59] In his own poems published as "Songs" (*pieśni*), Mickiewicz makes claims for the abstraction and universality of music. Art cannot be reduced to the heard utterance, but is positioned in a full sensory experience in which I hear resonances of Steve Goodman's emphasis upon dangerous vibes and "modulated moods," which he argues are constitutive of wartime hearing.[60] In a dedication of his *Sonnets* to a friend, for example, Mickiewicz figures this wartime hearing as heightened receptivity to foreboding sounds: "Oh poetry, you are not an art to be sung, because my sentiments do not have a voice that can be understood; they are nearer subterranean vibrations that no one can hear as murmurs."[61]

The literary identity of Włodzimierz Wolski stands in contrast to that of Adam Mickiewicz and clarifies the radicalism of Mickieiwcz's musical imaginary. Whereas Mickiewicz's epic poems articulate the mystical, ordained talent of a bard, Wolski collaborated and linked his writings to direct political action from within Poland. A generation younger than Mickiewicz, he moved to Warsaw after the January Uprising and was one of the twenty-somethings who reestablished Warsaw's literary haunts—cafes, salons—with politically engaged writings in *The Scholarly Review* (*Przegląd Naukowy*).[62] The scene extended across the arts with Wolski active in many of the collectives that together are known as Warsaw's Bohème (*cyganeria*) from 1840 through 1843. These poets and writers conceived of their work as an inflammatory collective, reading and critiquing one another's works so as to draw out and sharpen the politically insurgent. Wolski wrote across literary genres, though perhaps of the greatest interest to musicologists are his crafty renderings of folk dances (obereks, krakowiaks) and his lyrical fantasy on Chopin. Not all of Wolski's song-inflected verses were written to be set to music. His "Siberian March," for example, builds the drudgery of deportation out of the slogging sectional form of a march. The foreboding

[59] Adam Mickiewicz, *Konrad Wallenrod and Grażyna*, trans. Irene Suboczewski (Lanham, MD: University Press of America, 1989), pt. 6, line 233.

[60] Steve Goodman, *Sonic Warfare: Sound, Affect, and the Ecology of Fear* (Cambridge, MA: MIT Press, 2010), xvi.

[61] Quoted in Gomolicki, *Dziennik*, 168.

[62] "Wolski, Włodzimierz," *Literatura Polska: Przewodnik encyklopedyczny*, vol. 2, ed. Julian Krzyżanowski and Czesław Hernas (Warsaw: Państwowe Wydawnictwo Naukowe, 1984), 624.

Siberian landscape contrasts with the wonders of the Black Sea so prevalent in Mickiewicz's Crimean Sonnets.[63]

Mickiewicz nearly returned to Crimea—during the Crimean War—just before his death. He left France for Istanbul on what he called a "mystical-moral-military mission" to form a Jewish legion to fight alongside the "Sultan's Cossacks" with Ottoman forces in Crimea.[64] The boat journey and reconnaissance trips through Polish émigré circles brought him ever closer to the battlefield. His eldest son transcribed the poet's notes during this journey. These are packed with sonic detail and reveal a traveler who increasingly *listens* for the martial as he brainstorms rallying rhetoric. For example, upon leaving Paris Mickiewicz imagined himself marching with squadrons, listening to the "music of the horses' hooves" to keep his journey from becoming boring.[65] Already in Lyon, as he made his way toward the port in Marseille, his letters reveal a heightened interest in the battlefield. To his children back in Paris he wrote, "Crossing the bridge, we saw a couple of caravans loaded with wounded soldiers for the first time."[66] Upon arrival in Istanbul, he walked the streets, taking in the city's public spaces as a meeting point of cultures.

Sadyk Pasha invited the poet and his companion to Burgas, to the camp in which he trained his Polish Cossacks. Their collaboration was imbued with the potential to transform history. Mickiewicz's personal secretary, Armand Lévy, imagined a triumphant conclusion to the endeavor that would mean "changing the refrain in *Jeszcze Polska* [. . .] everyone would be able to sing: March, march Sadyk, from Turkish land to Poland."[67] At the military camps they were treated as generals to a feast and even presented with a spectacle—Cossack dances ("no Western ballet can compare!") as well as an equestrian play. Mickiewicz was touched that the foot soldiers sang together, teaching one another songs.[68] In his final account of the camp—one of his last writings, as he died one month later—we can at last listen in upon a Polish military band. This ensemble very likely later joined in the music making that punctuates accounts of the fighting and living in Crimea during war.[69] Satiated and motivated, Mickiewicz relishes the "lavish dinner . . . they sing, more music, until finally the real krakowiak

[63] Włodzimierz Wolski, "Marsz Sybirski," *Promyki: Nowe poezje liryczne* (Brussels: J.-H. Dehoue, 1869), 31–34. For a discussion of the trope Siberia as hell in Polish literature, see Stanisław Eile, *Literature and Nationalism in Partitioned Poland, 1795–1918* (New York: St. Martin's, 2000), 76–80.

[64] Quoted in Janion, "Mickiewicz's Jewish Legion," 184.

[65] W. Mickiewicz, *Żywot*, cxiv.

[66] Ibid.

[67] W. Mickiewicz, *Żywot*, lxxxvi.

[68] Ibid., civ.

[69] I think here, for example, of the panoramic attention to music making at the camps in William Russell, *The British Expedition to the Crimea* (London: Routledge and Sons, 1877), 140–42.

with violins, banduras, drums."[70] In this tantalizingly bare description, the "orient in miniature" returns, as does the multiethnic performance of Polishness in Wolski's "March of the Zouaves."

My goal has been to reorient the relationship between Poland and European geopolitics in the nineteenth century through poetics, framing it as a history of migration and military engagement rather than occupation and disenfranchisement. Over and against the silences of the archive, legion songs have allowed me to take the spiritual turn by Rousseau (and my poet-interlocutors) seriously. After all, it is almost exclusively in this form that the sounds of a then-minor power's participation in a global conflict may be recuperated now. The narrative of Polish Crimean history in Mickiewicz's notes brings us closer to Polish mediations of the war through music and sound, connecting the metaphysical vibrations of Mickiewicz's inspiration with the physical energy of soldiers making music, recasting the role of Polish struggles for independence in shaping the theaters of the Crimean War. The mobilizing gestures of the mazurka set this essay in motion, and Mickiewicz's project lands it squarely in a march toward Crimea—one cut short by his death.

[70] W. Mickiewicz, *Żywot*, cvi.

Appendix 5.1

Marsz Żuawów	**March of the Zouaves**
Nie masz to wiary, jak w naszym znaku!	You do not have the faith that we do, under this sign!
Na bakier fezy, do góry wąsy,	Our fezzes askew, mustaches to the heavens,
Śmiech i manierek brzęk na biwaku,	Our laughter and the clinking of flasks at camp,
W marszu się idzie jak gdyby w pląsy.	In the march we walk as if it were a dance.
Lecz gdy bój zawre, to nie na żarty,	But when the battle comes to an end, it's no joke,
Znak i karabin do ręki bierzem,	We take the target and rifle in our hands,
A Polak, w boju kiedy uparty	And when a Pole is stubborn in battle
Stanie od razu starym żołnierzem.	He immediately stands as an old soldier.
Marsz, marsz, żuawy!	March, march, Zouave!
Na bój, na krwawy,	In the battle, for the bloody,
Święty, a prawy,	Holy, and just,
Marsz, żuawy—marsz!	March, Zouave—march!
Pamięta Moskwa, co żuaw znaczy,	Remember Muscovy, what it means to the Zouave,
Drżąc sołdat jego wspomina imię;	The trembling "soldat" remembers his name;
Sporo bo nakłuł carskich siepaczy	Especially since it was he who stabbed a load of the Tsar's henchmen,
Brat nasz, francuski Żuawek w Krymie.	That brother of ours, the French Zouave in Crimea.
Miechów. Sosnówkę. Chrobrz. Grochowiska.	Miechów. Sosnówka. Chrobrz. Grochowiska.
Dzwoniąc też w zęby, wspomni zbój cara —	Also calling out baring teeth, he refers to the Tsar bandit —
Krwią garstka doszła mężnych nazwiska.	Blood flowed from a few brave names.
Garstka się biła, jak stara wiara.	A handful fought, as one seeks faith.

Kiedy rozsypiem się w tyraliery

Zabawnie z bronią pełzać, jak krety;
Lecz lepszy ogień—gęsty a szczery,
I lepszy rozkaz: "marsz na bagnety!"
Bo to sam bagnet w ręku, aż rośnie

Tak wzrasta zapał w dzielnym ataku.
"Hura" i "hura!" huczy żałośnie

Górą krzyż biały na czarnym znaku!

W śniegu i błocie mokre noclegi,
Choć się zasypia przy sosen szumie.
W ogniu rzednieją diablo szeregi.
Chociaż się zaraz szlusować umie.
A braciom ległym na polu chwały.

Mówimy: "Wkrótce nas zobaczycie.
Pierw za jednego z was pluton cały
Zbójów, nam odda marne swe życie."

When they are dislodged in their line formations
They comically crawl with their guns like moles;
Yet even better with fire—dense and in earnest
And better the command: "March with bayonets!"
For with that same bayonet in hand as things escalate
Such growing zeal in the valiant attack:

"Hurrah!" and "Hurrah!" they roar mournfully
To the heavens a white cross on a black shield!

Sleeping in wet snow and mud,
And falling asleep to the rustle of the pines.
They thin out the devils with their fire.
While they also immediately join them
And the brothers lying on the field of glory.
We say: "Soon you will see us.

First so that one of your entire platoons
Of thieves gives unto us his miserable life."

6

Who Sings the Song of the Russian Soldier?

Listening for the Sounds and Silence of War in Baltic Russia

KEVIN C. KARNES

The inscription below this "Song of a Russian Soldier" (Figure 6.1) tells us who sang it, when, and where: a rifleman named Fedor Filonin, waiting on the Gulf of Finland in April 1854. Here are the opening two stanzas, and the last stanza, loosely translated:

> As we stood in Reval
> Very close to the enemy,
> We waited for him
> On the Baltic shore.
>
> We scanned the sea swiftly,
> We waited hour after hour;
> We'll give you a good helping of woe—
> You're in for a real treat! [. . .]
>
> When the commanders call:
> "Forward, boys, for the Tsar!"
> There won't be a peep from the Mussalmans [Busurmane],
> When we're all barking hurrah!

Filonin, the singer, was encamped among some 20,000 men: a garrison stationed to guard against a British incursion by sea into the empire's northwest. That attack, however, was never to come. The experience of which he sang echoed that of many in the Baltic theater of the Crimean War: waiting along the coastline, passing the time in vague apprehension (and miserable conditions). But to

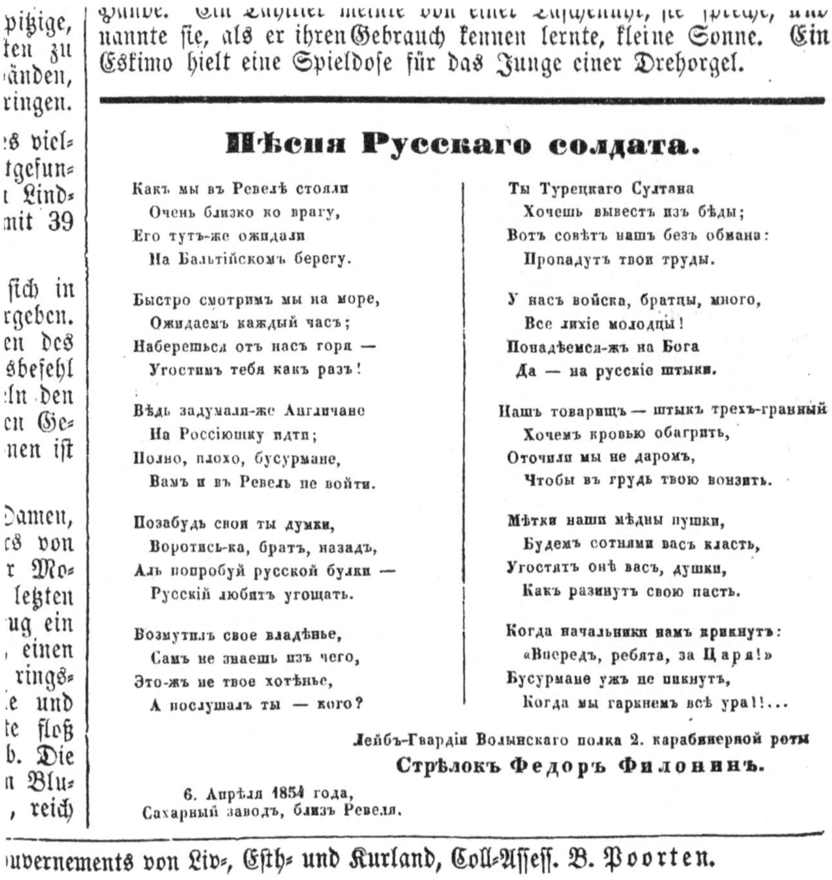

nannte sie, als er ihren Gebrauch kennen lernte, kleine Sonne. Ein Eskimo hielt eine Spieldose für das Junge einer Drehorgel.

Пѣсня Русскаго солдата.

Какъ мы въ Ревелѣ стояли
Очень близко ко врагу,
Его тутъ-же ожидали
На Бальтійскомъ берегу.

Быстро смотримъ мы на море,
Ожидаемъ каждый часъ;
Наберешься отъ насъ горя —
Угостимъ тебя какъ разъ!

Вѣдь задумали-же Англичане
На Россіюшку идти;
Полно, плохо, бусурмане,
Вамъ и въ Ревель не войти.

Позабудь свои ты думки,
Воротись-ка, братъ, назадъ,
Аль попробуй русской булки —
Русскій любитъ угощать.

Возмутилъ свое владѣнье,
Самъ не знаешь изъ чего,
Это-жъ не твое хотѣнье,
А послушалъ ты — кого?

Ты Турецкаго Султана
Хочешь вывестъ изъ бѣды;
Вотъ совѣтъ нашъ безъ обмана:
Пропадутъ твои труды.

У насъ войска, братцы, много,
Все лихіе молодцы!
Понадѣемся-жъ на Бога
Да — на русскіе штыки.

Нашъ товарищъ — штыкъ трехъ-гранный
Хочемъ кровью обагрить,
Оточили мы не даромъ,
Чтобы въ грудь твою вонзить.

Мѣтки наши мѣдны пушки,
Будемъ сотнями васъ класть,
Угостятъ онѣ васъ, душки,
Какъ разинутъ свою пасть.

Когда начальники намъ крикнутъ:
«Впередъ, ребята, за Царя!»
Бусурмане ужъ не пикнутъ,
Когда мы гаркнемъ всѣ ура!!...

Лейбъ-Гвардіи Волынскаго полка 2. карабинерной роты
Стрѣлокъ Федоръ Филонинъ.

6. Апрѣля 1854 года,
Сахарный заводъ, близъ Ревеля.

...uvernements von Liv-, Esth- und Kurland, Coll.-Assess. B. Poorten.

Figure 6.1 Fedor Filonin, "Pesnya Russkago soldata" [Song of the Russian Soldier]. *Rigasche Zeitung* (28 April 1854); in the public domain.

whom was his song addressed, published here, on the pages of a German newspaper in Riga, some two hundred miles to the south? And who was its singer, Filonin? Second carbine commander, the paper recorded, and possibly an ethnic Russian, since his song was identified as one of a *russkii* soldier (potentially an ethnic marker), rather than a *rossiiskii* one. But Filonin's regiment was from Volynsk, now Novohrad-Volynskyi, a Ukrainian city closer to Vienna than to his present station. And that station itself? To most of its inhabitants, it was known not as Reval but as Tallinn, spoken in the Estonian.

Aside from the title under which it was printed, Filonin's song is striking for the absence of unambiguously Russian peoples and spaces in its address. We have a Russian soldier who might be Ukrainian, singing in a largely Estonian space, published in the German press of a predominantly Latvian-speaking imperial province. Linguistic, national, and ethnic diversity is registered in virtually every aspect of his song—in its inscription of authorship, its place of publication, and the words it intones above all. But of course, even the most radical diversity might easily pass unseen. In that same spring of 1845, just as Filonin waited and sang, Sir Charles Napier, British commander of Victoria's Baltic Fleet, was anchored off the coast of Reval/Tallinn, gazing upon the shore. When he requested intelligence from the Board of Admiralty about those people he glimpsed on land, he was told of a historical "Esthonian" people, once "regarded as unconquerable but generous foes." However, Napier's counsel explained, within a handful of generations those people had mostly disappeared. "The majority have become Russian."[1]

In this anticipated engagement between Napier's men and the soldiers of Filonin's division, we encounter what the historian Larry Wolff calls the "operations of mental mapping" inherent to the project of "inventing" Eastern Europe—a transnational Enlightenment discourse that took on still familiar contours in the early nineteenth century. As Wolff reveals, the imaginative partitioning of Europe into eastern and western domains took place largely through the literary "operations" of "association and comparison" in work by writers from the continent's western reaches: "association among the lands of Eastern Europe, intellectually combining them into a coherent whole, and comparison with the lands of Western Europe, establishing the developmental division of the continent."[2] Tracing these operations of association and comparison through works of memoir, fiction, and geography, Wolff reveals what might be called the interchangeability of "Eastern" spaces in discourse within the European West. He catalogues what we might describe as modes of unseeing the diversity of easterly places and their inhabitants, modes that enabled Western observers to construct for themselves a fathomable whole of Eastern geographies and peoples, often with the purpose of vividly contrasting them with life at home. In its published form, Filonin's song testifies to the diversity of wartime experience in imperial Russia. But when regarded by the sailors aboard Napier's ship, Filonin and others who stood on the shore were Russians one and all.

In this essay, I take Wolff's arguments as my theoretical point of departure. But I will open them to critical engagement by highlighting two of their distinguishing features. First, Wolff's principal concern is a literary one. He studies what people wrote in prose and poetry, how they reported and how they fictionalized. Second, he is primarily concerned with what people *saw* or imagined

[1] G. Butler Earp, *The History of the Baltic Campaign of 1854: From Documents and Other Materials Furnished by Vice-Admiral Sir C. Napier, K.C.B.* (London: Richard Tentley, 1857), 101–4 (quoted at 104).

[2] Larry Wolff, *Inventing Eastern Europe: The Map of Civilization on the Mind of the Enlightenment* (Stanford, CA: Stanford University Press, 1994), 6.

that others might *see*, or with how a person might show on a map what there was to be seen when voyaging abroad. What I will try to do here is to suggest some ways in which we might understand things differently if we consider what was heard alongside what was seen on the Baltic coasts during and after the war. I am interested in how attending to experiences, practices, and discourses of listening can yield fresh historical understandings that unsettle those formed by attending primarily to acts of looking and seeing.[3] I wish to understand how audition structured encounters within Eastern Europe, just as its contours were being mapped by travelers from the continent's West. And I am concerned especially with easterly geographies renowned in the middle of the nineteenth century for the diversity of their peoples and the complexity of their contested histories. "A narrow strip of land," wrote a historian from Leipzig of Russia's Baltic provinces in 1868, "inhabited by small groups of peoples [*Völkersplittern*] of the most various kinds [...] whose highly exceptional circumstances are as little known in the western half of Europe as in the eastern."[4]

I take my historical bearings from the Crimean War, for that war afforded unprecedented opportunities for encounter among those hailing from westerly as well as easterly locales. To Napier's sailors, voyages through the Baltic Sea in 1854 and 1855 brought daily glimpses of, and occasional exchanges with, individuals living or stationed onshore. Although the British never made incursions on land, the coastlines were likewise sites of innumerable encounters, as thousands of soldiers from across imperial Russia encamped to defend against anticipated attack. In Reval and Riga, as in London and Manchester, families of soldiers scoured the papers for news from the Baltic theater, news that might now arrive in days rather than weeks, thanks to the new technology of the telegraph. In the Baltic provinces, these updates came not only in Russian but in the Estonian, Latvian, and German languages as well, creating a surge of vernacular discourse that owed to an unprecedented relaxation of imperial censorship in an attempt to cultivate popular support for the war effort.[5]

[3] I elaborate these theoretical concerns more fully in Kevin C. Karnes, "Inventing Eastern Europe in the Ear of the Enlightenment," *Journal of the American Musicological Society* 71 (2018): 75–108. In the present chapter, my work is inspired especially by Ana María Ochoa Gautier, *Aurality: Listening and Knowledge in Nineteenth-Century Columbia* (Durham, NC, and London: Duke University Press, 2014); Mark M. Smith, *Listening to Nineteenth-Century America* (Chapel Hill and London: University of North Carolina Press, 2001); and Alain Corbin, *Village Bells: Sound and Meaning in the Nineteenth-Century French Countryside*, trans. Martin Thom (London: Macmillan, 1998).

[4] Julius Eckardt, *Die baltischen Provinzen Rußlands: Politische und culturgeschichtliche Aufsätze* (Leipzig: Duncker & Humbolt, 1868), 1.

[5] On the telegraph and mediation of wartime experience, see Orlando Figes, *The Crimean War: A History* (New York: Picador, 2010), 305–6. An example of telegraphic dispatches on the pages of the Riga press is "Die letzten telegraphischen Nachrichten über die Bewegung der feindlichen Flotte," *Rigasche Zeitung* (25 June 1854), 1. The wartime relaxation of press censorship in the region

Ranging widely across historical sources, this essay examines traces of auditory experience in and of the Baltic theater of the Crimean War, traces that come down to us in the form of inscriptions (here largely textual ones) made by historical witnesses testifying to what they heard.[6] On the one hand, I will argue that some of these inscriptions confound the operations of association and comparison catalogued in Wolff's scholarship. They memorialize sites where or moments when those operations were undone, sometimes by listening and hearing rather than looking and seeing. On the other hand, I will point to inscriptions that testify to the presence of Wolff's operations where we might not expect them: as they structured the auditory encounters between individuals in easterly spaces and others in their midst. I follow Mark M. Smith in reading these inscriptions not as literal descriptions of sounding landscapes, but rather as "aural metaphors" shaped by individuals with "ears pricked by the day-to-day shape and defense of their worlds." At the same time, I heed his warning that such metaphors often seemed "real to those who did the [always] selective listening"—real enough to inspire some who conjured or read them to engage in physical action.[7]

While attending to sounds like the shouts given voice by Filonin's Russian soldiers, I join Mary Favret and Hillel Schwartz in pointing to war's defining silences alongside its famous din, silences like those that marked the advance of Filonin's imagined enemy.[8] And I will attend especially to the capacity of silence to function as an index of absence. In doing so, I follow Michel de Certeau, who observes, "the places people live in are like the presences of diverse absences. What can be seen"—and, I would argue, what can be *heard*—"designates what is no longer there: 'you *see*, here there used to be . . . ,' but it can no longer be seen." He continues, "There is no place that is not haunted by many different

is described in Vita Zelče, "Grāmatniecība un prese" [Book-publishing and the press], *Latvija 19. gadsimtā* [Latvia in the 19th century], ed. Jānis Bērziņš (Riga: Latvijas Universitātes Latvijas vēstures institūts, 2000), 363–64.

[6] Following Ochoa Gautier, I use the word *inscription* in a sense anticipated by Lisa Gitelman, to denote a "legible representation of aural experience." See Ochoa Gautier, *Aurality*, 7; and Gitelman, *Scripts, Grooves, and Writing Machines: Representing Technology in the Edison Era* (Stanford, CA: Stanford University Press, 1999), 15.

[7] Mark M. Smith, "Listening to the Heard Worlds of Antebellum America," *The Auditory Culture Reader*, ed. Michael Bull and Les Back (Oxford and New York: Berg, 2003), 159; and in *Hearing History: A Reader*, ed. Mark M. Smith (Athens and London: University of Georgia Press, 2004), 379–80. I borrow the image of the sounding landscape (*paysage sonore*) from Corbin, *Village Bells*, xi and xx, while departing from the translation by Thom, who renders it "auditory landscape."

[8] Mary A. Favret, *War at a Distance: Romanticism and the Making of Modern Wartime* (Princeton and Oxford: Princeton University Press, 2010); Hillel Schwartz, *Making Noise: From Babel to the Big Bang and Beyond* (New York: Zone, 2011).

spirits hidden there in silence, spirits that one can 'invoke' or not."[9] Absence, for Certeau, conditions experience just as powerfully as presence. Absence is often marked by silence, and it is everywhere linked to memory. For just as sound inscribed by historical listeners may record moments or conditions of encounter and presence, so may silence index the conditions of absence that shaped wartime experience in turn. This silence might stand for physical distance or emotional loss. It may also register failures of recognition or understanding, or the failure of the expected to materialize.

Noises of Encampment

In spring 1854, Filonin stood on the northern edge of a landscape of encampment by imperial soldiers, recruited from across the empire to safeguard its northwestern shores. By then, some 270,000 soldiers had been stationed along a curvy littoral extending westward from St. Petersburg to Reval/Tallinn before turning southward to pass through Riga and Memel (Klaipėda) in the Pale of Settlement. In 1863, Reval was found to have 20,680 inhabitants. Nine years earlier, when Filonin had sung, some 20,000 troops were stationed in its vicinity. In Kurland province to the south, some 40,000 were encamped.[10]

With this migration to the Baltic coasts came daily, often mundane encounters between soldiers, and also between residents and others who more regularly inhabited distant imperial locales. As the folklorist Fricis Brīvzemnieks recalled of his childhood on the Kurland coast in the 1850s, "the period of the Crimean War brought with it a great deal of noise. Soldiers traipsed around, took up residence in the tavern, and played wartime music." Unable to bear the din or the figures whose noise unsettled his formerly peaceful home, Brīvzemnieks remembered "sometimes fle[eing] to the nearby portage shores."[11] For another Kurland witness, the writer Aleksandrs Vēbers, the encampment of soldiers he remembered from childhood inspired recordings of noisy plurality and previously untold diversity. "The noise of war resounded not only on the Black Sea

[9] Michel de Certeau, *The Practice of Everyday Life*, trans. Steven Rendall (Berkeley and Los Angeles: University of California Press, 1984), 108 (emphasis and ellipses in original). See also Certeau, *The Writing of History*, trans. Tom Conley (New York: Columbia University Press, 1988 Conley's translation appeared in 1988, the original French is 1975), 1–2.

[10] Encampment figures are taken from *Latvijas PSR Vēsture* 1, 481. The population figure for Reval/Tallinn is from Toivo U. Raun, *Estonia and the Estonians*, 2nd ed. (Stanford, CA: Hoover Institution Press of Stanford University, 2001), 52.

[11] In Teodors Zeiferts, *Brīvzemnieks: Tautas darbinieka mūžs un laikmeta aina* [Brīvzemnieks: The life and times of a servant of the nation], vol. 1 (Riga: R. L. B. Brīvzemnieka rakstu komisija, 1929), 19. (An anticipated second volume was never published.)

but along the Baltic coast as well," Vēbers recalled. "At that time, there were substantial numbers of Russian soldiers camped at Rucava, as there were on the coastlines of Pūpe and Sventāja. Along with the soldiers there were Hussars, Cossacks and Bashkirs. The Cossacks, together with Kurland's forest guards, led by their forest masters, protected the Kurland coast."[12]

For Brīvzemnieks, the experience of encampment was principally one of disruption, and particularly disruption of the sonic environment: of the noisy music of foreign soldiers disturbing his childhood idyll. He was, after all, just seven years old in summer 1854. But for Vēbers, the undifferentiated noise of Russian soldiers dissolved with encounter into something else: a polyphony of Cossacks, Hussars, Bashkirs, and "Kurland's forest guards." Where Napier's gaze enfolded a multiplicity of individuals onshore into generic Russianness, Vēbers's acts of listening, together with his looking, worked in the other direction. He saw and heard a mass of Russian soldiers dissolve into a group of individual men and boys, each the member of a collectivity of unanticipated, surprising diversity.

A Silence of Unknowability

From the opening of hostilities between Ottoman and Romanov forces in October 1853, the pages of the *Rigasche Zeitung*, in which Filonin's song was published, had filled with reporting on events in the war's several theaters. Just as wartime had been a boon for newspaper printing in Britain at the turn of the nineteenth century, so too did the experience of the Crimean War sharply increase demand for printed news in western Russia.[13] Alongside accounts of mobilizations, casualties, and battles, the *Zeitung* printed patriotic poems and soldiers' songs like Filonin's, which served to rally its diverse readership in support of a unified imperial cause. The paper also published travelogues, essays on Ottoman history, and dispatches penned by visitors to the shores of the Danube, the Black Sea, and the Bosphorus, reflecting on expressions of local culture witnessed abroad.[14] These narrative essays typically functioned as a complement to items like Filonin's song. For while the printing of soldiers' songs might help to

[12] Vēbers, *Baltijas Westneşcha diwdesmitpeezu gadu jubilejai par peemiņu* [In memorial of the celebration for twenty-five years of *Baltijas Vēstnesis*] (Riga: B. Diriķa un beedru grahmatu drukatawa, 1893), 89.

[13] Zelče, "Grāmatniecība un prese," 363–64. On the parallel phenomenon in Britain at the turn of the nineteenth century, see Favret, *War at a Distance*, 61–65.

[14] For example: "Eine Donaufahrt zur Türkischen Grenze" (12, 22, and 26 Jan. 1855); Hubert v. Boehn, "Geschichtlicher Rückblick auf das Erscheinen, die Uebermacht und den Verfall des Osmanen-Reiches in Europa bis auf die neueste Zeit" (5 and 9 June 1854); "Tänze der Zigeuner in den Donauländern" (29 May 1854); "Das Innere eines Harem" (9 Mar. 1855).

meld the diversity of imperial experience into a coherent and common Russian one, these travelers' essays gestured to the constitutive outside of that ostensibly cohesive identity. They testified to ways of being that no Romanov subject could ever share.

One such essay, consisting of a series of anonymous missives entitled "Letters from Constantinople," presents an extreme manifestation of Wolff's operations of association and comparison.[15] The letters recapitulate the common Enlightenment trope of Turks as wholly other, as "beyond 'any idea of comparison'" with the peoples of Europe's west. And they extend that realm of incomparability eastward, to enfold Romanov spaces as well.[16] In these letters, the absolute difference of Ottoman peoples and spaces is registered in terms of auditory experience, as their anonymous author distinguished Constantinople from Riga not by its appearance but by its sound, and in particular by its silence. The silence described by the author was not unlike that which Philip V. Bohlman identifies in certain Brahminical texts: it was a silence of unknowability.[17]

> In the daytime, one can almost question whether Istanbul belongs to Europe or to Asia, but not for a moment at night. In the dark one perceives immediately that the Turkish city belongs to a foreign world with which we have nothing in common, not even in a dream. In Istanbul, the city of unmediated contrasts, nowhere do the troubles of the day or the pleasures of the evening carry over into the night, for the greatest relish of the Oriental is silence. For him, it is not a reward for work, recovery from strain, or the pleasure of rest. It is the silence of idleness, idleness too profound to be leisure. [. . .] Nowhere on European soil does one sleep more deeply, more soundly, or in such an untroubled way as on the Golden Horn. From the day's last prayer, an hour and a half after sunset, until sunrise nothing disturbs the quiet of the night. With the evening call of the muezzin from the minarets, life is extinguished.[18]

For the writer from Riga, the silence of the Turk was unknowable not only for reasons of cultural difference. Rather, the gap was ontological. It was a silence that entailed a cancellation of the very subject of Enlightenment discourse by which his readers could be assumed to have structured and comprehended

[15] "Briefe aus Konstantinopel," *Rigasche Zeitung* (12 June 1854), 6–7.

[16] Wolff, *Inventing Eastern Europe*, 48–49. (Wolff quotes from testimony of Charles-Marie, Marquis de Salaberry.)

[17] Philip V. Bohlman, "Analysing Aporia," *Twentieth-Century Music* 8 (2011), 133.

[18] "Briefe aus Konstantinopel," 6–7.

their worlds. In Jacques Attali's relatively recent formulation of this same Enlightenment perspective, "life is full of noise and [. . .] death alone is silent."[19] Much like the inhabitants of a noisy and industrializing American north who recoiled at the quiet of the southern plantation, so too the visitor from Riga "demarcated difference" with respect to another through inscriptions of auditory encounter.[20] The difference he demarcated could not have been starker:

> The Turks have something that reminds me of the Indian [*Indianer*] clans (one can hardly call them peoples) now gripped by extinction. Where they cannot stand up against the onrushing culture, where civilization penetrates among them, they gradually die out, as if they were compelled to live in an environment hostile to their inner nature, which they cannot bear forever. They retreat, they shrivel up, and they will not rise again. Like the Gypsies [*Zigeuner*] after centuries of wandering through Europe without pause or rest, they can only be exterminated [*kann man sie nur vernichten*], not assimilated. The time will come when, at the Bosphorus, if one asks about the people who once ruled this land and made Europe tremble, one will point in answer to the pines that adorn the hospital in Scutari. One will tell the stranger: like the Indians in the primal forests of America, so the Ottomans left nothing but graves as markers of their existence and of their extermination [*Vernichtung*]. This was something that the Turks had in common with those red races [*mit jenen rothen Stämmen*] like an omen: the cult of the dead. I do not wish to disturb their silence, for I trade with the living race.[21]

In Fedor Filonin's song from Reval, Russian soldiers cried "for the Tsar!" and barked together: "hurrah!" And in that way, with the vital noise of life, they awaited the arrival of their unknowable enemy, the Muslim who makes no sound. In these "Letters from Constantinople," an unnamed visitor brought his readers into the silent heart of unknowability itself, into imaginary encounter with the Ottoman Turks—barely living, he reported, and unworthy of life.

[19] Jacques Attali, *Noise: The Political Economy of Music*, trans. Brian Massumi (Minneapolis and London: University of Minnesota Press, 1985), 3.

[20] Mark M. Smith, "Listening to the Heard Worlds of Antebellum America," *The Auditory Culture Reader*, ed. Michael Bull and Les Back), 142; and in *Hearing History*, ed. Smith, 368.

[21] Bohlman, "Briefe aus Konstantinopel," 7.

Departures

Around the time when Filonin sang, one E. F. Schönberg, schoolteacher and organist in the Kurland town of Gramzda, composed his own memorial to the experience of wartime, a "Song of the Recruit."[22] The song was published in Latvian, and its narrative time antedates that of Filonin's song from Reval, for it addresses its sounds not to soldiers in the field but to their loved ones at home: mothers and daughters, brothers and fathers preparing the recruit for his departure, and readying themselves for the silence in the home that will soon attend his absence:

Now be with God, my loved ones,
Relatives, brothers, friends.
Now be with God, my parents,
Who are so good and lovely
Now be with God, my sweetheart,
Who has lately been my purpose.
I must go far away,
Where I have new brothers already! [. . .]

And if I were to perish
In some foreign place,
May you remember:
God awaits everyone!
Then I'll see my friends
In other blessed houses,
There where God's children
No longer fear death.[23]

Schönberg's "Song of the Recruit" was published in the provincial capital of Mitau, which the city's Latvian-speaking residents knew as Jelgava. It sang of an experience ubiquitous among Latvian households in the Baltic theater: the conscription of their men and boys into service with the imperial army, the transformation of their sons and fathers into Russian soldiers.

Little is known about how individuals were conscripted from households like the one from which Schönberg's soldier hailed. But the recruitment of which he

[22] Information on Schönberg taken from the cover page of his *Skohlas-dseeşmas pee rihta un wakkara luhgşchanas* [School songs for morning and evening prayer] (Jelgava, Latvia: J. W. Steffenhagen un dehls, 1870).

[23] E. F. Schönberg, "Rekruşcha-dseeşmiņa," *Latviešu Avīzes,* 18 Mar. 1854, 48.

sang was surely a euphemism; so-called peasants were enlisted for twenty-five years, which often amounted to a life sentence.[24] Local eligibility for the draft is not well understood, but it was a more complicated affair in the Baltic than elsewhere, since serfdom had been abolished there (anomalously) more than three decades earlier.[25] Of local Jewish men, we know that many were exempt: merchants, guild members, machinists, agricultural workers, rabbis, those attending school.[26] We know much less about how conscription touched the lives of their gentile, Latvian-speaking neighbors.

A week before Schönberg's song appeared, it was reported that 356,023 men and boys were eligible for the draft in the neighboring province of Livland, of whom 3,202 had been conscripted in the latest, eleventh round.[27] In November 1855, an "extraordinary" (*auserordentliche*) increase in conscription was reported, to 1% of all those eligible in the upcoming fourteenth call-up.[28] Literacy in the Baltic provinces was widespread even among those without formal education, so the ability to read is in itself unlikely to have signaled exemption from compulsory service. In print, Schönberg's song addressed the families of innumerable boys and men, and the silence it anticipated in their absence was one that all too many already knew. It was a silence that threatened daily to envelop the reader or her loved ones, either in the present moment or following a departure soon to come.

Where Are the Russian Soldiers?

In spring and summer 1854 and 1855, the fleet of Sir Charles Napier cruised the Baltic Sea, blockading ports, seizing cargoes, and carrying out reconnaissance. His sailors were sometimes joined by travelers, whose memoirs of adventures with the Baltic fleet catered to a substantial English market for war reportage. For those aboard the British ships, as for Filonin's comrades stationed onshore, the experience of wartime consisted largely in waiting: waiting for engagement by Russian warships that would never seriously threaten the British fleet, waiting for orders to sail to St. Petersburg, orders that never came.

[24] Andres Kasekamp, *A History of the Baltic States* (New York: Palgrave Macmillan, 2010), 72.

[25] Serfdom was abolished in 1816 in Estland, 1817 in Kurland, and 1819 in Livland. In most of the empire, the institution was abolished only in 1861.

[26] Mendels Bobe, *Ebreji Latvijā* [Jews in Latvia], trans. Rita Bogdanova, Eva Sinkēviča, and Ilmārs Zvirgzds (Riga: Shamir, 2006), 81–82.

[27] *Rigasche Stadtblätter* (11 Mar. 1854), 81.

[28] *Rigasche Stadtblätter* (3 Nov. 1855), 356.

At times, travelers and sailors alike reported feeling starved of hostile engagement, if only to break the tedium of months spent in plying the Baltic coasts. More often than not, when they approached Russian land, they found places emptied of their inhabitants, and a striking and often bewildering absence of Russian soldiers. One such traveler, William Henry, Duke of Grafton, reported accompanying Napier on what had seemed a daring move toward landing near Kronstadt, a fortress island situated at the mouth of St. Petersburg's harbor. "The Fleet first brought up near the island of Tasker," Henry recalled of the approach. "On going ashore there the island was found to be deserted. A pretty little octagon-shaped church stood near the place we landed at; the keys were in the door, giving the impression that the people had been scared away but a very short time before."[29] Another traveler, the Rev. Robert Edgar Hughes, fellow of Magdalene College, Cambridge, recalled accompanying Napier's fleet as it approached an island off the northwest coast of Estland. "We lay a whole day and night becalmed," Hughes wrote,

> and drifting towards the Russian lighthouse of Odensholm [in Estonian, Osmussaar]; at one time we were within a quarter of a mile of it. This certainly seemed too much of a good thing, and we got our boat out, and towed off to seaward. We afterwards heard that none but women were left upon the isle, and that we really were in considerable danger of being boarded by a bumboat laden with butter, eggs, and female Finns.[30]

In July and August 1854, Napier's fleet scored what might have been its only strategic victory in the campaign, the capture of the Russian-held Åland Islands, midway between Finland and Sweden. Hughes stood alongside Napier's sailors as they stormed an imperial fortress whose men offered no resistance. Inside he witnessed firsthand the tragedy of imperial policies of wartime conscription. The soldiers who manned the fort were hardly fit to fight. More surprisingly still, his encounter with the soldiers undid in an instant the associating and interchanging perspectives that had previously enabled Napier's crew to identify as Russian all who stood onshore. The field of difference Hughes encountered at Åland he registered by listening as well as by looking:

> We passed into the fort, and there we saw a strange sordid crowd of convict-looking wretches in long workhouse drab coats, scrambling and huddling together in all the attitudes of drunken, senseless

[29] William Henry, *The Baltic, 1854. The Crimea, 1854–1855* (London: Hatchards, 1884), 9.

[30] Robert Edgar Hughes, *Two Summer Cruises with the Baltic Fleet, in 1854–5; Being the Log of the "Pet" Yacht, 8 Tons, R.T.Y.C.* (London: Smith, Elder, 1855), 206–7.

> merriment. They tore off their uniform, they stamped on it and threw it in heaps, they sang, they laughed, and danced. One spoke to me in English; another, quite sober, asked me in good German, "Wo soll ich die Kleider hinlegen?" [*Where shall I put the clothes?*] Many were brawling together, and exchanging yells and heavy blows in their contests for a filthy sheepskin or a bundle of loathsome bed-clothes; among these were many Jews. One fine-looking fellow, with a black moustache, was a soldierlike and conspicuous figure; but the mass were not set up like soldiers, and showed no symptom of martial training; little squeezy bald-headed old men, or raw loose-spun boys, they looked more like a herd of half-starved emigrants than the imperial troops of a great military power.[31]

The encounter, for Hughes, was remarkable foremost for yielding an absence of Russian soldiers: an absence of men suited to soldierly duty, and an absence of conscripts speaking Russian. In their stead, he catalogued a diversity of individuals that rivals Vēbers's childhood memories of Cossacks, Bashkirs, and Hussars. There were Jews, a man speaking German, and another speaking English. With his eyes and ears suddenly and unexpectedly attuned to diversity, however, Hughes stood apart from Napier's men and the French sailors who accompanied them, whose response to the noise of the captured fort was to impose undistinguishing silence. He continued,

> "What do you think of our enemy" I said to a French sous-officer.
> "Canaille [*contemptables*]," he replied with a sneer.
> "What shall we do with them?" I asked again.
> "Fusiller [*shoot them*]," was his laconic answer.[32]

As Hughes ventured ever farther into the fort, he stepped into a room filled with bodies of the departed. "I looked around, and saw, on the floor before, behind, and beside me, the cold, clean, silent forms of the dead. The shock of the surprise was fearful; the light linen cloths that shrouded the stiffened figures waved and flickered in the draught, as if stirred by the breath of those that would breathe no more." For both Fedor Filonin and the anonymous author of the "Letters from Constantinople," the silence of the imagined enemy had marked his unliving, his unknowability. But for Hughes, it was precisely when the enemy was silent that he identified with him most fully, as he wondered at the silences

[31] Ibid., 118–19.
[32] Ibid., 119.

that the soldiers' departures had left in the homes they once knew. "What did these poor fellows know or care about the Turkish question?" Hughes asked himself. "And yet they had fought and trembled, they had writhed in agony, they had yielded up the breath of life, and now father and brother, maid and mother, were weeping and breaking their hearts for them, and all about the Danubian Principalities."[33]

When Hughes stepped out of the darkened fortress back into the light of day, the diversity he had heard in the soldiers' carousing and sensed in the silence of the dead dissolved back into what it had been before he set foot on the island. He gazed once more upon an undifferentiated mass of soldiers assembled before him. Those soldiers no longer spoke in a multitude of voices or carried private stories. Outside and together and all lined up, each of them sounded like every other, and they all sounded less than human.

> By this time the prisoners had been marched out of the fort, and were collected under a strong guard of English marines and French infantry. It was strange to see the three nations thus brought together.—The English, bold, sturdy, and strong, like bulls of Basan, staring and gaping on the foe; the French, small, active, and brisk, like horses of the desert; the Russians (I am unwilling to speak slightingly of a vanquished foe, but it is the truth) like unclean animals, grunting, wallowing swine.[34]

In Hughes's account of encounter at Åland, a fortress full of Russian soldiers exploded into a polyphony of languages, sounds, and faiths. But the polyphony he heard and remembered was fleeting, and it vanished in the daylight. At the end of his story, in the silence imposed by their murder at the hands of Napier's men, all of the people captured at Åland—the men and the boys, the Jews and the Germans, the soldierly and the invalid—became Russian soldiers once again.

Silent Country

The end of the war in 1856 was widely greeted with joy and relief. In Kurland's Latvian press, the peace accords were translated alongside a celebratory song: a "Song of Peace" by E. F. Schönberg, the organist who had earlier sung of the departure of the recruit for the theater.

[33] Ibid., 121.
[34] Ibid., 122.

Listen! The bells of peace are ringing everywhere,
From Paris to the banks of the Neva,
That the spilling of blood has ended;
What great joy, in which everyone can share.[35]

The treaty was signed on 30 March, and rejoicing continued into the summer. The departures of conscripts from farmsteads had ceased. Perhaps, it was hoped, the noise of life would return to soldiers' homes.

In fact, these early celebrations aside, the war had been a catastrophe for the Russian empire and its peoples, the effects of which would be felt for generations, and are arguably still felt today. The almost unfathomable costs of the conflict were registered socially and economically. They were felt in terms of dislocation and disillusion, as the empire's defeat "exposed the shortcomings of every institution in Russia": its infrastructure, its armed forces, its monarchy, its systems of regional governance, its peasant-based economy. They were registered foremost in the loss of over a quarter-million men and boys, killed on the Crimean Peninsula and in other theaters of engagement.[36]

Shortly after the accords were signed, imperial censorship of the press was restored, and published assessments of the war were to be silenced for decades. It was only after the Bolshevik Revolution and the consequent establishment of Latvia's republican independence that the losses were publicly tallied.[37] In a report of 1922 by the new republic's national office of statistics, the Crimean War was recalled:

> During the Crimean War, the dangers of war approached the city as well. [. . .] Riga's walls and ramparts were reinforced, new cannons were installed, and a flotilla of gunships was assembled on the Daugava [Dvina, Düna]. The English were confined to firing from shore, and Riga suffered only economically; the English blockade severed trade completely. [. . .] In 1852 there were 65,413 people living in Riga. In

[35] Schönberg, "Meera-dseeşma," *Latviešu Avīzes* (3 May 1856), 71. The peace accords were published on the front pages of the issues of 3 and 10 May.

[36] Figes, *The Crimean War*, 442–44 (quoted at 443) and 492–93.

[37] See Zelče, "Grāmatniecība un prese," 364–66. Published postwar assessments were cursory. For instance, "Of all the wars begun in the time of Nicholas I, the Crimean War was the only one in which victory was not wholly on Russia's side." (Matīss Kaudzīte, *Kreewu walsts attihstibas gahjums jeb Kreewijas wehstures pahrşkats* [Evolution of the Russian state, or survey of the history of Russia] [Riga: Schnakenburgs, n.d.; imperial censor cleared 19 Aug. 1887], 51.)

> connection with the Crimean War, the number of inhabitants fell to 60,463.[38]

Here, after nearly three quarters of a century, the departures that had defined wartime experience were finally quantified. The city of Riga was never directly attacked, but it had lost nearly 10% of its population.

Conscription undoubtedly accounted for a large share of the departures among men, many of which were to be permanent, as Schönberg's "Song of the Recruit" had foretold. But the emptying of spaces was likely felt more deeply in the countryside than in the city, since probable exemptions from the draft (attending school, membership in a guild, serving as a rabbi) were more widely attainable in urban locales.[39] While imperial censors effectively silenced public discussion of wartime losses, the postwar years saw the region's residents attesting to their costs in a striking way: by remarking on the silence of the recently emptied countryside. That silence registered in the conspicuous absence of musics of habitation, which indexed in turn the absence of individuals departed in the war.

In 1870, some fourteen years after hostilities had ended, one traveler observed, "Like the Russians, the Latvians and the Estonians are singing peoples." And yet, "I often happen upon schoolchildren while walking in Latvian and Estonian lands. To one, I remarked with amazement about the silence in the fields and villages."[40] Another, hiking the forests and fields of Livland that same year, wrote, "Here and there a person could be seen doing summer's work, but songs, folk songs, not one of them sang." He continued,

> One of the people traveling with me described for us a lovely scene from his youthful days, telling of how he used to live cuddled up so closely to nature's breast, when the flames of love flickered for the first time in his heart, and how happy he was to sing songs from early morning until late at night, by himself and together with others. Right after exiting the forest, we climbed aboard an oxcart that some shepherds were driving home. But the shepherds were as mute as the animals; not one of them sang any songs. [. . .] Growing up at my father's home, I heard shepherds singing in the early morning, in the afternoon, and in the evening while they herded their

[38] Marģeris Skujenieks, *Latvija: Zeme un eedzīvotaji* [Latvia: Land and inhabitants], 2nd ed. (Riga: Valsts statistiskās pārvalde, 1922), 94–95.

[39] See also Figes, *The Crimean War*, 317–18.

[40] D. M., "Kas dseed Baltijâ?" [Who sings in the Baltic?], *Baltijas Vēstnesis* (7 May 1870), 140.

animals. And how lovely were the echoes that resounded through the forest.[41]

This writer of these words was Atis Kronvalds, born in 1837. The song-filled childhood he recalled had unfolded just prior to the outbreak of the war, before the songful shepherds of his homeland had their lives transformed by distant events.

Reading these inscriptions, these testaments to silence, I am reminded of Certeau's distinction between places and spaces. Whereas a place, for Certeau, is a simple "configuration of positions," a space is a place made living by habitation; a space, he writes, "is a practiced place."[42] Specifically, he writes, the thing that "found[s] and articulat[es]" a space—what makes a place come alive—is the telling of its stories. He explains, "[W]here stories are disappearing [. . .] there is a loss of space: deprived of narrations [. . .] the group or the individual regresses toward the disquieting, fatalistic experience of a formless, indistinct, and nocturnal"—that is, *silent*—"totality."[43] If we exchange Certeau's *stories* for the Latvian writers' *songs*, then his words would seem to describe something similar to what they experienced of the Baltic countryside. With the absences wrought by war and its ravaging policy of mass conscription, sound itself—the musics of habitation—vanished from the landscape. With the loss of those sounds, there was lost a way of understanding and articulating spaces that had recently been practiced and alive. As it happened, this loss of song was noted amidst the first stirrings of discourse about Latvian cultural nationhood, a discourse that Kronvalds himself, at that very moment, was helping to direct.[44] For him, the silence that had befallen the countryside threatened to envelop the nation as a whole, a nation that he and others were only then beginning to imagine.

Re-sounding Space

Two years after Kronvalds wrote his "Excursion to the fatherland," the first volume of Latvian folk songs to include melodies along with texts was published: *A Garland of Songs*, compiled by the Livland schoolteacher Jānis Cimze. The very

[41] Atis Kronvalds, "Isbraukums pa tehwa-semmi" [Excursion in the fatherland], *Baltijas Vēstnesis* (30 July 1870), 237.

[42] Certeau, *The Practice of Everyday Life*, 117.

[43] Ibid., 123.

[44] See Otto Kronwald (Atis Kronvalds), *Nationale Bestrebungen* (Dorpat [Tartu, Estonia]: C. Mattiesen, 1872).

first song in Cimze's volume was yet another "Song of the Recruit," transcribed from a performance by an unknown singer and harmonized by one "J. S.":

Girl of my soul,
soon I will leave you;
but wherever I am,
I will remember you. [...]

But when fate ordains,
I have to depart,
to face the fierce Turks,
to shed their blood. [...]

And when a bullet
rakes me to the ground,
my last words will be:
You, you are mine; I'm yours.[45]

This song was one of three in Cimze's book to declaim a history of conscription in the Crimean War. It was one of many such songs that were sung for years and even decades after the conflict ended.[46]

To his fellow Latvian-speaking Russian subjects, Cimze described his collection as inscribing "the ancient and only inheritance of your grandfathers," which "reveals to you their joys and sorrows." His hope, he explained, was that "you may now come to behold it as something belonging to yourself, so that you may come to say: this is bone from my bones, this is flesh from my flesh."[47] In transcribing songs from oral tradition and publishing them so as to enable their broad distribution and amateur performance, Cimze hoped to give voice, and help others give voice, to the sounds of history itself, of the histories of countless lives led on Russia's Baltic shores. By imploring his readers to sing—to articulate in sound the memories they might share, so that others might hear and respond, and even take them as their own—he sought to cultivate what David Lowenthal once described as a "sense of history [that] goes beyond knowledge

[45] Jānis Cimze, *Dseeşmu rohta jaunekļeem un wihreem* [A garland of songs for youths and men], vol. 2, *Lauka puķķes: Latweeşchu tautas-dseeşmas* [Field flowers: Latvian folk songs] (Leipzig: G. Kreysing, 1872), 1.

[46] In addition to the three recruit songs in Cimze's volume (nos. 1, 4a, and 4b), these songs include six out of sixty-seven folk songs preserved in the unpublished collection of 1865 by the pastor Carl Christian Ulmann (Latvian National Library [Riga], Rare Book and Manuscript Division, Rx57, 2, 41).

[47] Jānis Cimze, "Dseeşmu rohta jaunekļeem un wihreem (Preekşchrunnas weetâ) [A garland of songs for youths and men (in place of a prologue)]," *Mājas Viesis* (1 May 1871), 141.

to empathetic involvement."[48] Cimze undertook his project, he explained, in response to the silence of the Baltic countryside. The task he set for his *Garland of Songs* was to return singing, the sounds of habitation, to spaces silenced during the war.

Given this, it is no coincidence that Cimze launched his project with a song that commemorates the wartime departures that had silenced Baltic spaces in the first place. For as Kronvalds had attested and subsequent generations later documented with statistical precision, the Crimean War had emptied cities, and the silence of the countryside was unyielding. By the 1870s, wartime absence and postwar silence constituted the pervasive background of public history in the region. But here, in Cimze's collection of songs, the experience of absence was reframed. His volume promised that memories of absence, however diversely registered, just might, if recalled and sounded with others, enable the founding of a new practiced space, a new space of living and being together. Mary Favret writes of wartime as "the experience of an undoing or damaging of rational sense," which "works to dismantle the forms that prop up our sense of the world and our place in it." But as she also acknowledges, "wartime may establish something that war would otherwise destroy, namely a culture."[49] The establishing of a culture—the founding of a new lived space—was what Cimze imagined for his work. That culture was to be grounded in the sounding of inscriptions attesting to histories of wartime experience, of the noise of encounter and the silences that indexed departure, unseeing, and loss.

In the first decade of Latvia's republican independence, the experience of the Crimean War was widely recalled. It emerged as a crucial component of cultural memory that attended the so-called national awakening of the 1850s, 1860s, and 1870s—the very movement that Kronvalds and Cimze helped to shape. In the 1920s, the beginnings of the movement were widely framed in terms of collective responses: to opportunities presented by the wartime opening of public discourse, to collective traumas experienced in the war. As one text recalled in 1928, "[i]n order to activate the strengths that had been gathered, the right moment needed to arrive. That time came in the second half of the nineteenth century, when Russia, after the loss of the Crimean War, opened itself to ideas from western Europe, and we Latvians began to sing: 'We have awoken to a new, lovely life.' "[50] In such tellings, the national awakening coalesced in writings by

[48] David Lowenthal, *The Past Is a Foreign Country* (Cambridge: Cambridge University Press, 1985), 212.

[49] Favret, *War at a Distance*, 15 and 18.

[50] *Rīgas Latviešu Biedrība sešos gadu desmitos 1868–1928* [The Riga Latvian Association across six decades, 1868–1928] (Riga: Rīgas Latviešu Biedrība, 1928), 6.

Kronvalds, Cimze, and their contemporaries. But the work of those figures was enabled in turn by a complex of external forces and twists of fate—by the kinds of unexpected events that Michel Foucault once described as "the details and accidents that accompany every beginning."[51] The experience of the war, its affordances and losses, comprised a crucial "accident," perhaps *the* crucial accident, that enabled the national movement to begin.

And yet, the sounds and silences of the war were soon put aside amid later turns in the region's history. With new traumas in the 1930s—most acutely, a coup led by the nationalist president Kārlis Ulmanis in 1934—it became newly important for many Latvians to recall the national awakening not as a response to accidents and twists of fate but as a forceful, even inevitable reaction to conditions of foreign (Russian, Romanov) rule. Amid extensive meditations on the injustices of serfdom and its lingering effects, the experiences and affordances of the Crimean War receded and eventually vanished from discourse on Latvian history.[52] This trend continued in narratives unfolded by émigré historians following Latvia's annexation by the USSR in 1940.[53] By the time independence was regained in 1991, the experience of the Crimean War was all but forgotten. In a foundational history of Latvian peoples produced in 1995, for example, the war is never mentioned. In another from 2011, it is referenced only in passing. In a monumental history produced by a team of Latvian historians in 2000, the war is treated only in relation to its impact upon the press.[54]

I close this essay by noting the absence of the Crimean War in national historiography of the present, for this present absence stands as mirror and postlude to the absences that defined the experience of wartime for many. To a remarkable extent, those who fought, observed, and remembered the war inscribed their responses in terms of auditory experience. Their inscriptions attest to the songs and sounds that accompanied and shaped encounters, and also to the silences that indexed departure, loss, and failures of imagination or understanding. Sometimes, their inscriptions attended the undoing of Wolff's operations of association and comparison. Other times, they reframed those

[51] Michel Foucault, "Nietzsche, Genealogy, History," *Language, Counter-Memory, Practice: Selected Essays and Interviews*, ed. Donald F. Bouchard; trans. Donald F. Bouchard and Sherry Simon (Ithaca, NY: Cornell University Press, 1977), 144.

[52] For instance, *Latvijas Vēstures Avoti* [Sources of Latvia's history], vol. 5, *Dokumenti par tautas atmodas laikmetu 1856–1867. g.* [Documents on the period of national awakening, 1856–1867], ed. Augusts Tentelis (Riga: Latvijas Vēstures Instituta apgādiens, 1939).

[53] For instance, Arvēds Švābe, *Latvijas vēsture 1800–1914* [History of Latvia, 1800–1914] (Stockholm: Daugava, 1958).

[54] Andrejs Plakans, *The Latvians: A Short History* (Stanford, CA: Hoover Institution Press of Stanford University, 1995); Plakans, *A Concise History of the Baltic States* (Cambridge: Cambridge University Press, 2011), 215; Zelče, "Grāmatniecība un prese," 352–76.

operations, as individuals and groups demarcated difference with respect to others in terms of the sounds they made or the silences that engulfed them. A century and a half after the end of the war, we no longer attend to the sounds of the Baltic theater. Its defining silences await us amid a greater silence over the war itself.

7

A voice that carries

DELIA CASADEI

Few elements in Italian nineteenth-century history were as unheroic as the Crimean War. Italy's modest participation in the war alongside the Western allies was spearheaded by the Kingdom of Sardinia, the northwestern territory governed (in the years before national unification) by the House of Savoy. The Kingdom was at the time under the political leadership of Camillo Benso di Cavour, the canny statesman who sought to unify the Italian peninsula under the House of Savoy, and indeed succeeded in doing so in 1861. Cavour masterminded Italy's involvement in the Crimean War as a strategic move to gain favor with France, whose military support was needed to assume control of the northeastern parts of the country still under Austrian rule. The small contingent—18,058 men and 3,496 horses—left Genoa on 25 April 1855, eighteen months into the war. They arrived in early May and fought little, the only exception being the battle on the Tchernaja River in August 1855—an attack launched by the Russian army outside the main theater of war at Sevastopol—which the Sardinian and French troops successfully repelled. Thousands of Italian soldiers perished nevertheless, many of them victims of rampant cholera. The disease remained entirely untreated in the Italian camp because of the burning of the *Croesus* (a ship loaded with medical supplies destined for Crimea) shortly after it departed from Genoa.[1]

[1] Within Italian historical scholarship, the Crimean War is treated as part of larger Risorgimento histories written by and large between the late 1930s and the late 1960s, and reflecting therefore the shifting ideological aims of the passage from Fascism to democratic governance, across World War Two. See, for instance, Franco Valsecchi, *L'unificazione italiana e la politica europea dalla guerra di Crimea alla guerra di Lombardia, 1854–1859* (Rome: Istituto per gli studi di politica internazionale, 1939) as well as his *L'alleanza di Crimea: Il Risorgimento e l'Europa* (Milan: Mondadori, 1948); a sizable section on Crimea is also found in Piero Pieri, *Storia militare del Risorgimento* (Turin: Einaudi, 1962). Monographs on specific aspects of the Crimean expedition have since been published, such as on military correspondence, the role of the Catholic Church, and the impact on Ligurian mercantile

Unheroic it was. Yet as a tactical move in international diplomacy, Italy's participation in the war was to prove momentous. Cavour's participation at the ensuing peace proceedings—an occasion on which Italy's grievances against Austria were finally heard by an international assembly—opened the way for an alliance with France that eventually made possible the Savoyard appropriation of Austrian territories, and, ultimately, the unification of the peninsula under the Piedmontese crown. This imbalance—between the war as a military event and its weight as a tool in ensuring international support for national unification—persists in the historiography of the Italian Crimea War. It may even explain why sustained literary accounts—consisting, between 1858 and 1896, almost exclusively of memoirs written by high-ranking officers—began to appear only two years after the war had ended. Indeed, this post facto literary archive thrives on the war's relative obscurity and geographical remoteness. For, in contrast to more famous highlights of the Risorgimento that took place around this time, the Italian Crimean War offered those who reminisced a key tool of myth telling—a "floating signifier," an insignificant item whose meaning could be respelled again and again according to poetic and political purpose.[2] Such myth telling is, as we will see, riven with currents of anxiety regarding national belonging within the Savoyard monarchy and Italy's geopolitical relationship—as southern periphery—to Central Europe, and, simultaneously, to the Orient.[3] These geopolitical anxieties were incarnated, I want to suggest, by a particular aural attitude to sound, and especially to voice.

trade. The only attempts at exhaustive historical accounts of the Italian Crimean War are rare and of relatively recent vintage, and are to chalk up the combined recurrence of the 150th anniversary of Italian unification (2011) and the rising tensions between Ukraine and Russia over Crimea, which culminated in Russia's occupation of the peninsula in early 2014. See Franco Rebagliati, Furio Ciciliot, and Liliana Betruzzi, eds., *La spedizione d'Oriente: Volontari italiani ed esercito sardo alla Guerra di Crimea, 1855–1856* (Savona: L. Editrice, 2011); Alberto Caminiti, *La guerra di Crimea, 1853–1856* (Genoa: Edizioni Liberodiscrivere, 2013).

[2] This expression, which has since been appropriated by thinkers as disparate as Jacques Lacan and Ernesto Laclau, was first coined by Claude Lévi-Strauss in "Introduction à l'oeuvre de Marcel Mauss," *Sociologie et Anthropologie*, by Marcel Mauss (Paris: Presses universitaires de France, 1950), vii–lii.

[3] The many problematic tendencies involved in representing the Orient have given rise to a critical tradition, departing from Edward Said's seminal *Orientalism* (London: Penguin, [1978] 2003). (On the topic of orientalism in the Crimean War see, in this volume, Andrea F. Bohlman's "Orienting the Marshal: Polish Legion Songs on the Map" 118–23 and Kevin C. Karnes' "Who Sings the Song of the Russian Soldier" 135–37.) However, my use of "Central Europe" in this chapter may require some clarification: it roughly covers an aggregate of the Prussian-Austrian-Hungarian Mittel Europa of the mid nineteenth century, Switzerland, and France. It stands for a "center" that is viewed from a position of southern peripherality, and specifically, of Italian peripherality to that which lies beyond the Alps.

We can begin with a particular voice: that of would-be Italian king Vittorio Emanuele II. As his troops prepared to embark for Crimea, he spurred them on their way with a rousing address—one that subsequently echoed down the whispering gallery of Italian memoirs of the Crimean War. Near the close of the nineteenth-century, officer Antonio Ricci recalled the event:

> On that occasion, His Majesty the King delivered to the troops the order of the day that became so famous, in which, among other things, these words were said: "You will see distant territories where the name of Italy is not unknown" ["Vedrete lontane terre ove il nome d'Italia non è ignoto"]. I said that the order remained famous, and indeed you often heard soldiers repeat the words I have written above. They repeated them especially in the painful circumstances in which they found themselves on arrival in Crimea, adding, in truth, a few glosses of their own, but these were innocent annotations that harmed no-one, least of all His Majesty the King, whose lips had first uttered the words.[4]

Mishearing is the name of the game in Ricci's account of Crimea, especially in this passage. Note, too, how the king's words create a curious pivot between physical spaces: from their utterance on the Piedmontese shores of Alessandria, to the soldiers' misremembering the words as they settle into the Italian camp at Mount Hasford, near Balaklava. Ricci's apologetic downplaying of the modifications wrought on the hallowed royal voice has the predictable effect of drawing attention to them. And yet, the mishearing and distortion in the camp points to a more fundamental distortion: one effected by Ricci himself in recounting the king's words. The sentence, annotated in historical documents of the time, had been "you will see distant lands where the cross of Savoy is not unknown" ("Vedrete lontane terre, dove la croce di Savoia non è ignota")—a much more

[4] "In quella occasione [. . .] S. M. Il Re diede alle truppe pazienti quell'ordine del giorno che poi è rimasto così famoso, in cui fra le altre cose si diceva: 'Vedrete lontane terre ove il nome d'Italia non è ignoto.' Ho detto che l'ordine rimase famoso, e difatti si udivano spesso i soldati a ripetere le parole che ho poc'anzi riportate. E le ripetevano specialmente nelle dolorose circostanze per le quali dovettero passare appena giunti in Crimea, aggiungendovi veramente qualche chiosa del proprio, ma erano chiose innocenti che non facevano torto ad alcuno e tantomeno a S. M. il Re, dalle cui labbra quelle parole erano state pronunziate. [. . .] Una volta [il generale La Marmora] udì un soldato, il quale con voce stentorea chiosava l'ordine reale in questo modo: vedrete lontane pietre, colle parole che vengono appresso. [. . .] Il soldato che era in buona fede, spiegò senza esitazione ciò che voleva dire, ed aggiunse come complemento queste parole: 'Le terre le potranno vedere i battaglioni che sono in basso, ma noi che siamo in alto presso l'osservatorio non vediamo proprio che delle pietre, e che pietre!' " Antonio Ricci, *In Crimea* (Turin: Roux Frassati e Compagnia, Editori, [1885] 1896), 12–13.

likely coinage at a time when national unification was some way off, and on the occasion (as Ricci had previously signaled) of the King's handing of Savoyard flags to the departing troops at Alessandria.[5]

Ricci's own mishearing adapts the sentence for the purposes of post-unification Italy. He does so by switching the anticipated moment of national recognition by outsiders from a visual to an auditory mode, from visual recognition of the Savoy Cross to auditory recognition of the name of "Italy." He then compounds this turn from the visual to the aural by recounting the soldiers' verbal repetition of the sentence, driving toward a key moment of mishearing: a soldier who, "with a stentorian tone," delivers the line, "'you will see distant rocks,' with the words that followed" ("'vedrete lontane pietre,' colle parole che vengono appresso").[6] The modification, which in Italian is effected through aural substitution—"terre" (lands) become "pietre" (rocks)—is picked up by General La Marmora, who demands that the soldier explain himself. The soldier good-humoredly tells him that his comrades on the observatory see nothing but rocks all day, and are thus a long way away from the populated lands imagined by the King. La Marmora, amused, presses a coin into the soldier's hand—"so that the rocks of Crimea won't be too tough on you"—and walks away.[7]

The attention that Ricci's recounting lavishes on the circulation of a phrase is loaded with meanings that are entangled in the politics of speech: a high literary pronouncement from a king is distorted by a foot soldier whose indifference toward the bombast is a mixture of political and linguistic alienation from a literary language that has no place in his speech. The senselessness of the expedition translates into the obscurity of the mangled edict. Ricci orientates the reader's senses toward listening and voice only to stage a linguistic malfunction, all the while ostensibly reassuring us that this is in good fun. The odd monetary

[5] The phrase "Vedrete lontane terre, dove la croce di Savoia non è ignota" is reported along with Vittorio Emanuele's transcribed address to the troops at Alessandria in *Ricordo pittorico militare della spedizione sarda in Oriente, Pubblicato d'ordine del ministero di guerra per conto dello stato maggiore* (Turin, Italy: Ministero di guerra 1857), and was engraved on memorial medals coined by the Savoy monarchy for distribution to Crimea veterans in 1855. See <http://numismatica-italiana.lamoneta.it/moneta/W-ME51M7/4>, accessed 5 Jan. 2016. It is also featured in a book on the history of Crimea published in Italy during the Crimean War; see Michele Giuseppe Canale, *Della Crimea, del suo commercio, e dei suoi dominatori, dalle origini fino ai dì nostri* (Genoa, Tipografia del Regio Istituto dei Sordomuti, 1855), 215.

[6] See n2 for the full Italian text of the anecdote.

[7] "Il generale Lamarmora che era rispettoso della disciplina, ma al tempo stesso aveva il cuore più buono del mondo, fece osservare al soldato che non era bene la traduzione libera da lui fatta delle parole reali, ma al tempo stesso, non vedendoci nulla di cattivo intenzionalmente, estratto un pezzo da cinque franchi lo regalò al soldato, dicendogli: Prendi questo onde le pietre non ti paiano tanto dure." Ricci, *In Crimea*, 13–14.

resolution of the incident, meanwhile, bespeaks condescension of a high-ranking officer toward his inferior—swiftly silencing an act of harmless insubordination.

Voices, and particularly "Italian" voices, frequently occasion strange actions in the stories told about the Italian Crimean War. In Ricci's memoirs, these range from oddly humble brags about being requested—by his French or English superiors, although crucially, never by the Turkish ones—to provide entertainment by singing an operatic aria; to overblown tales of the "Orphic power" of the Italian *bersaglieri*'s military band, whose arrangement of operatic arias predictably bewitched the armies of other nations; to moments of hearing and (more frequently) mishearing snippets of foreign language conversations, and also spoken Savoyard edicts (as we have seen).[8] These kinds of anecdotes are in and of themselves unsurprising: by the mid-nineteenth century, thanks to the successful export of *bel canto* and to the long-established philosophical discourse around Italian vocality, the aural recognition of gorgeous excess was routinely invoked as a key cultural asset—almost a natural resource—of the emergent nation state. Yet rarely are the voices recorded by Italian historiographers of the war straightforwardly presented; they are more often laced with doubt and undercurrents of violence, and are in constant peril of being misheard, or of falling on the wrong ears.

The theme of *bella voce* in the Crimean War conjures darker aspects of the geopolitics of Italy and the construction of Europe during the second half of the nineteenth century. Thanks to the unusual perspective provided by the journey south, and then east, from Piemonte to Crimea, Italian vocality becomes transformed. As well as being the familiar cultural asset ripe for international (usually operatic) export, it is now also the sound of a lack: a debt toward Central Europe's enlightened republican nation states, which Italy strived to emulate. I am here evoking what Roberto Esposito calls the *munus*, a gift that is also an inextinguishable debt.[9] From the perspective of European Enlightenment, Italy's blessed vocality may be just such a *munus*. If we imagine a broad spectrum of discourse around the political value of Italy's voice, we would have to begin beyond the Alps, with Rousseau's enthusiasms for Italy as the portal to the operatic land

[8] For more on Italian opera singers and bands performing operatic hits within the international wartime soundscape, see Flora Willson, this volume, 175–95.

[9] The concept of the *munus* as the "giving [of] something that one can *not* keep for oneself and over which, therefore, one is not completely master" lies at the heart of Roberto Esposito's idiosyncratic etymology for the word and concept of "community," which is the departure point for his *Communitas: The Origin and Destiny of Community* (1998) (Stanford, CA: Stanford University Press, 2009). The key passage with regard to "munus" is the discussion on pp. 4–5, and the passage just quoted is from p. 5.

before *logos*.[10] Only then might we witness this discourse move into Italian-speaking lands, and observe as vocality comes to be incorporated into Italian intellectual production. Take, for example, Giuseppe Mazzini's *Filosofia della musica* (1836), a study of the inadequacies of Italian opera composers as composers, quite literally, of the sound of a functional republican Italian state. In other words, discourse on the voice in Italian political thought was alive and well by the second half of the nineteenth century, but it simultaneously aped and challenged earlier, toxic perceptions of Italy as heard from the outside: thus Mazzini's (and his successors') appropriation of a tone that is, essentially, both dismissive of Italy's ability to partake of Enlightenment reason and suspiciously celebratory of her power to surpass it.[11]

Monarchic Italy will organize this same discourse around an ideology of opera as a symbol of unequivocal national belonging—an ideology that will be further compounded during Fascism. The Italian gift for voice has to function—in the lead-up to unification and all the more in its aftermath—as a marker of a national belonging within a territory divided into competing monarchies, and into different languages, for centuries. It is easy for us now to discern the catch-22 at play in such an ideology: how can a nation state, whose primary asset is a common language, be united (or recognized by others) in a voice that both transcends and fails to be language? This is the critique heralded by Antonio Gramsci, whose *Prison Diaries* became something of a textbook for the intelligentsia of Italy's first republic. Selections from Gramsci's *Diaries* have long been known among musicologists for their damning thoughts about opera as a "pestiferous" disabler of language, and thus of democratic politics.[12] Gramsci decried the poverty of *logos*

[10] The text referenced here is Jean-Jacques Rousseau's *Essai sur l'origine des langues* (1781). To date, the most exhaustive critique of Rousseau's romantic approach to prelinguistic vocality—and an important text for the critique of Italian vocality presented here—is Jacques Derrida's *De la grammatologie* (Paris: Les Éditions de Minuit, 1967).

[11] Similar themes regarding music's relationship to reason and language in Romanticism have been treated in John Hamilton's excellent *Music, Madness, and the Unworking of Language* (New York: Columbia University Press, 2008). Hamilton works on a Germanic tradition of thought in which, I would argue, the discourse about the limits of language as revealed by music (a discourse that both explores and performs various kinds of mental disturbance) is part of the ebullient and self-assured literary culture of late-eighteenth- and early-nineteenth-century Germany. My work here could be seen as relating to Hamilton's work within a more geopolitical perspective. To my mind, only a literary culture highly confident of its worth and of the historical relevance of its language could dwell with such sensuousness upon the limits of language. An emergent nation-state peripheral to the core of European political affairs—such as Italy was—could not have developed such a highly aestheticized literary take on the failures of language. Thus my literary examples are unquestionably of a much lower aesthetic and philosophical rank than anything in Hamilton's book, but they are also riddled with a more palpable undercurrent of concrete political anxiety.

[12] Gramsci's reflections on music are rare, and embedded into his much more frequent but less than systematic reflections on language. Musicology has not taken up Gramsci in any systematic way,

in Italy, and was intensely suspicious of the bourgeoisie's appropriation of the literature and tastes of Central Europe. But ultimately his *Diaries* fall within the constitutive contradiction of the Italian *bella voce,* whose riches necessarily stem from, and reproduce, a poverty of *logos* tantamount to faulty internal politics and—thus—a lowly rank within the European geopolitical hierarchy.

With the historiography of Crimea we are dealing with a novel facet, or perhaps even a modification, of the traditional post-unification musical tropes of the Risorgimento—the rousing choruses picked up by supposedly riot-prone audiences during operatic performances in Milan and elsewhere. In the materials I am about to discuss we are at once mired in Risorgimento history and geographically displaced from it: we have moved both outside prospective national boundaries, and beyond the domain of the cosmopolitan Europe north that Italy wished to be recognized by. As mentioned a moment ago, the trajectory to Crimea from Italy moves south, and then east. Shiploads of soldiers set sail from Genoa down the Tirrenian Sea, past the Strait of Messina, across the Ionian and then the Aegean Sea and toward the Bosphorus. The southeastern trajectory by water—recounted at a time in which both Italy's role in Europe and its very existence as a nation state—will make audible Italy's own accursed southernness, bringing out the politically charged *lack* sounding in the *bella voce* that, even if it overcomes *logos,* must always fall short.

Masking the Uncommon Tongue

And so to Crimea. The earliest travelogue I consider here is *I Piemontesi in Crimea,* published three years before unification by Mariano D'Ayala (1808–1877), an erudite statesman working for the House of Savoy. Unlike many accounts that came in the wake of unification, D'Ayala's is not a personal memoir but rather an attempt at narrative history, probably based on the military accounts to which D'Ayala had access in Turin, as well as official "ricordi pittorici

with the exception of Mary Ann Smart's "Liberty on (and off) the Barricades: Verdi's Risorgimento Fantasies," *Making and Remaking Italy: The Formation of Cultural Identity around the Risorgimento,* ed. Albert Ascoli and Krystyna von Henneberg (Oxford: Berg, 2001), 103–118. Smart's essay is in many ways a musicological response to Gramsci's suspicions on opera. Yet to dwell excessively on Gramsci's ideas about the linguistic failures on opera is to overlook his broader reflections on mishearings and respellings of hegemonic languages on behalf of the subaltern; language, and listening to language, is always as much a site of resistance as it is of oppression. Among musicological takes on Gramsci, perhaps the most striking to date is Aaron Fox's *Real Country: Music and Language in Working-Class Culture* (Durham, NC: Duke University Press, 2004), which uses Gramscian thought to analyze the modes in which the working-class community of Lockhart, Texas constructs speaking patterns and modes of subjectivity, thanks to the practice of country music.

militari"—oversized illustrated albums released on behalf of the Ministro di Guerra to document the war.[13] The history is a predictable glorification of the King of Sardinia and his prime minister, Cavour, as the noble minds behind the territorial unification of a spiritually bonded nation.

Sound comes into D'Ayala's narration mostly through the question of language and linguistic identity. In what was later identified—at least since Gramsci—as a trope of Risorgimento cultural production, D'Ayala pushed the idea of a nation united by a common tongue:

> [One cannot] say that some of the people of Italy are more or less, or even not at all Italian, just because there are differences between them. A different accent in speaking the same tongue, cool rather than ardent courage, that is the full extent of the differences between soldiers of the Italian fatherland.[14]

In the intensely fragmented peninsula of 1858, there was little more than a literary Italian whose use mitigated regional linguistic differences among the higher classes. D'Ayala's idea of a common spoken tongue among the majority of the population was, in short, blatantly an ideological construct. We could put this even more strongly: D'Ayala—born in Messina (Sicily) and educated in Naples, going on pointedly to avoid republican riots in 1848 in Naples, Florence, and Turin—probably experienced this linguistic disunity more vividly than many less well-traveled functionaries of the monarchy. His dismissal of linguistic differences may have gained urgency through the friction it produced with his own experience. D'Ayala wrote with the strained authority of a southerner acting as proxy for—speaking on behalf of—the Savoy kingdom. As mentioned earlier, the ideology of a common tongue was dismantled, perhaps for the first time by Gramsci—who, not by coincidence, also hailed from the Italian south,

[13] The book, referenced in n4 but worth bringing up again, is the *Ricordo pittorico militare della spedizione sarda in Oriente*. Parts of this publication were cribbed and reworked into a subsequent *Ricordo pittorico militare della spedizione sarda in Oriente* published in 1884 by the *Consiglio direttivo della Società dei Reduci dalla Crimea* on the occasion of the Esposizione generale Italiana of 1884 in Turin.

[14] "Né perché i popoli d'Italia abbiano fra loro alcune differenze, potranno dirsi alcuni più ed altri meno, anzi nulla Italiani. [. . .] Un accento diverso nel parlare la lingua comune, un coraggio più freddo o più ardente, ecco tutta la varietà che corre fra i soldati della medesima patria Italiana." Mariano D'Ayala, *I Piemontesi in Crimea: Narrazione storica* (Florence: Tipografia Barbera, 1858), 13. It's important to note that D'Ayala's text was published by a private press as part of a series entitled "biblioteca civile dell'Italiano" (civic library of the Italian language), devised and financed by a group of noblemen who later became high-ranking statesmen in unified Italy. With regard to the question of language, note how D'Ayala had previously described Austria in exactly opposite terms, as "an unnatural and violent aggregate of twenty peoples different in race, culture and language," ibid., xiii.

Sardinia. Yet even without invoking Gramsci, we might detect uncertainty behind D'Ayala's protestations over language in some of his subsequent statements about the uses and abuses of national "voices." A common tongue was not enough once in Crimea:

> It is not enough that the fatherland's armed forces be commanded by voices from the fatherland, nor that orders and regulations be formulated in an appropriate and solemn language; it is also necessary that Italian soldiers be of one mind and one heart, and that, shedding all forms from the outside, they take on pure national form.[15]

D'Ayala is here referring to the edicts of Vittorio Emanuele II, King of Sardinia, to his troops in Crimea, which he quotes in full and comments on positively, but is clearly unconvinced by. The bombast of the edicts—no matter how apt and elevated the content—was simply not sufficient to assert radical nationhood. Neither were military commands delivered in Italian on site. It is almost as if D'Ayala were warning against the precise incident Ricci was to describe some forty years later—the distortion of the King's speech. D'Ayala's solution to this insufficiency of the national language is little more than a surge of rhetoric urging primal tuning among bodies in which *logos* plays no part. Voice is transformed into a form of physical collective fusion—the primal formation of the body politic as visceral national identity. One mind, one heart, pure form. Voice—the sound that issues from the body that utters it—is no longer even in the picture, so violently inward is the wiring of nationhood into flesh.

The condition for this overcoming of language is the exportation of the nation outside its boundaries, but also—and this is important—outside the Western scheme of powers into which it wishes to enter. It is the journey southeast that peels away regional differences and uncovers the idea of voice as collective tuning. The journey to Constantinople on the way to Crimea is thus endowed with mythical markers. It is a journey by sea, and one that involves traversing that most Greek of mythical passages: for D'Ayala—and, as we shall see, for others after him—the Strait of Messina was a particularly charged site. It is here that,

[15] "Non basta che le armi patrie sieno con voci patrie comandate, né che le provvisioni e i regolamenti abbian lingua propria e solenne; è necessario altresì che i soldati italiani abbiano una mente e un cuore: e che, deposte tutte le forme altrui, rivestano le pure forme nazionali." D'Ayala, *I Piemontesi in Crimea*, 2–3.

on the way east, the southern lands can be contemplated as part of a prospective whole. Voice is, again, of critical importance:

> [The ships] greeted the islands of Capraia and Gorgona, no longer invoked as the "hedgerow of the Arno's mouth," and the Aeolian islands with Stromboli and its smoke, and then Ponza and Capri, then Messina and Reggio; and on the shores of the Strait, passionate youths and an unhappy people greeted the passing of those glorious flags with their gaze and by waving little white cloths, barely stifling in their heart the cry forming on their lips: "Viva L'Italia!"[16]

The journey along the western coast of Italy and then through the Strait of Messina is full of Homeric reminders: the Sirens near the Aeolian islands, Scylla and Charybdis at the Strait. D'Ayala added to these a Tuscan boundary—the islands at the mouth of the Arno in Tuscany, which Dante had dubbed the hedgerow separating the Arno from the sea. Again, this seemingly innocent quotation was politically astute. D'Ayala probably knew that Florentine dialect—which Dante had, as was well known, consecrated into high literature in the thirteenth century—was in these very years being proposed by Alessandro Manzoni as the future national language of united Italy.[17] By the time the Strait had been reached, the linguistic impulse toward nationhood had been engendered; at the edges of the peninsula, it was already a potential voice, a common impulse in the minds and hearts of the people. But this potentiality never translated into voice. No sound came from the lips of the forlorn onlookers at the Strait. The desired common tongue must exist only as a spasm, well away from the vagaries of intersubjectivity. It is as if D'Ayala had placed this extreme pressure on the voice so as to thrust it back into the depths of a flesh where it could neither be heard, nor—crucially—misheard.

[16] "[Le navi] salutavan la Capraia e la Gorgona, non più invocate a far 'siepe all'Arno in su la foce,' e l'Elba e le Isole Eolie con la Stromboli fumante, e Ponza e Capri, poi Messina e Reggio; sulle cui rive i giovani ardenti del Faro e il popolo infelice salutavan col guardo e con lo sventolare di bianchi lini il passaggio di quelle gloriose bandiere, appena soffogando nel cuore il grido che correa sulle labbra di 'Viva l'Italia.'" D'Ayala, *I Piemontesi in Crimea*, 56.

[17] Manzoni revised his novel *I promessi sposi* (1827) so that its language conformed to Florentine literary dialect, and republished it in 1840. D'Ayala might even have been familiar with the phrase widely used to describe Manzoni's linguistic retooling of his novel: the "risciacquatura in Arno," the rinsing [of language] in the Arno river. After the unification, Manzoni became part of the unified monarchy's committee on linguistic unification, and in 1868 published a celebrated brief essay specifically on the subject of linguistic reform: "Dell'unità della lingua e dei mezzi di diffonderla."

Hearing the Italian Voice

In order to understand the shriveled physicality, the engendering and yet willful retention of voice in D'Ayala—the extent to which voice existed under a pressure too great to be released into the realm of the audible—we might look at Marx's treatment of Italy in his journalism for the *New York Herald Tribune*. Marx's work of this period includes, notoriously, a formidably cosmopolitan account of the war. Within his account Italy is, as one might expect, a footnote—but a telling one. To Marx, the failures of 1848 were to be examined carefully in light of the subsequent revolutionary insurgence. In this context he offered a withering commentary on the last embers of the failed 1848 riots against Austrian occupation. He singled out 6 February 1853 in Milan, an uprising organized *in absentia* by an exiled leader of 1848, Giuseppe Mazzini. The uprising—which has an odd, posthumous literary relationship to Crimea—consisted of a series of spectacular miscalculations and miscommunications: firearms meant to be sent from Genoa and Switzerland were never delivered; the plan to storm the Austrian headquarters and persuade Hungarian soldiers to defect in favor of the rioters was also unsuccessful; street protests failed to gain traction; and eventually the uprising was swiftly and bloodily quelled by Austrian troops. Marx concludes,

> Let us hope that henceforth there will be an end of *révolutions improvisées*, as the French call them. Has one ever heard of great improvisators being also great poets? They are the same in politics as in poetry. Revolutions are never made to order. After the terrible experience of '48 and '49, it needs something more than paper summonses from distant leaders to evoke national revolutions.[18]

The play on the figure of the improvisator here is an odd, tantalizing detail to a reader suspicious of vocal metaphors in relation to Risorgimento politics. Marx offers a complex simile by way of path between the aesthetic and the political ("they are the same in politics as in poetry"). The anatomy of the simile goes something like this: "improvisation" is to "poetry" what "(ineffectual) paper summonses" are to "revolution." This is a rather sophisticated diagnosis. Marx did not write—as he easily could have—that, in politics as in poetry, Italians are mere improvisators, stuck in a primitive oral phase that has not yet blossomed into literacy. Rather, he seemed to imply that as politicians, Italians have literacy (Mazzini issues paper summonses) but not *good* literacy. Their being stuck in an

[18] Karl Marx, article in the *New York Daily Tribune* (7 Mar. 1853).

undeveloped orality is manifested through the fact that they use the written letter badly, as a flimsy substitute for, rather than a fulfillment of, voice.

Thus the leader who conceives of a revolution spontaneously and without adequate planning is an improviser whose prowess, while impressive, is tied to the time and place of performance; the leader who plans everything away from the site of political revolt is like a letter that, however well intentioned, may be misdelivered, misread, and then badly acted on, if at all. In order to produce lasting political change, revolution has to be like a voice that not only is powerful and masterful, but that also *carries*. The balance between orality and literacy is key to politics as it is to literature: the best revolutionary is the poet, whose command of the rhythms of spoken utterance translates into literature that retains the resonance and immediacy of voice, but transcends the limits of the *hic et nunc*.

This is not to say that Marx was a kind of political theorist of the Italian voice—far from it. I am, however, suggesting that Marx's seemingly inconsequential choice of metaphors to describe the Italian political condition reflects a broadly hegemonic discourse on the Italian voice in the nineteenth century. Whether or not Marx is a witting participant in this discourse is hard to discern. Could Marx, for instance, have known that his political target in this excerpt—Mazzini—had published a *Filosofia della musica* in 1836 in which the genre of opera—and the writing for voice in opera—was taken to be allegorical of Italy's potential as a democratic republic?[19] Perhaps not. He may, however, have been aware of the appraisal of Italian culture delivered by Central European writers earlier in the century, in particular the famously incendiary essay published by Madame de Staël on the subject of translation in 1816.[20] De Staël's basic argument was that Italy did not have a literature fit to compete or even converse with that of Central Europe, because it was based on classical, outdated models whose subjects were no longer contemporary. One of the causes of this problem was the sheer beauty of Italian as a language: when spoken or sung, it could cover up even the dullest literary material, thus facilitating the country's cultural conservatism.[21] Not much could be done about this: the phonetic riches of Italian amounted to a vessel that could not aspire to adequate intellectual content. It might, however, serve to carry and appropriate literary advances of the Central European countries.

[19] Giuseppe Mazzini, "Filosofia della musica" (1836), *Scritti letterari* (Milan: Bietti, 1933), vol. 2, 36–73.

[20] The essay was first published in Italian as "Sulla maniera e l'ultilità delle traduzioni" in *Biblioteca italiana* 1/1 (1816), 9–18. The translator of the essay and editor of the journal was Pietro Giordani.

[21] I am borrowing the insights on Madame de Staël's commentary on Italian literature and theater from Gary Tomlinson's article on the affinities of Italian Romantic Opera and Romantic Literature; see his "Italian Romanticism and Italian Opera: An Essay in Their Affinities," *19th-Century Music*, 10/1 (Summer 1986), 43–60.

And perhaps Marx might have heard of de Staël's novelistic accounts of her travels in Italy, one of which—*Corinne, ou l'Italie* (1807)—amply predated the essay on translation and focused exclusively on a female vocal improviser (the titular Corinne) whose performances she notated in her novel, effectively carrying out a work of translation complementary to the one she exhorted Italians to take on in 1816.[22] Whether Marx was conscious of this or not, improvisers, or "improvisators"—in Marx's nomenclature—were not a vague, neutral term in nineteenth-century literary discourse, but one of the ciphers of the encounter between Central European literary erudition and the sensuous riches of the European South. Through the literary account of the *improvvisatrici,* de Staël performed the act of capturing Italian vocal performances: she notated improvisations into French language poetry that could thus be carried, quite literally, across the Alps. As it was for Marx, for de Staël the voice was a powerful technology for both art and politics, but one whose ideally literary use escaped Italians themselves. Italians are improvisers, not poets. Only by appropriating Central European styles could they achieve an optimal, cosmopolitan literature. Only then would their voice truly carry.

The rarefied sensibilities of a de Staël have ostensibly little to do with Marx's wry musings on the political causes of the failure of 1848. Yet Marx is a participant—a fleeting and perhaps unwitting one, but a participant nonetheless—in a discourse that de Staël actively shaped in the early nineteenth century. And it is in Marx's very high-profile account of the months leading up to the beginning of one of the nineteenth century's deadliest international conflicts that the aestheticizing rhetoric of de Staël begins to display its full political weight. Italians cannot make lasting, internationally acclaimed use of either speech or literature in the same way that they cannot work revolutions. Again, they are improvisers, not poets. By the same token they are, in Marx's view, agitators, not political leaders.

Italian vocality thus expands into something of a geopolitical economy. As Italian literary scholar Roberto Dainotto has boldly argued, the fashioning of the southern corner of Enlightened Europe was the result of decades of work stemming, initially, from the pinnacles of French and German literature. From Montesquieu to Rousseau, from Hegel to Madame de Staël, Southern Europe—and Italy especially—served as a means of maintaining symbolic ties, and yet also substantially warding off, the southeastern Mediterranean, at once understood as a point of origin, and as an embarrassing premodernity to be overcome. For Dainotto, Montesquieu crystallized the thought that "as colonies of the Oriental

[22] A twenty-first-century reflection on the practice of female improvisation in early nineteenth-century Italy and its representation in both literature and opera is Melina Esse's "Encountering the *improvvisatrice* in Italian Opera," *Journal of the American Musicological Society* 66/3 (Fall 2013), 709–70.

world of Islam, the civilizations of Spain and Italy did not constitute an integral part of Europe but were its negative south."[23] Yet it was crucial that they were included—as the aestheticizing flair for voice that runs from Rousseau to de Staël shows—precisely because they served to render Europe *immune* to the Orient, by folding elements of it within itself.

The political economy of voice lies with the geopolitics of the European South, as an excess which bears an inextinguishable debt toward an enlightened Europe that begrudgingly, but necessarily, includes it. Rousseau and de Staël are thinkers whose names haunt opera scholars' bibliographies to this day. Often overlooked is that the supposed beautiful orality of Italians, as heard from France and Germany, amounts to their eschewal of, but also subjugation to, the *lettres*: the literate thought that defines the Republic of European states. As Dainotto argues, literature—the very idea of belles lettres—comes to embody not simply the act of writing, but the very form of (French) Enlightenment *logos* as well.[24] Taking Dainotto's thought one step toward the aural—and with a sidelong glance toward Derrida—we might argue that the Italian voice is primarily the by-product, the lack/excess produced by this notion of the literate.[25] It is the gift of the Italian peninsula, its contribution to the Republic of letters—but only insofar as it is also the sonic embodiment of that which Italy does not have, its *debt* to the superior literature, ability for revolution, and democracy of its core-European siblings.

Inhuman Voice, Inscribed Ground

An anti-monarchic counter-narrative to accounts like that of D'Ayala was produced almost simultaneously in the literary movement known as Scapigliatura. A mix of French bohemianism and watered-down socialism and anarchism, the Scapigliati were young members, mostly of the haute bourgeoisie, who opposed monarchic unity and upheld radical republican values—Mazzini's lost cause. Their style involved a plethora of images of physical decay and festering wounds,

[23] Roberto M. Dainotto, *Europe (in Theory)* (Durham, NC: Duke University Press, 2007), 6.

[24] Dainotto works through Voltaire and Diderot to come up with a definition of letters as "not literature as erudition, but literature as a key to practical knowledge; not literature as a cult of the past, but as praxis on the present and creation of a progressive future; not literature as knowledge for knowledge's sake, in the end, but literature as the formation of citizens—of a society of polished spirits, perfect taste, and graceful sciences. This is literature, in sum, understood as the basis for the transnational Republic of Letters of poets, doctors, and mathematicians already praised in le siècle de Louis XIV." *Europe (in Theory)*, 90.

[25] The deconstruction of the binary of writing and an imagined original orality is the subject of Derrida's reflections on Rousseau's *Essay on the Origin of Languages* (1781) in *De la grammatologie*.

as well as sensory malfunctions of various kinds, all of which could be broadly conceived as allegorical of the corrupt moral and political order (Catholicism, the Savoy Monarchy, capitalist expansion in the northern cities).[26] Crimea would not, by any stretch of the imagination, become a key theme for this group, but it did play some part. Few note that it appeared in the Scapigliatura's literary manifesto, the novel *La Scapigliatura e il 6 febbraio (un dramma in famiglia)* (1862) by Milanese writer Cletto Arrighi, whose subject was the uprising of 6 February 1853 that had spurred Marx's aversion to Mazzini. The Crimean War is referenced in the epilogue; the context is, again, a journey by water. Two years after the Milanese uprising, two of the novel's minor characters meet on a ship sailing across Lake Maggiore, in Piedmont. Both of them are now *émigrés*, but are traveling back to Italy from their homes abroad in Switzerland and Paris. One of them, a young rioter in 1853, is on his way to fight in the Crimean War; the other, an elderly doctor, is on his way back to Milan, where an outbreak of cholera is decimating the population. "Cholera is my Sevastopol" says the old doctor, thus linking in one sinister sweep the sick Italian body politic and the exportation of Italian nationalist grievances.[27]

Years later, Crimea became the high-profile subject of a novel by one of the Scapigliatura's most eccentric writers. Iginio Ugo Tarchetti's *Una nobile follia* was published in Milan in 1866, five years after national unification. Tarchetti's account is neither—like D'Ayala's book—a history nor a personal memoir; it is, however, written *as though* it were a memoir, from the perspective of a first-person narrator. It charts the Crimean War through the eyes and ears of Vincenzo D., a fictional conscript. In response to the horrors of the war, Vincenzo suffers

[26] The Italian scholarship on Scapigliatura (taken as a movement involving literature, visual arts, music, and theater) is predictably vast. A good recent general survey is Giuseppe Farinelli's *La Scapigliatura: Profilo storico, protagonisti, documenti* (Rome: Carrocci, 2003), and on the literary Milanese/Lombard Scapigliatura it is important to mention Enrico Ghidetti, *Tarchetti e la Scapigliatura lombarda* (Naples, Italy: Libreria scientifica editrice, 1968) and Massimo Arcangeli's *La Scapigliatura poetica milanese e la poesia italiana fra Otto e Novecento: Capitoli di lingua e di stile* (Rome: Aracne, 2003). English language work on Scapigliatura is rarer; a key text is David Del Principe's *Rebellion, Death and Aesthetics in Italy: The Demons of Scapigliatura* (Madison, WI: Fairleigh Dickinson University Press, 1996). It is also worth noting that Tarchetti's most renowned novel, *Fosca* (1869), was adapted into the musical *Passion* by Stephen Sondheim in 1994, although Sondheim approached Tarchetti's subject through a film adaptation by Ettore Scola, *Passione d'amore* (1981).

[27] Cletto Arrighi, *La Scapigliatura e il 6 febbraio (un dramma in famiglia)* (Milan: Francesco Sanvito, 1862), esp. 304–15. The passage just quoted is found at 312: "Io sono un soldato della salute pubblica né più né meno di questo giovanotto che lo è della civiltà e dell'indipendenza. Il cholera è la mia Sebastopoli. Che diresti di un soldato che il giorno dell'assalto non corresse sotto la bandiera? Il mio posto è dove si muore. È a Milano." [I am a warrior for public health no more and no less than this youth who defends civilization and independence. Cholera is my Sebastopol. What would you think of a soldier who doesn't fight for his flag on the day of the attack? My place is where people are dying. It is in Milan.]

a mental breakdown that eventually results in his deserting the front, smuggling himself back to Milan, and isolating himself in a small apartment entirely populated (like a modern day's Noah's Ark) by animals and insects.

The cultural and political networks within which Tarchetti operated could be described as a blow-by-blow inversion of D'Ayala's background. If D'Ayala was a nationalist and a monarchist, Tarchetti was both anti-unification and anti-military. Born some forty-five miles east of Turin—the seat of the House of Savoy's power—Tarchetti moved to Milan and became part of Scapigliatura; D'Ayala, on the other hand, was a southerner who had trained in a military academy in Naples and then integrated himself into Piedmont as a statesman and scholar of the military arts. Tarchetti had volunteered to join unified Italy's new army in 1862, and his first—and, it turns out, last—assignment was in the southern region of Puglia, where troops were engaged in the bloody suppression of anti-unification rioters (a phenomenon broadly known as *brigantaggio*). Horrified, he began to conceive of the monarchy as a colonial project of annexation of the peninsula's South and abandoned the army in 1863.[28]

Tarchetti's anti-militarism permeates his Crimean novel—so much so that its second edition, which came out in 1869, was publicly burned in Milan's military barracks. It takes a radically anti-identitarian stance, avoiding all mention of Italy or indeed of the reason that Italian soldiers were sent to Crimea. This undoing of identity is one of the novel's most innovative formal features: it is written in the first person, but there are in fact three narrators, two of whom are homonymous. The lead character—who is the last "I" to speak in the novel—defects from the battlefield of Crimea by taking the uniform of a dead Polish soldier on the battlefield; he changes his name to Vincenzo D. after arriving safely back in Genoa, and eventually ends his life so that his homonym dear friend—Vincenzo D.—may take on his identity, escape his creditors, and live as a free man. This overwrought splintering of authorial voice produces a semiotic surcharge, making the "I" into a taunting, opaque recurring sign, and the novel into a litany of personal pronouns whose referent is always elusive.[29] The multiplication is, strikingly, matched by an almost complete obliteration of human voices within the diegesis. Not only is there little dialogue, but a studied absence of hearing and speaking as well. The battlefield is devoid of human sound, whether musical

[28] Tarchetti's background, political convictions, and literary ambitions with regard to *Una nobile follia* are explored in Roberto Carnero's preface to the 2004 edition by Mondadori. See Iginio Ugo Tarchetti, *Una nobile follia: Drammi della vita militare* (Milan: Mondadori, 2004), i–xxv.

[29] On the themes both of material decay and of the doubling/splitting of the self, both very prominent in the novel, it is important to mention Alberto Carli's *Anatomie scapigliate: L'estetica della morte tra letteratura, arte e scienza* (Novara, Italy: Interlinea, 2004), and especially Tommaso Pomilio's *Asimmetrie del due: Di alcuni motivi scapigliati* (Lecce, Italy: Manni, 2002).

or linguistic—despite being thickly populated with soldiers. What does emerge, though, almost incessantly, is the landscape of the battlefield, a landscape that seems—far more than the warfare waged upon it—responsible for thousands of deaths. Bodies fall in clusters from precipices, are swallowed by floods, or buffeted by wind or land configuration into the thrall of the enemy, regardless of nationality or allegiance.[30]

The figure of the ground, the bare earth, becomes for Tarchetti not only the material substratum that warring armies trample, but also a metaphor for a natural order of peaceful coexistence among living things that is breached by war. Retribution issues from it as if from an angry deity. The glorified body politic of nationalism is thus unsettled by then common ground, which gives the lie to war as a nefarious artifice, an artifice that creates a violating inscription, a gash in the earth. On the day of the battle of the Tchernaja, Vincenzo D. notes,

> It is but seven in the morning, and everything writhes and lives [. . .] blue flies buzz over maple leaves, dragonflies hover in clusters over the river, clouds of gnats dance in the sunlight [. . .] and cuckoos make the whole plain resound with their drawn-out, monotonous song. Upon this natural idyll, men are about to inscribe an epic poem drenched in blood.[31]

Mocking heroic epic poetry, Tarchetti equates war with writing. Paradoxically, the only vocal sound in Tarchetti's battlefield is an inhuman scream produced precisely by the wounded earth. As the earth gathers up the fallen in battle, Vincenzo recalls,

[30] We might wonder where, if at all, Tarchetti drew his account of the Crimean battlefield. In my research, I have found his descriptions to be remarkably similar in style and tone to those of a French lieutenant whose memoir of his brief sojourn in Crimea was translated into Italian in 1855. See Barone de Bazancourt, *Cinque mesi al campo di Sebastopoli*, trans. Giuliano Landucci (Florence: Tipografia Giacomo Terni, 1855). De Bazancourt's account is striking because it relies heavily on the notion of natural "spectacle," using inflated literary language to describe a battlefield that de Bazancourt clearly observed from a position of mastery—the safety of the French observatory—as a horrific but thrilling visual performance. In linguistic terms, Tarchetti's account could be understood as an appropriation of a Central European account into the Italian language, a cosmopolitan counterattack on nationalist narratives brought forth within the boundaries and language of the united monarchy.

[31] Tarchetti, *Una nobile follia*, 138: "Non sono in fatto che le sette ore del mattino, e tutto si agita o vive: le melonte saltellano a migliaia sui prati, le mosche azzurro si posano ronzando sopra le foglie degli aceri, le libellule aleggiano a stuoli sul fiume, nubi di moncherini volteggiano nei raggi del sole, le lucertole verdi si affacciano alle screpolature dei massi e s'inseguono, e i cuculi fanno risuonare tutta la valle dei loro canti monotoni e prolungati. Dinanzi a questo idillio della natura, gli uomini si apparecchiano a scrivere un'epopea di sangue."

> We could hear neither the cries of the victors, nor the wails of the wounded and dying; but there was something in the air, something that seemed to weep, that seemed to ache; there was that great voice, that great emanation of grief that matter lets out as it dies.[32]

It is as if the violent inscription of war carved a screaming mouth into the earth, opening a space for voice as the remainder—rather than the premise—of the written trace.

South by Southeast

Personal memoirs of Crimea flourished some twenty-three years into Unification, partly as a commemorative tribute to Vittorio Emanuele II. Indeed, Crimea-related activities and documentation enjoyed an upswing in the wake of his death in 1878. The Società dei Reduci della Crimea (Society of the Veterans of Crimea) was founded in Turin that year, with branches opening simultaneously in Milan and Genoa. The Society was active in maintaining living memory of the war, by collecting funds for war veterans, celebrating yearly the battle of the Tchernaja, and even organizing pilgrimages to Cavour's grave in Santena. It also successfully promoted the building of an ossuary in Crimea for the collection of the remains of Piedmontese fighters, holding a ceremony there on its inauguration in August 1882. At the national exhibition (Esposizione Generale Italiana) of 1884, which took place in Turin, personal memoirs by participants in the war were encouraged, and war correspondence was further gathered. The reason for this surge in activity was not, of course, just a matter of commemoration, but had a broader political purpose: in the early 1880s Italy was gearing up for colonial occupation of Eritrea, which it partially accomplished in 1885.[33] The export of the nation, the aural prestige of Italy's voice, began to be couched in more aggressive terms. Portraits of Crimean military leaders enjoyed a surge in demand, and were matched by a fresh stream of commemorative literature.

It is at this time, it seems, that memories of the musical prowess of the Italian military band in Crimea began to circulate. One such memoir was penned by Carlo Osvaldo Pagani in 1880. Offered as a heroic memoir of Alfonso La Marmora, the leader of the Crimean expedition, the memoir dwells on the

[32] Tarchetti, *Una nobile follia*, 145: "Non udivamo le grida dei vincitori nè i gemiti dei feriti e dei morenti, ma vi era nell'aria qualche cosa che sembrava piangere, che sembrava soffrire; vi era quella gran voce, quella grande emanazione di dolore che la materia emette morendo."

[33] This information is given in the preface to Ettore Bertolè Viale, *Lettere dalla Crimea, 1855–56*, ed. Umberto Levra. (Turin: Comitato dell'Istituto per la Storia del Risorgimento Italiano; Carrocci, 2006).

band's sonic assertion of nationhood. But this time the ideal audience is both the powerful allied troops and the Turkish contingent, the latter often treated with palpable colonial disdain:

> The music of the first regiment of grenadiers band—the only one to accompany the Sardinian expedition—had become famous. The band played well, and it played good Italian stuff that went right to the heart even of the English and the Turks, the most unmusical men on earth. Our local melodies were enjoyed best by the French, already taught by Rossini to appreciate *bel canto*. By the others it was absorbed mysteriously, with intense pleasure. It produced in them the inebriation of opium, opened up new patches of sky to their eyes. Sometimes as many as thirty- or even forty-thousand allies gathered around the music, and that strange mixture of types, of languages, of cultures, lent something strange and fantastical to the echoes of the sound waves resounding in the farthest corners of the camp.[34]

As it hovers over the camp's disparate nationalities, the grenadiers' music—which is repeatedly indicated to be operatic in provenance—enfolds sounding nation and cosmopolitan *passepartout*. It is worth noting that Pagani also recalls the language used by the allied troops to communicate with one another: Sabir, a pidgin used by mercantile communities in the Mediterranean basin between the eleventh and nineteenth centuries.[35] It is doubtful whether the strange mixture described by Pagani amounts to any certified language, but his decision to

[34] "La musica del primo reggimento granatieri—la sola che accompagnasse il corpo sardo di spedizione—era diventata celebre addirittura. Suonava bene, eppoi roba Italiana che andava dritto al cuore perfino degli Inglesi e dei Turchi, gli uomini più antimusicali della terra. La melodia nostra, paesana, gustata meglio dai Francesi abituati già da Rossini al bel canto, veniva dagli altri come assorbita, misteriosamente, voluttuosamente; produceva in essi un'ebbrezza come di oppio, apriva ai loro occhi lembi di cielo; In alcuni giorni trenta e perfino quarantamila alleati facevano circolo intorno alla musica, e quella strana mescolanza di tipi, di linguaggio, di costumi, aggiungeva qualcosa di strano, di fantastico, al ripercuotersi delle onde sonore negli echi più lontani dell'accampamento." Carlo Osvaldo Pagani, *Alfonso La Marmora: Pagine nuove; Ricordi storici della campagna di Crimea* (Rome: Carlo Voghera, 1880), 413.

[35] Pagani, *Alfonso La Marmora*, 409: "Ma come faceva tutta questa gente ad intendersi? La cosa, infatti, era difficile, e si può dire che non ci sarebbero riusciti se non fosse stato per quello strano idioma (il sabir), nato sulle coste del Mediterraneo, miscuglio d'italiano, di spagnuolo, d'arabo, di turco, ridotto d'altronde a pochissimi vocaboli ed aiutato energicamente dall'eloquenza del gesto." [But how did all these people manage to understand one another? It was, indeed, a difficult thing, and one might say they wouldn't have managed it without that strange language (sabir), born on the coast of the Mediterranean, a mix of Italian, Spanish, Arabic, Turkish, ultimately reduced to very few words and energetically enhanced by the eloquence of gesture.]

record it as Sabir is striking, evoking as it does the centuries of Genoese and Venetian mercantile glory, indeed the very period in which Genoa and Venice had set up coastal colonies in Crimea. On Crimean ground, pidgin mercantile tongues and band arrangements of bel canto arias now share the same symbolic purpose: to bind a variety of bodies through a cosmopolitanism laced with colonial ambition.

Such optimism was not to last. By the turn of the century memoirs of Crimea were taking on a darker tinge as Italy renewed its attempts at colonial expansion. The result was their devastating, bloody defeat in the Italo-Ethiopian War of 1896, which caused a backlash of anti-colonial sentiment: one that spilled into the imagery of travel away from home and onto the battlefield.[36] One of the most startling musical episodes recounted in later narratives of the Crimean campaign is found in an account by the previously mentioned Agostino Ricci, staff officer in the Crimean expedition who published his memoir in 1896, more than forty years after the war and shortly before his death. It concerns his journey by sea via the Strait of Messina. Ricci stages the passage dramatically. Having sailed all night from Genoa, the Italian troops' ship reaches the Strait in the morning. Those aboard wake up as they glide past the lighthouses at either side of the Strait. But the episode has an odd musical soundtrack: a dissonant, badly performed fanfare. Ricci rushes over to the first trumpeter of the military band aboard to inquire as to the origin of the sound; the trumpeter promptly informs him that he is hearing the royal fanfare, which

> is to be played on solemn occasions, and I wanted to let it be heard by those on either side of the Strait [. . .] at the moment we are leaving Italy and may never see it again. After all, both those in Messina and in Reggio Calabria are Italian, and should listen with joy to the fanfare of their gentleman King.[37]

But there is a dislocation between the trumpeter's pompous aim and its aural result. It would have been unlikely for lands that were under a monarchy separate from (and rival to) that of Sardinia to have known the royal fanfare—or have recognized it as a national hymn of any kind. Indeed, Ricci muses that

> someone would have had to explain to the people on the mainland that they were hearing the royal fanfare; this is, assuming that they would

[36] See, for example, Luigi Guarnieri, *La battaglia di Adua e il popolo italiano* (Turin: Roux Frassati, 1897).

[37] Ricci, *In Crimea*, 17.

> have been able to hear it at all, which, given the distance, was very unlikely.[38]

It is unclear whether Ricci takes the unintelligibility of the fanfare to be a matter of political alienation or a consequence of the poor performance. The Strait—the passage between the inside and outside of the nation—engenders a warp in which sounds lose their intelligibility, in which one reaches a disagreeable state of hovering at the threshold. Called upon to signify unity beyond political and linguistic boundaries, music falls into the same vagaries as language. It too fails to be recognized, to signify, to unite. Instead of serving as valediction to the imagined national territory, it illustrates the nation's undoing as the ship passes by the last stretch of known land. The journey south, past the recalcitrant regions of Sicily and Calabria, brings the ship closer to Africa than to Piedmont. As it continues eastward toward Constantinople, Italy's perilous belonging to Europe is in turn uncovered, awakening the memory of the territory's historical ties with the Ottoman Empire. The Italian South's inability to hear itself as part of the unified nation-state thus overlaps with the Enlightened European's wary listening to Italian voices. Sensuous southernness morphs into incoherence.

The misheard fanfare echoes the Savoyard foot soldier's distortion of the royal edict, Ricci's other aural anecdote, recounted at the beginning of this chapter. The fanfare, in turn, is echoed by the incipit to one of the most important recent philosophical accounts of voice, Mladen Dolar's *A Voice and Nothing More*:

> There is a story which goes like this: in the middle of a battle there is a company of Italian soldiers in the trenches, and an Italian commander who issues the command "Soldiers, attack!" He cries out in a loud and clear voice to make himself heard in the midst of the tumult, but nothing happens, nobody moves. So the commander gets angry and shouts louder: "Soldiers, attack!" Still nobody moves. [. . .] He yells even louder "Soldiers, attack!" At which point there is a response, a tiny voice rising from the trenches, saying appreciatively "che bella voce!," what a beautiful voice![39]

Dolar memorably invokes the commander's failed interpellation, along with the soldier's aestheticizing impulse, and leads us a third way, toward a voice object that "does not go up in smoke in the conveyance of meaning, and does not solidify

[38] Ibid.

[39] Mladen Dolar, *A Voice and Nothing More* (Cambridge, MA: MIT Press, 2006), 6–7.

in an object of fetish reverence, but [is] an object which functions as a blind spot in the call and as a disturbance of aesthetic appreciation."[40] I, too, have pursued this third way, this voice object. Yet, unlike Dolar, I have tried not to hurry past the specific implications for Italian nationhood with regard to the voice. For geopolitics run all the way down in the constitution of the voice object. The *bella voce* is produced by a celebratory discourse that constitutes it as the object, and never the speaking subject, of literary theory as well as of European-scale political shifts. Probing unloved archives, such as that of the Italian Crimean War, can thus bring out the disquiet and rupture that emerges as Italian vocality is squared, in post-unification historiography, with a Eurocentric aural vantage point. Such archives allow us to observe a voice object taking shape in distorted sentences whose falling short of sense provides the occasion (and the potential) for infinite linguistic renewal: a game of "telephone" in which phone and *logos* are each other's constant remainders.

[40] Dolar, *A Voice and Nothing More*, 4.

PART III

WARTIME AS HEARD

8

Operatic Battlefields, Theater of War

FLORA WILLSON

As the allied bombardment of Sebastopol persisted through spring 1855, British, French and Turkish forces were joined by General La Marmora's troops from Piedmont and Sardinia.[1] Dispatches from the Crimean front continued to appear in the international media, but for the most part there was little progress to report. Undeterred by the lull, the satirical press offered readers a seemingly endless stream of lampoons and baroque absurdities. In April, the serious-minded, Boston-based *Dwight's Journal of Music* jettisoned its conventional tone and overcame its geographical and political distance from the action to present its own contribution to wartime silliness. *Dwight's* reported that the Sardinian tenor Giovanni Mario—celebrated as a lion of *bel canto* and as partner of the still more famous soprano Giulia Grisi—was thinking of joining his native army in Crimea. Evidently unsure whether the louder laugh was to be had from believing the claim in the first place, or from the absurdity of Mario's military ambitions, the article settled for the latter. If Mario did indeed take to the battlefield,

> the world will have to lament the spoiling of a good singer to make a bad soldier; for Signor Mario will find the bona fide soldier's life in the field very different from the sham soldier's life of the opera. The little squad of thirty or forty basses and tenors that make a grand army on the stage, is a much more manageable force than a division in the field, and the best martial air with which the prince of tenors ever rallied his forces in an opera, would be of little avail in the trenches of Sebastopol.[2]

[1] For more on the participation of Italian military forces in the war, and its subsequent place in Risorgimento historiography, see Delia Casadei, "A voice that carries," in this volume, 150–51, 163–71.

[2] "Musical Chit-Chat," *Dwight's Journal of Music* (14 Apr. 1855), 12–13.

Having drawn up this basic distinction between operatic playacting and the realities of military life, *Dwight's* produced a litany of Mario's "greatest hits" that would perhaps be heard no more. From "Una furtiva lagrima" (Donizetti's *L'elisir d'amore*) to "Spirto gentil" (the same composer's *La favorita*), these extracts—slow *romanze*—seem to have been chosen for their obviously sentimental qualities: they are vehicles for the tenor voice at its least aggressive, least forceful. It was precisely for this manner of singing that Mario was famous, his demeanor far removed from that of the military man as idealized at the mid-nineteenth century.

Yet Mario was not the real target of *Dwight's* satire. The article goes on to consider the best course of action for "Miss Coutts"—a mysterious figure famed for attending every single Mario performance, no matter where in the world he sang. For her, *Dwight's* recommended

> immediate enlistment in male apparel, in the Sardinian contingent, or a purchase of a choice loop-hole in the fortification of Sebastopol, from whence, with a hundred-horse power opera glass, she may inspect the movements of General Mario.[3]

Here the comedy is more complicated. Coutts is considered automatically amusing, the scenario into which she is inserted redolent of many an opera plot. But there is also a sideswipe at the so-called purchase system, by which officers tendered family money for their commissions, regardless of qualification or experience on the battlefield. It is ultimately at this vilified practice (one abolished in the wake of the Crimean War), rather than at the familiar satirical double act of Mario and Coutts, that *Dwight's* takes aim.

I start with this item of journalistic ephemera because it foregrounds a central feature of Crimean War historiography. The idea that the Crimean War was to an unprecedented extent a *mediated* conflict has received considerable attention in recent scholarship, and has frequently been advanced as evidence of its status as the first truly modern war.[4] Mary Favret's work on the emergence of what she calls "modern wartime" makes this connection axiomatic, arguing that "the epistemology of modern warfare is an epistemology of mediation."[5] Favret's focus is the Napoleonic Wars, but her contribution neatly encapsulates the ongoing

[3] Ibid., 13.

[4] See Ulrich Keller, *The Ultimate Spectacle: A Visual History of the Crimean War* (Australia: Gordon & Breach, 2001); elaboration of Keller's "first media war" thesis can be found in *Der Krimkrieg als erster europäischer Mediankrieg*, eds. Georg Maag, Wolfram Pyta, and Martin Windisch (Berlin: LIT, 2010).

[5] Mary Favret, *War at a Distance: Romanticism and the Making of Modern Wartime* (Princeton, NJ: Princeton University Press, 2010), 12.

debate about the relative modernity or conservatism of nineteenth-century warfare in general. In the Crimean context in particular, it is W. H. Russell's frontline dispatches in *The Times*, Roger Fenton's *in medias res* photographs, and the battles staged as a regular attraction of London's mid-century nightlife that are repeatedly invoked in the accumulated evidence for the war's status as—for the British, at least—fundamentally mediatized.[6]

In each of these instances of a distant war mediated for a civilian population at home, we might also note a complementary reverse process: what we might call the militarization of the media forms themselves. With the electric telegraph and postal service between Britain and Crimea still unreliable, the reprinting of soldiers' letters in newspapers became a primary conduit of information for local and national readerships. Fenton, we might recall, rose to prominence as a photographer largely through his documentary images of events and places in Crimea—the first time the medium was used in a military environment. And numerous theatrical shows in London that addressed the war brought invalided soldiers onto the theatrical stage as "living" proof. Not only did media forms become martial; members of the military increasingly became consumers of militarized media productions in their own right. The effects of mediatization on those at the frontline have not generally been addressed by Crimean War historians. Nor have military contributions to what was clearly a two-way process, in which events at the front were mediated and consumed, and subsequently formed a frame through which to view future events.

There is no shortage of examples of such mediations of the war within the military itself. We might, for example, consider the various dramatic troupes that emerged from among the British forces stationed in Crimea and put on military-themed entertainments—in one case in a specially constructed 250-person theater; or the fact that officers' accounts of "their" war (in letters or diaries, later published) are routinely punctuated by the arrival of newspapers from home; or that it was apparently conventional for officers and visitors lucky enough to be accommodated in huts rather than tents to paste those newspapers over the wall boards, as both insulation and decoration. Colonel Edward Cooper Hodge, for instance, reported in a letter from October 1855 that he had "various scenes from *Punch* and the *Illustrated London News* depicted on the walls of my

[6] In this essay, I use the term "mediatization" to denote the proliferation of technological, literary, and theatrical forms that conditioned the knowledge and experience of war in the nineteenth century. However, the word is also meant to suggest its overtones in twentieth-century media discourse, in which it has come to signify the revaluation of "truth" brought about by mass media. See, for example, Jean Baudrillard, *For a Critique of the Political Economy of the Sign*, trans. Charles Levin (St. Louis: Telos, [1972] 1981), 175.

domicile"—a revelation that complicates any attempt to read those publications solely as records of wartime sentiment and public opinion back in London.[7]

I raise these moments in which life at the front can be shown to be always already mediated—indeed, in which attempts to separate war and its mediations inevitably collapse—since they bring to light contiguities and continuities between spaces often considered at an almost total remove: between the men on the front line and the civilian population. There are various explanations that might be summoned as to why this new proximity of Britons at home and at war came about around the middle of the nineteenth century. The increased number of literate common soldiers in the British army must have played a part, as must the more rapid, more widely established communication and travel networks open to a growing proportion of the Western European bourgeois populace. These social catalysts may have worked to blur boundaries between battlefield and domestic spaces, their differentiation "at a distance" being replaced by a gray area characterized by cycles of remediation. It is this gray area that concerns me above all in this chapter and that is central to my project in two ways. First, my principal geographic focus is on Constantinople, capital of the Ottoman Empire and the city through which all allied troops, observers, and tourists passed between 1854 and 1856 en route to the Crimean front. It was an urban switchboard intimately connected with the war, many of its British wartime visitors and commentators eliding it with Crimea in a single, topographically vague sense of exotic Otherness. Yet Constantinople was simultaneously characterized by its distance from the battlefield and was seized on as the nearest metropolis—as an outpost of relatively "modern" urban life—for those with the means to come and go from the Crimean front.

Second, I hope to offer a new perspective on the war as a site of mediation by examining the presence of opera (which mostly meant Italian opera) in Constantinople, in Crimea, and ultimately on the battlefield itself. Unlike the war poetry read by Favret, or the visual art produced to record or monumentalize battles since time immemorial, music not only *reacted* to war.[8] True, composers and publishers scrambled to produce Crimea-themed sheet music for the British musical public—and enough "Sebastopol" quadrilles appeared in 1855 to merit an article about these "Dances of Death" in *Punch*.[9] But music was also played and heard on the battlefield: in the form of the trumpet-and-drum signals

[7] *"Little Hodge": Being Extracts from the Diaries and Letters of Colonel Edward Cooper Hodge Written during the Crimean War, 1854–1856*, ed. Marquess of Anglesey (London: Leo Cooper, 1971), 131.

[8] In this sense, although my project builds to some extent on Ulrich Keller's rich and wide-ranging account of the war's visual traces, I do not share his aim to examine the war's mediation simply as a set of representations (however subtly excavated); see Keller, *The Ultimate Spectacle*.

[9] "Dances of Death," *Punch, or the London Charivari* 28 (1855), 73.

still employed to communicate with the troops; and, more significantly for my purposes, in the performances of military bands in the evenings, on the march, or during funerals and other military ceremonies. On these occasions, as we will see, the repertoire played was often dominated by operatic excerpts. This habitual remediation of opera on the Crimean frontline might channel our attention to larger historiographical themes, perhaps reflecting the newly elevated status accorded of band music in general around the time of the Crimean War.[10] More importantly, the routine nature of opera's presence might also prompt us to interrogate the kinds of auditory attention that the military band could demand. After all, it was in only the recent past that audiences of elite European music had, according to one highly influential narrative, "stopped talking and started listening."[11] And there is little question—even when we allow for geographical variation and for a more gradual process of change than such watershed rhetoric might suggest—that elite music performed in certain metropolitan spaces had, by the middle of the nineteenth century, begun to be thought to demand new modes of attention from its listeners.[12] The Crimean frontline, and its bands' remediation of operatic hits, may therefore provide the occasion to test the social diffusion and geographical reach of "modernized" listening habits at the mid-century.[13] We might even ask how such shifts in listening practices in the years before the Crimean War might also have produced a widespread perception of the Crimean battlefield as one inundated by sound of all kinds; but reaching any conclusion (insofar as one might be possible) falls outside the remit of this project.

[10] It may be no coincidence that, as Trevor Herbert and Helen Barlow have noted, there was a distinct change in the status of military bands at mid-century. In 1854 the Royal Artillery Band made a British concert tour—the first ever undertaken by a military band. It was above all a symptom of a nascent understanding of such outfits as musical ensembles as well as military accessories; and of an attendant desire to listen to what could thus be considered a performance rather than a colorful byproduct of military activity. Trevor Herbert and Helen Barlow, *Music and the British Military in the Long Nineteenth Century* (Oxford: Oxford University Press, 2013), 196–99. See also Henry George Farmer, *Military Music* (London: Max Parrish, 1950), 48.

[11] This central problematic was identified by James Johnson in his groundbreaking study, *Listening in Paris: A Cultural History* (Berkeley: University of California Press, 1995), 1.

[12] For an overview of the larger context in which this epistemological shift took place, see William Weber, *The Great Transformation of Musical Taste: Concert Programming from Haydn to Brahms* (Cambridge: Cambridge University Press, 2008).

[13] That opera in this context indicates almost exclusively the early nineteenth-century Italian repertoire (from Rossini to middle-period Verdi) says much about that tradition's hegemonic status in elite European circles at this time; but that ubiquity is also symptomatic of its increasing *global* spread—an international dissemination that both predated the Crimean War and was perhaps bolstered by it. See Benjamin Walton, "Italian Operatic Fantasies in Latin America," *Modern Italian Studies* 17/4 (2012), 460–71.

Nevertheless, by focusing on the war as it was saturated—however unexpectedly—by opera, this chapter attempts to recover a crucial aspect of the sensory experience it afforded. On the one hand, then, such an approach encourages closer consideration of the rhetoric and social circumstances of the Crimean narratives available to us, pushing us to address the contingencies of particular elite references and experiences. On the other (and in some sense arguing from the opposite corner), this focus serves to question our assumptions about the physical and discursive reach of opera at the mid-nineteenth century: to illuminate its presence in spaces far removed from its original social and geographical milieux. In the pages that follow, the very notion of "opera" stabilizes according to the three essential modes in which it was produced, consumed, and made meaningful in the Crimea. First, opera as staged performances: an assemblage of performers, works and conventions imported from Western European urban centers; a luxury item to be bought and displayed alongside other mobile commodities that signaled and were afforded by nineteenth-century globalization. Second, opera as a sonic metonym, represented (and, for many of its less socially elevated consumers, overshadowed) by a looser collection of associated musical experiences: excerpts and arrangements played by itinerant musicians or on abandoned pianos, or overheard on the streets or, most regularly, performed by military bands for the amusement of soldiers and civilians alike. Finally, opera in a third sense: as it was invoked by observers of the Crimean War—whether tourists, reporters, or commanding officers, all of whom participated in the war at a distance—as a metaphorical or symbolic point of reference.[14] For these Crimean actors, opera was a filter through which to see and hear the war and its surroundings, a shorthand for plush, gilded world inhabited by its military elite. The twenty-first-century reader of officers' memoirs and Crimean travelogues is thus confronted with a secondary battlefield: one in which observers' and soldiers' perspectives might be productively set in opposition.

Cosmopolitan Constantinople (or, Between the Battlefields)

In 1837, when R. T. Claridge published his *Guide along the Danube, from Vienna to Constantinople* in the wake of a trip made the previous year, he marveled at how, in the Ottoman capital, "The total absence of carriage-wheels, clocks, bells,

[14] It is worth noting that this third sense is thus simultaneously furthest removed from the first, "literal" definition and most predicated on its assumed truth—that opera was ultimately a staged performance and a mode of consumption.

and all sonorous occupations, leaves the whole city wrapt in almost unbroken silence; while the people appear to be mute, and desirous of passing along the streets without being seen."[15] Claridge lingered over Constantinople's reported sonic void in order to draw a starker contrast with its magnificent Byzantine past.[16] Yet he might equally have been comparing its soundscape to that of his native London, the "world city" of the 1830s, where noise seemed an inevitable byproduct of industrialization—where hubbub and modernity went hand in hand.[17] As heard by Claridge, Constantinople's silence was not the peaceful idyll so often mourned by Londoners. Rather, it constituted a series of striking absences: a dearth of modern technologies of transport and timekeeping, of any "theater or public place of resort,"[18] and ultimately of life itself, as if the inhabitants of such a backward city could be little else than muted and invisible.

Insofar as this Constantinople ever existed, it was vanishing in a wave of urban transformation by the late 1830s. When Sultan Abdülmecid succeeded his father, Mahmud II, in 1839, in a change of power that historian Philip Mansel identifies as inaugurating Constantinople's "third golden age," Mahmud had already begun rapid modernization of the city.[19] The first official Ottoman newspaper, *Takvim-i Vekayi*, was launched in 1831; a new postal system was established in 1834;[20] and, from the mid-1830s, regular steamer services ran to Odessa, Izmir, and Marseille—reducing the journey time to the last of these cities from six weeks to six days.[21] Abdülmecid shared his father's desire to modernize the Ottoman Empire; he led it into an era that was, according to Mansel's account, "torn between contradictory forces: between dynasticism and nationalism; capitalism and the pre-industrial state; Islam and Christianity; the Russian army and the Royal Navy."[22] Most important for my purposes, however, is the city's equivocation between traditional Ottoman culture and its Western counterpart, with the

[15] R.T. Claridge, *A Guide along the Danube, from Vienna to Constantinople, Smyrna, Athens, the Morea, the Ionian Islands, and Venice, from Notes of a Journey Made in the Year 1836* (London: R. C. Westley, 1837), 88.

[16] The orientalist trope of the Ottoman city's "silence" it seems, was to continue throughout the nineteenth century and later; see, for example, Edmondo de Amicis, *Constantinople*, trans. Caroline Tilton ([1878] New York and London: Putman's, 1888), 241. See also Kevin Karnes's discussion of the orientalizing of Turkish subjects through urban silence, in this volume, 135–38.

[17] See James Q. Davies, "A Musical Souvenir: London in 1829" (PhD thesis, University of Cambridge, 2005), 1–16.

[18] Claridge, *Guide along the Danube,* 106.

[19] Philip Mansel, *Constantinople: City of the World's Desire 1453–1924* (London: John Murray, 1995), 261.

[20] See Emre Aracı, "A Levantine Life: Giuseppe Donizetti at the Ottoman Court," *The Musical Times* 1880 (Autumn 2002), 51.

[21] Mansel, *Constantinople,* 262.

[22] Ibid., 261.

latter represented most obviously by Italian opera, nineteenth-century Europe's most prestigious cultural export.

The decisive moment took place in 1826, when Mahmud II abolished the Janissary corps, his increasingly resented (and feared) elite infantry. He thereby did away with the Janissary bands, which had been the principal performers of traditional Ottoman ceremonial and military music. In their place, Mahmud recruited European musicians to establish a new Ottoman military band based on the European model. His first recruit, a French conductor called Manguel, was short lived. His second, Giuseppe Donizetti—the elder brother of the famous composer—was a great success: he remained in his post from 1828 until his death in 1856, by which time he had been promoted to the lofty status of brigadier general and had been conferred the honorific Donizetti Pasha. It was through this "other" Donizetti (the composer of *Lucia di Lammermoor* called Giuseppe his "fratello turco") that Italian opera was officially introduced to the Ottoman Court. Under Donizetti, the band's repertoire was not limited to military music: it also included operatic extracts. The band frequently performed arias by Rossini or Donizetti Jr. in the city streets, even while accompanying the Sultan on his weekly processions to the mosque.[23] Such was the popularity of this newly imported repertoire that word reached Paris that, as a result of the Sultan's love of Italian opera, "the ancient Turkish music has died in agony."[24]

There are complex dynamics of cultural politics and domination at work here, symptomatic of Constantinople's presence in an increasingly worldwide circuit traveled by European (and above all Italian) operatic works and performers. It was not only the new Ottoman military band that played operatic extracts in the city: entire works were performed at theaters established in the Pera district of Constantinople (now Beyoğlu), a neighborhood that was populated largely by foreign emigrants and diplomats. Pera was so closely associated with the "Franks" (as all Western foreigners were known) that in 1840 Murray's first *Handbook* to the city reported sourly that "This suburb [. . .] is devoid of any Oriental character, and bears much resemblance to a second-rate Italian town."[25] It can be no

[23] See Aracı, "From Napoleon to Mahmud: The Chequered Career of the Other Donizetti," *Giuseppe Donizetti Pascìa: Traiettorie musicali e storiche tra Italia e Turchia*, ed. Federico Spinetti (Bergamo, Italy: Fondazione Donizetti, 2010), 11; Bülent Aksoy, "Musical Relationships between Italy and Turkey through Turkish Eyes," *Giuseppe Donizetti Pascìa*, Spinetti, 69.

[24] Aracı, "Levantine Life," 51. Even Pers Tuglaci's late-twentieth-century account of the history of Turkish military bands proclaims that "[i]n 1831, with the official establishment of the Imperial Orchestra, the centuries-old *mehterhane* [the traditional music played by the Janissary bands] entered a dark period in its history." Pers Tuğlacı, *Turkish Bands of the Past and Present* (Istanbul: n.p., 1986), 18.

[25] *A Handbook for Travellers in the Ionian Islands, Greece, Turkey, Asia Minor, and Constantinople* (London: John Murray, 1840), 155.

coincidence that, in the same year, Constantinople's first Italian opera house was opened in Pera (its inaugural production was apparently of Bellini's *Norma*).[26] The venue remained open until 1842, at which point it changed hands and was renamed the Naum Theater after its new owners, two Syrian brothers. Under their directorship Italian opera became established in Constantinople, with new works—those of Verdi a case in point—increasingly receiving their first Turkish performances after a time lapse comparable to that achieved by major operatic centers such as London and Paris.[27]

More difficult to trace than *when* individual works first reached Constantinople is *how* audiences at the Naum Theater reacted to the imported tradition. Special leaflets were produced by the theater on its opening, not only to serve as publicity but also to furnish advice on how one should behave in the new establishment. Prospective audience members were requested, among other instructions, not to stand during performances and were informed that there should be no smoking, no fighting over seats, and—above all—no noise.[28] This final injunction is the most striking, revealing as it does something crucial about the Naum brothers' ambition for their theater. In 1840, elements of the audience at Her Majesty's Theatre in London—the major opera house in a city that fashioned itself as a cultural as well as an industrial world leader—had descended into riot in response to the establishment's refusal to hire the star baritone Antonio Tamburini for the coming season. Newspaper reports of that occasion make it clear that, while rioting was unusual and generally considered unacceptable, talking during performances remained a conventional (if increasingly criticized) aspect of theater attendance even in this prestigious venue. The shift charted by James Johnson by which operatic audiences fell silent and started to listen attentively was, in other words, still ongoing at this point. That around the same time, Constantinople's new operatic public could be informed that silence was a prerequisite thus suggests a remarkable degree of respect for the medium as a form of instruction as well as entertainment.

This respect was shared, significantly, by opera's other new market: London's growing middle classes, whose attentiveness was matched by its desire to better itself. That the importation of Italian opera to Constantinople was anything but politically neutral is clear; indeed, it must be seen as one manifestation of

[26] Mehmet Baltacan, "The Relationship between Turkish and Armenia Regarding the Ottoman Empire and Contributions of Armenian Artists to the Turkish Opera," *International Journal of Social, Business, Psychological, Human Science and Engineering* 8/5 (2014), 1149–55; here, 1151.

[27] *Ernani* and *Nabucco* both had their Turkish premieres in 1846 and were followed by *Macbeth* (1848), *I Lombardi* (1850), *I masnadieri* (1847), *Il trovatore* (1853), *Rigoletto* (1854), and *La traviata* (1856). I take these figures from Baltacan, "The Relationship between Turkish and Armenian," 1150.

[28] Baltacan, "The Relationship between Turkish and Armenian," 1151.

the globalizing dynamics governing international cultural capital and participation in a "universal" elite culture. The comparison between Pera and London is instructive, not least because mid-century operatic cultures in cities across the globe developed in increasingly active dialogue with one another. In London audiences were marked by hierarchical divisions, with the top occupied by the established, aristocratic "fashionables," for whom operagoing remained a largely social occasion, central to everyday life; on another level were the middle classes, growing gradually in number and political weight, whose new, markedly attentive presence in the opera house drew media attention (and constituted an emerging market for specialist music journals); on another level still were those in the lower classes who could never hope to attend an operatic performance, but for whom opera may nevertheless have constituted an important element in the auditory experience of urban life, whether mediated via domestic arrangements, open-air band performances, or street musicians. What is striking about the Crimean War's establishment of a wartime experienced in Constantinople, as in London or in Sevastopol, was that it brought constituents of each of these groups of London's operatic consumers into contact with Pera's operatic culture. Moreover, as members of the allied forces regularly attended the Naum Theater during the war, they wrought significant changes to Constantinople's operatic scene. As the *United Service Magazine* observed ruefully toward the end of the war, "[w]hat a difference there was between the Pera opera-house of 1853 to 1854 and 1854 to 1855! I am not alluding to the cantanti, but to the audience."[29]

Traces of operagoing abound in Crimean War memoirs penned by officers and hangers-on alike. Lord George Paget, Colonel of the Fourth Light Dragoons (who rode in the Charge of the Light Brigade but whose reputation suffered from his decision to take home leave with the war still ongoing) makes passing mention of regular visits to Pera's "fairish" opera between accounts of other excursions and occasional overwrought commentary on the conflict itself.[30] Colonel Hodge (commander of the Fourth Dragoon Guards, the man whose hut was decorated with copies of *Punch* and the *Illustrated London News*) records a night at the opera in late December 1855—the work was Verdi's *Il trovatore*, "the singing and acting very respectable."[31] Edwin Galt, one of the war's many battlefield tourists, reported on the "capital Italian opera" where he twice attended *Lucrezia Borgia*;[32] but a self-styled "Roving Englishman" warned

[29] "Society at Pera since the Western Invasion," *United Service Magazine* (Jan. 1856), 35.

[30] George Paget, *The Light Cavalry Brigade in the Crimea: Extracts from the Letters and Journal of the Late Gen. Lord George Paget, K.C.B., during the Crimean War* (London: John Murray, 1881), 154.

[31] Hodge, *"Little Hodge,"* 140.

[32] Edwin Galt, *The Camp and the Cutter; or, a Cruise to the Crimea* (London: Thomas Hodgson, 1856), 50.

"the enlightened traveller" against an operatic excursion to Pera: "There is an unhealthy smell of dead rats about it—a prevailing dampness and dinginess—a curious fog, a loudness, a dirtiness."[33] Such terms were regularly employed by Western visitors critical of Constantinople, who diagnosed the city as filthy, disease ridden, and overrun with packs of wild dogs.[34] It is interesting, then, that the same author goes on to relate how, before attending the opera one night, he ducked into the theater's coffee house, where British sailors and French soldiers were simultaneously singing native melodies. The author confesses to

> a keen enjoyment of their songs. There is a fine raciness about those of the British tar, which it is positively invigorating to hear. I shall not have half so much fun in the theatre, where Mademoiselle Squallini, an autumnal *prima donna* from Islington, is tearing one of prolific Verdi's operas into shreds, and screaming in a manner which is inconceivably ear-piercing.[35]

What is more, he ends his account of Pera's operatic offerings by explaining that the theater's elite clientele—"Highly-connected young gentlemen, mostly from the neighbourhood of Sloane Street or Putney, and belonging to her Majesty's commissariat"—are in the habit of going "behind the scenes" during performances to display their knowledge of the "elegant dissipations of London and Paris."[36] The author does not approve. For him, it seems, opera could represent upper-class debauchery in Pera just as it did in the art form's longer established urban centers. Furthermore, in comparison with the "invigorating" songs of the soldiers and sailors, an opera by Verdi is imagined to be no more than noise—as if the degeneracy of its consumers had rubbed off onto the art form itself.

There is little doubt that by the mid-nineteenth century, in Constantinople as elsewhere, Italian opera was understood and even celebrated as an international

[33] *Pictures from the Battle Fields*, by "The Roving Englishman" [Eustace Clare Grenville Murray] 2nd ed. (London: Routledge, 1855), 48.

[34] One interesting example is that of Albert Smith, who visited the city before the Crimean War and drew a contrast (strikingly reminiscent of Michel de Certeau's writings on walking in the city) between the view of Constantinople from afar, and the impression from close up: his first glimpse of the city from the Golden Horn was a "gorgeous panorama" comparable only with the experience of looking down on London by night from a balloon. Once disembarked, however, he "felt that I had been taken behind the scenes of a great 'effect.' The Constantinople of Vauxhall Gardens, a few years ago, did not differ more, when viewed, in front from the gallery, and behind, from the dirty little alleys bordering the river"; Albert Smith, *A Month at Constantinople*, 2nd ed. (London: David Bogue, 1850), 42, 47.

[35] Murray, *Pictures from the Battle Fields*, 49.

[36] Ibid., 51.

art. Purveyed by itinerant troupes of singers, it had an audience governed by behavioral conventions established in London or Paris but, as we have seen, reproduced to great symbolic effect further afield. Yet for "The Roving Englishman" attending the opera in Pera during the Crimean War, opera's fundamental mobility was coupled to another international force: that of the aristocracy and other elite members of any European society, whose wartime leisure was partly based—as in peacetime at home—in the more or less luxurious surroundings of the opera house. What is more, the appearance of this foreign elite at the Naum Theater itself seemed to constitute a military sortie, albeit one directed against Britain's and France's Ottoman allies. The *United Service Magazine*—a British military publication whose regular coverage of theatrical events at home, in Constantinople, and in Crimea is itself significant—reported how, following this friendly invasion of foreign officers,

> [t]he native fair of Pera were no longer allowed to frequent the scene of these reforms by their jealous owners, and the only females to be seen in the edifice were a few English ladies, who were sharing the campaign with their husbands, and the fair vocalists on the stage.[37]

Taking suitability for female consumption as an international measure of respectability, the *United Service Magazine* charts a striking reversal: here it is the residents of Pera (whether Ottoman or Frank) who are understood to represent civilization, while the behavior of the recently arrived, Western European social elite is judged to be little short of barbarian, apparently "creating much disorder and confusion, to the inexpressible disgust of the native audience."[38] In the hallowed space of the opera house, the officers' claim to social elevation was trumped by their military status: they brought with them the aggression and strife of the battlefield, rendering the opera house—like the theater of war—an arena largely off limits to women of any nation.

Intermezzo: Operatic Maneuvers

There were, however, exceptions to this rule of female exclusion, as the *United Service Magazine*'s mention of military wives makes clear. In addition to the small number of spouses who had traveled with their officer husbands to Crimea (most remained in Constantinople or in Scutari, a nearby Black Sea resort town;

[37] "Society at Pera since the Western Invasion," 35.

[38] Ibid.

a handful insisted on being accommodated in the military camp itself), the war zone and its environs became a destination for British tourists as the bombardment of Sebastopol continued through 1855 and allied victory was awaited with ever greater confidence. These visitors appear to have formed a new operatic market in the Crimea: according to Elizabeth Grey, "regimental bands were detailed to play 'appropriate airs' to entertain these spectators during lulls in the fighting," as though (as for the officers visiting Pera) opera constituted a natural interlude for battle.[39] Yet allied military bands had long been regularly performing operatic excerpts, eliciting mixed responses from those at the front. The band of the Sardinian army—a military force generally looked on with patronizing amusement by their allies—was particularly well known for its operatic renditions.[40] Colonel Hodge, for instance, attended a review of the Sardinian army in November 1855 and reported (in a peculiar conflation of Italian stereotypes) that "[t]here was first a kind of High Mass, opera music playing the whole time," as if it were only natural that Catholic ceremonial should mingle with operatic performance.[41]

More surprising to visitors to the Crimea, whether military or civilian, and apparently more comment-worthy than opera's penetration into Constantinople, was the fact that the Ottoman forces also boasted a military band whose repertoire was centered on Italian opera. The officer Frederick Robinson, for example, visited the barracks of the Turkish artillery at Pera and found "an amusing burlesque on our English bands." He heard a selection of excerpts from Bellini's *I puritani,* in which "the high notes appeared to me particularly defective. Two of the performers carried 'trees' of small bells, which they rotated slowly."[42] The instrument to which Robinson refers is presumably the *chaghana,* or "Jingling Johnnie," whose popularity far outlived the Janissary bands in which it originally appeared. Still more unexpected than this performance of Bellini's opera (which was almost two decades old and its highlights unequivocally world famous) is that in April 1855 an officer's wife heard the same band play a medley from Verdi's *Rigoletto*. The opera had premiered in Venice less than four years earlier,

[39] Elizabeth Grey, *The Noise of Drums and Trumpets: W. H. Russell Reports from the Crimea* (London: Longman, 1971), 44.

[40] Mrs Duberly, the seemingly ubiquitous wife of Captain Henry Duberly (paymaster of the 8th Royal Irish Hussars), whose presence at the front was sufficiently famed that *Punch* could satirize her as Mrs Jubilee, was typical in her view that "[t]here never was such a pretty little army sent into the field as that of the Sardinians"; Frances Isabella Duberly, *Mrs Duberly's War: Journal and Letters from the Crimea, 1854–6,* ed. and intro., Christine Kelly (Oxford: Oxford University Press, 2007), 259. See also Delia Casadei's discussion of what she terms the Sardinian band's "sonic assertion of nationhood" in this volume, 167–68.

[41] Hodge, "*Little Hodge,*" 136.

[42] Frederick Robinson, *Diary of the Crimean War* (London: Richard Bentley, 1856), 48.

and its first performance at the Naum Theater had taken place only the previous year.[43] By the mid-century, opera evidently traveled at least as quickly in military band arrangements as it did in its staged form.

The mechanisms by which such musical circulation took place remain, as so often, elusive. Yet we might call attention to two prestigious bands that may have served as institutional nodes in the proliferation of Italian operatic music among the Turkish troops. First, there was the official Ottoman military band trained by Giuseppe Donizetti: the band that, as we have seen, had undergone explicit westernization (under the flag of modernization) in the years before the Crimean War. Second, the commander of the Ottoman army, Omer Pasha, possessed his own private, uniformed band, which was reported to play both traditional Ottoman and operatic repertoire at the Crimean front. Writing his memoirs in 1915, Thomas Buzzard, a member of the British medical staff with the Ottoman army (and Crimea correspondent for the *Daily News*), attributed the extravagance of maintaining a personal band to Omer Pasha's newfound "appreciation of luxury which we associate with the Oriental," the commander having been born in Croatia and only later converted to Islam. Yet Buzzard also reveals that the musicians in Omer Pasha's band were German rather than Turkish; and he recalls how "[i]t was just about this time that the *Trovatore* of Verdi had recently appeared, and we used to listen to the strains of 'Ah che la morte' under a brilliant sky in as lovely a landscape as the eye could rest upon."[44]

For all that he saw the German band as a sign of "Oriental" excess (overlooking the fascinating politics implied by the employment of German musicians by an Ottoman commander whether in the service of extravagance or anything else), it is clear that Buzzard was more than happy to enjoy its renditions as fitting accompaniment to his own exoticized experience of the Crimea. We might safely assume that Buzzard's memory of hearing a famous extract from act 4 of *Il trovatore* was calculated to give a certain tragic coloring to his account of the war. Yet this particular passage not only forms a strange juxtaposition with the "brilliant sky" of the Crimean landscape in which he heard it. It is also a moment in which Verdi's incarcerated hero is heard from offstage, his reflections on death alternating with a *Miserere* sung by a similarly off-stage chorus and interjections from his heartbroken lover—the one figure actually seen by the audience. The operatic excerpt thus brings its own complex sense of multidimensional space to the scene Buzzard describes. Indeed, to insert the sound of Verdi's music into that landscape—one sketched in distinctly picturesque terms—is almost to imagine the scene to be theatrical; to render Crimea itself operatic.

[43] Duberly, *Mrs Duberly's War*, 167.

[44] Thomas Buzzard, *With the Turkish Army in the Crimea and Asia Minor: A Personal Narrative* (London: John Murray, 1915), 122.

War through Opera Glasses

Buzzard was not alone in embellishing his memories of the Crimean War with an operatic accompaniment. The notorious Mrs. Duberly recalled a horse ride inland from the Crimean port of Eupatoria, during which she and her companions had entered one of the many houses abandoned by the retreating Russians and found a grand piano. Coming upon the intact instrument was, she gushed, "like meeting a dear and long absent friend." Duberly played various popular mid-century songs before declaring,

> One more song and I must hasten back, to be on board my ship by twilight. Heavy guns are pouring their dull, broadsides on our straining ears. What shall the song be, sad and low, or a wild outburst of desperate courage? I have it:
>
> Non curiamo l'incerto domani.
> Se quest'oggi n'è dato goder.[45]

Like Buzzard, Duberly refers to a specific operatic piece at a moment of heightened theatricality, posing her question against a melodramatic backdrop of nightfall and pounding guns. It might thus seem in some sense fitting that she should provide an answer in the form of the act 2 *brindisi* from Donizetti's *Lucrezia Borgia*, with its energetic exhortations to live in the present rather than worrying about the future. Yet, as in Buzzard's operatic turn, Duberly's reference fails to ring true to a reader with knowledge of the opera beyond the Italian snippet that she provides. Donizetti's drinking song is, after all, saturated with dramatic irony, its music markedly in contrast with the dark hues that pervade the opera; at its end, the work's eponymous villain will appear to announce that she has poisoned the wine of all those present.

As understood in its staged context, this is hardly the cheerful leave-taking Duberly seems to have intended (nor can it be wholly described as a "wild outburst of desperate courage"). There are, however, two connected explanations as to why both Duberly and Buzzard might have embedded such operatic meaning in their Crimean narratives. First, we might conclude that both authors were keen to demonstrate their fluency in the cultural language of the upper classes; but that neither was sufficiently literate in operatic terms to alight on examples that retained their intended meaning and original illustrative purpose under more expert scrutiny. Second, we might decide that scrutinizing these individual references is to miss the more important point: that "opera" (whether represented

[45] Duberly, *Mrs Duberly's War*, 68–69.

by an extract from *Il trovatore* or *Lucrezia Borgia*, or any other work) was most significant as an index of a particular system of cultural values; that "opera" was being invoked symbolically, rather than literally. Both explanations are probably true; but it is the broad symbolic meaningfulness of "opera" in the Crimean War that is important above all here, and that has the further-reaching implications.

Perhaps the best-known instance of such symbolic invocation of opera appears in the most famous British account of the Crimean War. *Times* correspondent William Howard Russell described watching the 1854 Battle of Balaklava alongside the commanding officers of the British army:

> The instant they [the Russians] came in sight, the trumpets of our cavalry gave out the warning blast which told us all that in another moment we should see the shock of battle beneath our very eyes. Lord Raglan, all his staff and escort, and groups of officers, the Zouaves, French generals and officers, and bodies of French infantry on the height, were spectators of the scene as though they were looking on the stage from the boxes of a theatre. Nearly every one dismounted and sat down, and not a word was said.[46]

Ulrich Keller has discussed at some length the association between military events and theatrical spectacle that is encapsulated here.[47] I would add that the "theatre" Russell imagined would almost certainly have been an opera house. Opera was, as already discussed, the most prestigious and most widely disseminated theatrical medium during the nineteenth century, not to mention the art form most automatically associated with the elevated social milieu of Russell's military co-viewers. In light of this intuition, we might usefully return to the final sentence—that, met with the spectacle of battle opening out below them, "[n]early every one dismounted and sat down, and not a word was said." Such a detailed account of the behavior of this "audience" is striking—not least because Russell paints a scene not only in which a battle is rendered broadly theatrical, but also in which its spectators behave according to the newly silent norms of modern operatic listening.

What is more, these operatic ears were matched by operatic eyes. During the bombardment of Sebastopol in June 1855, in which Colonel Hodge effused that

> [s]hot, shell and rockets were going all this time. The sight was awful, particularly where I knew that friends were in the midst. It was grand,

[46] William Howard Russell, *Russell's Despatches from the Crimea, 1854–1856*, ed. Nicolas Bentley (London: Andre Deutsch, 1966), 124.

[47] Keller, *The Ultimate Spectacle*, 5.

however, as a spectacle. [. . .] Much thunder today, mixed with the firing, the effect was very grand.[48]

Crucial here is the contrast between Hodge's friends, "in the midst," and his own position, at a sufficient distance that the events could be rendered spectacular—and pleasurably so. Such distance was nothing if not a mark of social elevation: a privilege accorded only those with the social pedigree to occupy the higher levels of military command, or those noncombatants who had the means to travel to the front as observers. It is no coincidence that these two groups—one civilian and one military—shared an essential piece of equipment: opera glasses. Ubiquitous in accounts of the war by tourists and officers alike, these literally provided their users with a perspective on the unfolding events. Small wonder in the circumstances that everything seen by these elite spectators seemed to be imbued "with the essence of opera," as Russell described a review of the French troops.[49]

There is little doubt that the mid-century British military was itself "a theatrical institution," as Scott Hughes Myerly has argued; or that, as Paul Fussell explored with great sensitivity in his study of the Great War, we now *expect* modern war and theatricality to be intimately connected, precisely because life at the front was so often felt to be "unreal" in comparison to news from back home.[50] But these readings of war's theatricality tend to address only soldierly experiences and perceptions, as if theatricality can emerge in the absence of an audience. On the contrary, the distinctly *operatic* theatricality of the Crimean War constantly throws our attention back onto the physical presence of audiences, and onto the observations of individual spectators, above all. At least among those whose experiences of Crimea we can most easily gain access to, there is nothing short of an operatic mode of perception. This mode had, for sure, primarily been learned in the opera house, but once established could reach far beyond its walls, to constitute a much broader experiential frame of reference.[51]

What might give us pause at this point is that recourse to operatic frames of reference was to some extent an established feature of orientalist travelogues by the time the war began. Yet among the endless references to "panoramas" and "backdrops" beloved of visitors to the neatly exoticized foreign landscape,

[48] Hodge, *"Little Hodge,"* 112.

[49] Russell, *Russell's Despatches from the Crimea*, 192.

[50] See Scott Hughes Myerly, *British Military Spectacle: From the Napoleonic Wars through the Crimea* (Cambridge, MA: Harvard University Press, 1996), 8; Paul Fussell, *The Great War and Modern Memory* (Oxford: Oxford University Press, 1975).

[51] By contrast, I suspect that those hearing "La donna è mobile" played on a barrel organ on a London street could equally have been hearing a ballad or street cry, so far as their mode of listening or degree of attentiveness was concerned.

the bazaars of Constantinople stand out, for they drew surprisingly specific operatic references. John Harwood reported that the Ottoman traders tended to assume "as supercilious and *goguenard* an air at the casual purchaser, as the regular frequenters of the Parisian opera assume when an uninitiated mortal ventures to intrude into the hallowed precincts of the *foyer*."[52] Murray's *Handbook* of 1854 warned prospective visitors that to pass through the city's bazaars, "more ceremony is required than amongst the well-dressed mob of an opera-house."[53] Again, the operagoers and their behavior, rather than the operatic performance itself, are the subject of the implied comparison—and both rely on the reader's familiarity with such social conventions. Indeed, insofar as the Crimean War might be understood to be "operatic" (in a way the writer for *Dwight's* with whom I began could only have found comic), it was so both because Constantinople and the Crimean peninsula already had its own operatic culture; and because these places were already understood in operatic terms by their Western European visitors.

Opera and "Modern War"

In approaching the Crimean War through an operatic lens, as did so many of its elite participants and hangers-on, I have attempted to reveal an important sense of continuity—however class-contingent—between wartime as experienced from Crimea and as lived on the British domestic front. I have thus been concerned with the relationship between military events and "opera"—understood multiply as performance, re-mediated sound and cipher—as figured by the war's many published witnesses. Perhaps inevitably, the majority of those narratives (whether civilian or military), have come from members of the British social elite—people, that is, for whom "opera" would indeed have been part of everyday life. At the very least, such an approach to the war via its operatic elements might—as Favret (rightly) recommends—remind us of the particular social origins of many of the war's eye- and earwitnesses. I hope, though, that I have also offered some sense of opera's sheer pervasiveness in and around the Crimea: as performed by military bands at the front, staged in Constantinople, and resorted to as a symbol or frame of reference. I have attempted to plot the outline of an argument about the art form's ability in the mid-nineteenth century to cross or even collapse geographical and social distance, and even about its centrality, for some, to the experience of war. Understood thus, opera demands inclusion in

[52] John Harwood, *Stamboul and the Sea of Gems* (London: Richard Bentley, 1852), 54.

[53] *A Handbook for Travellers in Turkey: Describing Constantinople, European Turkey, Asia Minor, Armenia, and Mesopotamia*, 3rd ed. (London: John Murray, 1854), 92.

any account of events in Crimea—at least for those who seek to address war as a sensory experience, or for those interested in the war's mediatization.

Taking opera's presence in Crimea more seriously may also enable us to gain a more socially sensitive perspective on the continuing debates over the war's claims to being either "the first modern war" or "the last of the old-style wars."[54] Earlier I quoted Philip Mansel's description of mid-century Constantinople as a city "torn between contradictory forces," in which rapid modernization could be seen alongside the persistence in the city of much older traditions. As discussed, one could indeed see the arrival of Italian opera in the city as a symptom of this process of modernization; yet although its advent was of course enabled by modern networks of transportation and communication, opera might also be understood as intimately connected with older technologies. The Crimean tourist Edwin Galt could report that "[t]he European population here use sedan chairs at nights, and in the dark streets, returning from the opera, almost every one carries a long white round paper lantern, purchased at the shops for a piastre (2*d*)": operagoing here was every bit as reliant on the sedan chair and the lantern as on the steamer or the telegraph.[55] We might also recall Frederick Robinson's account of a rendition of excerpts from *I puritani* by the (famously modernized) Ottoman military band, which nevertheless included two *chaghanas*: instruments from the Janissary bands that had supposedly been superseded almost thirty years earlier.

In one of the few studies dedicated to music produced during and in relation to the Crimean War, Didier Francfort makes his position on the conflict's relationship to modernity clear from his article's title, which heralds the war as "the founding moment of European military music."[56] He goes on to pose the question as to whether the war did not, surely, constitute the first European musical expression of the "concert of nations," ahead of the redrawing of the map of Europe in the 1870s.[57] It is certainly tempting to trace acts of musical diplomacy between the Crimean allies, and thus to see the conflict (as Francfort suggests we should) as an instance of musical universality at a time more famous for growing nationalist tendencies. But in contrast to Francfort, who is concerned above all with the production of new works in relation to events in the Crimea, I have largely been addressing the presence there of already established,

[54] For the first quotation, see Keller, *The Ultimate Spectacle*, 251; for the second, see Ian F. W. Beckett, *The Victorians at War* (London: Hambledon Continuum, 2003), 162–63.

[55] Galt, *The Camp and the Cutter*, 45.

[56] Didier Francfort, "La Guerre de Crimée, moment fondateur des musiques militaires européennes," *Der Krimkrieg als erster europäischer Medienkrieg*, ed. Maag, et al., 163–72.

[57] "[L]a Guerre de Crimée ne fut-elle pas, avant le remodelage de l'Europe dans les années 1870, une première expression musicale européenne du 'concert des nations?'," "La Guerre de Crimée," 163.

even canonized pieces from the Italian operatic repertoire. As such, any claim to international operatic diplomacy being enacted beyond the a priori fact that the prized status of canonic works rested on their supposed universality would also need to account for the successful mid-war staging of a season of Italian opera in St Petersburg; or for the fact that prewar Sebastopol itself boasted an opera house, where one "Lady Resident near the Alma" reported watching a performance of *Norma* in which a "motley group of Druids" could be seen from close up to be wearing the boots of Russian soldiers.[58]

Insofar as the Crimean War can be celebrated as an instance of musical universalism, in other words, it must be understood as such within the larger context of a musical culture increasingly centered on a canon of revered works, which were reproduced and disseminated across a vast network of urban centers. In this context as in so many others, the war must be seen as a single, midway stopping point on a much longer trajectory. As military historian David Edgerton writes,

> [t]he military, and war itself, have often been seen as left-overs from the past. War was not something which modern, democratic free-trading nations did. Soldiers, particularly officers, were relics of an older agricultural and warlike society, which like chivalry, would disappear as modernity marched on. Modern war was a tragic clash of the old and new.[59]

Edgerton's principal concern is with twentieth-century conflicts, in which this "tragic clash" was to come into ever sharper focus; the point is nonetheless still a useful one when applied to the Crimean War, offering a sobering alternative to the choice between old-fashioned war and modern war, old technology and new. Instead, Edgerton suggests that the crux of "modern" war is located precisely in this combination, this constant tussle, between the old, established ways and those that claim to be produced by "progress," forging the route to a better future.

With this in mind, I want to end with a final comic take on opera's role in events in Crimea—this time from *Punch*, the war's ultimate satirical opponent. The article in question combines an attack on what was becoming known as London's "organ grinding plague"—the musicians, often Italians, who played incessantly on the city's streets—with an account of a fantastical new system by which music could be piped directly into the homes of the capital's music-loving inhabitants. Consumers could apparently choose to be supplied from

[58] *The Crimea: Its Towns, Inhabitants, and Social Customs; By a Lady Resident near the Alma* (London: Partridge, Oakey, 1855), 74.

[59] David Edgerton, *The Shock of the Old: Technology and Global History since 1900* (London: Profile, 2006), 139.

the "Grand Mozart and Beethoven Junction," the "South Donizetti Milk-and-Water Works," or "the Great Hydraulic Processes of Verdi," according to individual preference. The article's relevance to this chapter, however, is revealed in its parting shot, aimed directly at the organ grinders. The author suggests that they might volunteer to join the allied forces in Crimea (perhaps even uniting with Miss Coutts, Mario, and the Sardinians?) where, "if they were properly organised with their own frightful organs, and brought well into play, every Russian would give his ears, rather than to endure the cruel infliction to which they would be subjected."[60] In a conflict famous for its showcasing of the new, and above all for its deep media imprint, *Punch*'s suggestion of a resolution to the continuing bombardment of Sebastopol is striking indeed: the deployment of a troupe of organ grinders playing extracts drawn from an art form in which longevity was increasingly venerated (and whose own prestige rested in part on its ancient pedigree) as a means to repulse an enemy equally sensitive to such offensive operatic reworkings. The true conflict here is neither of old and new, nor even that between nations: it is one waged by the international social elite against a changing world order, in which opera could be consumed by all—but only once removed from its gilded cradle and remediated irrevocably as noise.

[60] "Music Really for the Million," *Punch, or the London Charivari* 28 (1855), 251.

9

Earwitness

Sound and Sense-Making in Tolstoy's Sevastopol Stories

ALYSON TAPP

War demands representation, yet struggles to find adequate depiction. It may yield readily to verbal form that controls or eases through aestheticization, abstraction, or generalization. In this, "all wars are different and also the same."[1] And so, let us begin not with Leo Tolstoy's account of the Crimean War, but with Virginia Woolf's writing some sixty years later during the First World War. Tolstoy (1828–1910) and Woolf (1882–1941) are joined by their search to capture the flux of human consciousness, and for both wartime lent new urgency to this task.[2] Underlying the matter of artistic representation for both was, at root, the sense-making work of consciousness, the ways we perceive, mediate, and represent the world to ourselves. Woolf famously brought to fullest expression the technique of stream of consciousness, arguably invented by Tolstoy in the dying soldier's interior monologue in "Sevastopol in May" (and, notably, in the scenes preceding Anna's death in *Anna Karenina*). For Tolstoy, a serving officer, personal experience of the Crimean War sharpened the imperative to narrative representation of individual experience and the way such narrative involves externally received sensory impressions, including sounds, in making sense of the self and the world.[3] In pursuit of a form that is adequate to the truth of experience, Tolstoy moved beyond representations of war that are clothed in

[1] Kate McLoughlin, "Introduction" and "War and Words," *The Cambridge Companion to War Writing*, ed. Kate McLoughlin (Cambridge: Cambridge University Press, 2009), 1, 15.

[2] Tolstoy's presence is felt directly in *To the Lighthouse*, in which *Anna Karenina* becomes a subject of dinnertime conversation. In the draft, *War and Peace* is mentioned. For more on this, see Emily Dalgarno, "A British *War and Peace*? Virginia Woolf Reads Tolstoy," *Modern Fiction Studies* 50/1 (2004), 129–50.

[3] On the broader historical context of the literary mediation of sensation, see Dina Gusjenova's chapter, this volume, 6–9, 13–23

language that glorifies, abstracts, sentimentalizes, or distances. Woolf too recognized war's demands for representation, as well as its susceptibility to being troped in ways that suppress immediacy and actuality. "Heard on the Downs," her essay written after a summer in Sussex in August 1916, significantly bore the subtitle "The Genesis of Myth." It begins,

> Two well-known writers were describing the sound of the guns in France, as they heard it from the top of the South Downs. One likened it to "the hammer stroke of Fate"; the other heard in it "the pulse of Destiny."
>
> More prosaically, it sounds like the beating of gigantic carpets by gigantic women at a distance. You may almost see them holding the carpets in their strong arms by the four corners, tossing them into the air, and bringing them down with a thud while the dust rises in a cloud about their heads.[4]

The chalk-and-flint hills of the Downs double the landscape of northern France; the calcareous swells of land behind the English and French shores share a geological origin but are separated by the Channel, which in wartime provided a natural barrier of civilian immunity—though one seemingly permeable to sound: the landscape receives echoes of distant sounds from across the Channel.[5] Yet the male walkers' lofty abstractions keep the reality of the battlefield at a remove, its sounds heard as hammer strokes of Fate, pulses of Destiny. For her part, Woolf the essayist translates these sounds into something no less metaphorical, though concrete and "prosaic," admitting battlefield experience into the domestic realm. She goes on,

> All walks on the Downs this summer are accompanied by this sinister sound of far-off beating, which is sometimes as faint as the ghost of an echo, and sometimes rises almost from the next fold of grey land. At all times strange volumes of sound roll across the bare uplands, and reverberate in those hollows in the Downside which seem to await the

[4] Virginia Woolf, "Heard on the Downs: The Genesis of Myth," *The Essays of Virginia Woolf*, ed. Andrew McNeillie (London: Hogarth Press, 1986–2011), 3: 40.

[5] Robert Macfarlane writes of the ghostly doubling in wartime of the landscapes of the South Downs and northern France. See his *The Old Ways: A Journey on Foot* (London: Penguin, 2012), 329, 355. Paul Fussell also discusses the audibility of gunfire from across the Channel in the southern counties of England, and writes, of Thomas Hardy and his poem "Channel Firing," that "[o]ne reason modern English poetry can be said to begin with Hardy is that he is the first to invite into poems the sound of ominous gunfire heard across water." Paul Fussell, *The Great War and Modern Memory* (Oxford: Oxford University Press, 1975), 68.

> spectators of some Titanic drama. Often walking alone, with neither man nor animal in sight, you turn sharply to see who it is that gallops behind you. But there is no one. The phantom horseman dashes by with a thunder of hoofs, and suddenly his ride is over and the sound lapses, and you hear only the grasshoppers and the larks in the sky.[6]

Sound travels, moving across the landscape: its mobility conjures a phantom horseman—mobile, elusive in meaning, beyond visual verification. The sonic evokes spectral presences that signal the overlapping of multiple worlds: of the distant and the proximate, of the living and the dead. Behind these efforts to characterize wartime sound are bids to confront—or evade—questions of death and destruction.

The sounds on the Downs, and their manifold interpretations, are emblematic of the condition that Mary Favret has drawn our attention to: "war at a distance"—or how, in settings removed from combat or occupation, the experience of war "becomes part of the barely registered experience of the everyday."[7] For those experiencing it at a distance, war returns home through various kinds of mediation—mediations that sometimes take an unexpected form. Woolf's essay goes on to describe how, in a chicken coop not far from the Downs, the daughter of an elderly villager came to discover

> evidence of the supernatural state of things now existing without going farther than the shed in which her hens are sitting. When she came to hatch out her eggs, she will tell you, only five of the dozen had live chicks in them, and the rest were addled. This she attributes unhesitatingly to vibrations in the earth caused by the shock of the great guns in Flanders.[8]

For the hen keeper, the sound's meaning is not metaphorical, but wholly material: sensory perception slides from the audible to the tangible as the sounds transmit harmful vibratory forces.[9] In the age of heavy artillery—even within the

[6] Woolf, "Heard on the Downs."

[7] Mary A. Favret, *War at a Distance: Romanticism and the Making of Modern Wartime* (Princeton, NJ: Princeton University Press, 2010), 9. Favret invokes Woolf's writing of 1940, from before the Battle of Britain, where Woolf considers (counter to Favret's own thesis) that the wars of 1815 were entirely absent from the experience and literature of Austen, Scott, and others.

[8] Woolf, "Heard on the Downs."

[9] In his study of the relation between sound, affect, and violence, Steve Goodman makes a call to expand the study of sound beyond the humanly perceptible: "An ontology of vibrational force delves below a philosophy of sound and the physics of acoustics toward the basic processes of entities affecting other entities. Sound is merely a thin slice, the vibrations audible to humans or animals. Such an orientation therefore should be differentiated from a phenomenology of sonic effects centered on

relative safety of places of civilian immunity—the natural world is not untouched by war; "nature" does not provide a refuge set apart from the scene of fighting. In fact, even when the battle zone is far removed from immediate perception, the earth itself comes to function as a "medium" through which the experience of war might be made available. In winter 1915, ornithologist T. A. Coward noted the surprising behavior of birds in response to frequencies exceeding the range audible to humans, behavior brought about by the vibratory force emitted by distant gunfire.

> It is well known, that thunder, an explosion, or other loud noise will start the Pheasant's crow, and during the war, air raids before audible to human ears were responded to by the agitation of the sensitive birds. My most remarkable personal experience was on the morning of January 24th, 1915, when, in a Cheshire wood, I was struck by the frequent crowing of the cocks; in my notebook I wrote—"Cock Pheasants crowing constantly, and wing-flapping after each crow." Two days later I commented on the fact in the *Manchester Guardian,* but attributed it to the mildness of the weather, and it was only when reports were received of similar disturbances in Norfolk, Lincoln, and Cumberland that I realised that the Cheshire birds had also been influenced by the air vibrations of the heavy guns in the North Sea battle, some 400 miles away.[10]

As Woolf scholar Mark Hussey contends, "*all* Woolf's work is deeply concerned with war."[11] Nine years later, in her great novel of war at a distance *To the Lighthouse* (1927), in the highly experimental prose elegy section titled "Time Passes," Woolf was to return to the image of the tactile, vibratory force of the sounds of war, now felt not by egg shells, but by insensate objects: crockery and glassware—which nonetheless participate in a kind of sympathetic vibration, registering the death of a son in the Ramsay family.[12] For Woolf, the acquaintance

the perceptions of a human subject, as a ready-made, interiorized human center of being and feeling." Steve Goodman, *Sonic Warfare: Sound, Affect and the Ecology of Fear* (Cambridge, MA: MIT Press, 2009), 81.

[10] T. A. Coward, *The Birds of the British Isles and Their Eggs: Second Series Comprising the Families Anatidae to Phasianidae* (London and New York: Frederick Warne, [1920] 1950), 361. I am indebted to Robert Burton for pointing me to this reference.

[11] Mark Hussey, "Living in a War Zone," *Virginia Woolf and War: Fiction, Reality, and Myth,* ed. Mark Hussey (Syracuse, NY: Syracuse University Press, 1991), 3.

[12] The distant death of Andrew Ramsay is felt in the objects whose contours carry the narratives of domestic life and human relations in which they participate. The square-bracketed text toward the end of the following passage shocks us with its contrasting numb registering of fact and underscoring of spatial disjuncture. "[T]here came later in the summer ominous sounds like the measured blows of

with war at a distance and the generational experience of the First World War gives impetus to her experimentation with the novel, to the representation of fragile consciousness and modernity's new forms of mourning. Yet the memory of the Crimean War is present in *To the Lighthouse*, too. In its early chapters, Mr Ramsay—a man of the late Victorian age—comes to be associated with a refrain of circulating lines from Tennyson's "Charge of the Light Brigade." Its sound and rhythm propel him forward, as he re-enacts the charge on the front lawn, and any meaning the words retain associate him only with a foolish and impersonal insistence on valor and heroism. During the First World War, the Crimean War loomed large in the British imagination, partly because it was within reach of an older generation's memories.[13] But there is perhaps another, subtler reason for Crimea resurfacing at this moment. "Wartime has trouble measuring its distance from other times of war," Favret observes—and with this assertion justifies the anachronistic movements of her own study of British romanticism and the Napoleonic wars, which finds, among twentieth-century writers, recapitulations and inflections of phenomena that originated in the previous century.[14]

As for Woolf, for Tolstoy too the experience of war—at a distance and firsthand—was a formative impetus to narrative experimentation. He brought to bear a double generational perspective on the writing of war: a serving officer in the Crimean War, he took the Napoleonic Wars experienced by his parents' generation as the subject of his great novel *War and Peace*, written during the 1860s. In the coupling of Tolstoy's and Woolf's works, then, there is the accretion of a century's experience of war, from the Napoleonic Wars to the First World War, with the Crimean War at the midpoint. By moving now to read Tolstoy after Woolf, we may encounter a proto-modernist Tolstoy, whose ongoing experiment to capture the self in writing poses questions about the narrative

hammers dulled on felt, which, with their repeated shocks still further loosened the shawl and cracked the tea-cups. Now and again some glass tinkled in the cupboard as if a giant voice had shrieked so loud in its agony that tumblers stood inside a cupboard vibrated too. [...] [A shell exploded. Twenty or thirty young men were blown up in France, among them Andrew Ramsay, whose death, mercifully, was instantaneous.]" Virginia Woolf, *To the Lighthouse* (San Diego: Harcourt, 1927), 133.

[13] Virginia and Leonard Woolf's library contained the first volumes of Alexander William Kinglake's eight-volume *The Invasion of the Crimea: Its Origin and an Account of Its Progress Down to the Death of Lord Raglan* (Edinburgh and London: Blackwood, 1863–87). The volumes had been inherited from Woolf's father, Leslie Stephen, who referred to Kinglake's work and the possible agency of the *Times* in causing the Crimean War in the opening of his 1865 pamphlet, "The *Times* on the American War." Karen Levenback comments on the relationship between Tennyson and Stephen (and Woolf's knowledge of it): Stephen could not share Tennyson's support for the war, but the two men "were undoubtedly agreed on the misrepresentations by newspapers [...] in reporting on the Battle of Balaclava." Karen L. Levenback, *Virginia Woolf and the Great War* (Syracuse, NY: Syracuse University Press, 1999), 91, n17.

[14] Favret, *War at a Distance*, 30.

representation of consciousness and authentic experience that sounded again and acutely when modernist writers sought to remake a system of representation that had ruptured in the age of the First World War. For Tolstoy in the *Sevastopol Stories,* sound became a cipher for unmediated reality, and ultimately for truth—a means of gesturing toward authentic battlefield experiences prior to their narrative retelling.

Making Sense of War: The *Sevastopol Stories*

Tolstoy's *Sevastopol Stories* (1855) comprise a trilogy, united by their subject matter but varied in narrative form.[15] The *Stories*—which were to provide Tolstoy with numerous kernels for future development in *War and Peace*—are a crucible of narrative experimentation. Taken together, the stories stage a transition: the first, "Sevastopol in December," is a documentary sketch employing the unusual second-person narrative voice; the second, "Sevastopol in May," "discovers" and affirms the power of a third-person omniscient narrator; and the last, "Sevastopol in August" passes as a fictional novella in the more conventional third-person voice.[16] The stories "exist on the liminal boundary between reportage and fiction," portraying fictional characters, yet stamped, the reader knows, with the authenticity that "Tolstoy was there."[17]

Originally published in separate monthly editions of Russia's leading journal, the Petersburg-based *The Contemporary (Sovremennik)*, the stories appeared as dispatches from the front lines, and were received with all the keenness that greets bad news: "Sevastopol in December" appeared in the June 1855 edition of *The Contemporary,* and the following month's volume paints a picture of the thirst for news from the Crimea that had beset the Imperial capital of Petersburg.[18] A prominent example of the "thick journal," the key medium and institution for the development of nineteenth-century Russian literary culture, *The Contemporary* contained (in the 1850s) a mixture of a native and translated fiction and nonfiction, poetry, literary criticism, book reviews, news of the

[15] The individually titled stories were published together in a revised 1856 edition under the title *Military Tales* (*Voennye rasskazy*), and subsequently became known as the *Sevastopol Stories* (*Sevastopol'skie rasskazy*). The common English translation of *Sebastopol Sketches* is itself mildly indicative of the works' slippage between the documentary and the fictional.

[16] Kathryn B. Miller, *Tolstoy and the Genesis of War and Peace,* ed. Robin Feuer Miller and Donna Tussing Orwin (Ithaca, NY, and London: Cornell University Press, 1996), 16.

[17] Andrew Wachtel, "History and Autobiography," *The Cambridge Companion to Tolstoy,* ed. Donna Tussing Orwin (Cambridge: Cambridge University Press, 2002), 176.

[18] *Sovremennik* 52 (July 1855), 119.

Petersburg and Parisian cultural scenes, and more.[19] In addition to the individually titled pieces that later became known as the *Sevastopol Stories*, editions from the Crimean period contained aggregated eyewitness reports received as letters from Sevastopol, as well as formal, impersonal military digests of news from the conflict. This mixture of fiction and nonfiction could readily accommodate the generic ambiguity of the *Sevastopol Stories*.

The dynamic interplay of nonfictional and fictional modes was, however, also an operation motivated by the internal imperatives of Tolstoy's own evolving art. Growing out of his diaries, Tolstoy's narrative practice began with the perceiving consciousness and the desire to capture the self in writing. In an early narrative experiment from 1851 titled "A History of Yesterday," Tolstoy formulated this task:[20]

> I am writing a history of yesterday not because yesterday was extraordinary in any way, for it might rather be called ordinary, but because I have long wished to trace the intimate side of life through an entire day. Only God knows how many diverse and diverting impressions, together with the thoughts awakened by them, occur in a single day. Obscure and confused they may be, but they are nevertheless comprehensible to our minds.[21]

The *Sevastopol Stories* continued this project of making sense of sensory impressions and exploring how experience could attain truthful narrative representation.

For Tolstoy, the Crimean War sharpened the imperative to the narrative representation of individual experience: to represent the experience of battle posed anew questions about the limits of narrative to capture both the flow of thoughts and sensory impressions—and, what is more, the coherence they may take on, but only in retrospect.[22] Any retrospective conferral of meaning or attempt at

[19] For an overview of the mid-nineteenth-century institution of the Russian thick journal, see Robert L. Belknap, "Survey of Russian Journals, 1840–1880," *Literary Journals in Imperial Russia*, ed. Deborah A. Martinsen (Cambridge: Cambridge University Press, 1997), 91–116

[20] To gain an insight into Tolstoy's and Woolf's common task of giving narrative expression to the flux of human consciousness, one might compare the young Tolstoy's formulation to Woolf's command to the would-be writers of a new, modern fiction in 1919: "Examine for a moment an ordinary mind on an ordinary day. The mind receives a myriad impressions—trivial, fantastic, evanescent, or engraved with the sharpness of steel. From all sides they come, an incessant shower of innumerable atoms; and as they fall, as they shape themselves into the life of Monday or Tuesday." Woolf, "Modern Fiction," *The Common Reader: Volume 1*, ed. Andrew McNellie (London: Vintage, 2003), 150.

[21] Leo Tolstoy, "A History of Yesterday" (1851), trans. George L. Kline, *Russian Review* 8/2 (1949), 142.

[22] As Jan Mieszkowski notes, with reference to Tolstoy, Stendhal, Chateaubriand, and others, "The notion that a battlefield might serve as a sort of perceptual laboratory was common in nineteenth-century efforts to make sense of the challenges that Napoleonic combat presented to

sense-making of experience is a falsification, an aberration of Tolstoyan truth, which strives to capture an unmediated present. In *War and Peace*, this same epistemological quandary was to be wedded to the discussion of the nature of history in general; in the *Sevastopol Stories*, the same issue was being worked out for the sake of what one might call an ideology of narrative form: how to represent the truth of perceptual processes and individual consciousness. This was a project Tolstoy had begun in his diaries (from 1847) and continued, in his first published work, in the semiautobiographical *Childhood* (*Detstvo*, 1852). In the context of Tolstoy's oeuvre, the *Sevastopol Stories* effected a transition from the first-person narrative voice of his diaries and the shifting semiautobiographical voice of *Childhood* toward the stable, third-person (omniscient) narrative voice that characterized his major fiction from then on. This is the ethical discovery that Tolstoy revealed when, at the end of "Sevastopol in May" (of the three stories the most ambiguously poised between documentary and fiction), he declared: "No, the hero of my story, whom I love with all my heart and soul, whom I have attempted to portray in all his beauty and who has always been, is now and will always be supremely magnificent, is truth."[23]

The battlefield provides a limit case for exploring the process of sense-making, and the *Stories* foreground some of the problems of attaining this Tolstoyan truth in narrative representation without the means of an impersonal third-person voice. For example, "truth" is lost in the act of transmission of a story taken from life because of the spatial and temporal disjuncture that necessarily marks an act of personal narration. Tolstoy calls attention to the fact that stories are told in a space different from that in which the events occurred: the receipt of letters from the front in the fashionable drawing rooms of Petersburg society exposes the incongruity of milieux. Even where the disjunction is less acute, Tolstoy is nevertheless keen to emphasize the time lag between happening and telling—a delay that necessarily entails a prismatic shift in the narrator's perspective. So it is that when a cadet named Pest tells of how "the company commander had been killed and he [Pest] had slain a Frenchman" the

> principal elements of this story [. . .] were factually true but in recounting its details he boasted and made things up. He found himself boasting in spite of himself, and the reason for this was during the whole of the action he had been lost in a fog of oblivion, to such a degree that all that had happened had seemed to be taking place somewhere else,

the coherence of historical experience." *Watching War* (Stanford, CA: Stanford University Press, 2012), 84.

[23] Leo Tolstoy, *The Sevastopol Sketches*, trans. David McDuff (London: Penguin, 1986), 109. Subsequent references to the work will be given in parentheses in the text.

> at some other time and to some other person. It was, therefore, natural that he should now reproduce these details so that he came out of the affair with some credit. (92)

This moment serves the emerging critique of heroism in "Sevastopol in May": there are forces that act on the scene of the telling—such as ambition or the pursuit of recognition—that become pressed onto and distort the original experience and complicate the traditional commonplace of noble, pure heroism.

Yet it is the narration of death that ultimately bestows confirmation of the impersonal third-person voice for Tolstoy's realist art. It is only a third-person voice that can capture the simultaneous experiences of the two officers, Praskukhin and Mikhailov, as a shell explodes, killing the former, and give us insight into the individuals' consciousness at that moment:

> "But perhaps it will only be Mikhailov who's killed. Then I'll be able to tell the story of how we were walking side by side when he was suddenly killed and spurted blood all over me. No it's closer to me—I'm the one who's for it."
>
> It was at this point that he remembered the twelve rubles he owed Mikhailov, as well as another debt he owed to someone in St Petersburg, one he should have paid a long time ago; the gypsy melody he had sung earlier that evening came into his head; the woman he loved appeared in his thoughts, wearing a hood adorned with lilac ribbons; he remembered a man who five years earlier had insulted him and on whom he had never got his own back. (96)

The impersonal third-person voice is not limited to the external and retrospective point of view, but assumes a privileged perspective, situated in an unfolding present, privy to the flow of individual consciousness. As an instrument of truth, Tolstoyan realist narrative strives to capture experiences before they become mediated: to expose the falsity of retrospectively conferred coherence and generalization and convey instead (or alongside it) the very processes that coordinate sensory impression and sense-making. The *Stories'* attention to sound, as I will show, share a special affinity with this goal.

The Sounds of War in the *Sevastopol Stories*

Most commentators on the *Sevastopol Stories* focus on their visual qualities. They draw attention to the ways in which the unusual second-person ("you") narration of "Sevastopol in December" confronts its reader with sights, constructing—and instructing—its reader through a technology of vision appropriated from the tour guide.[24] (The December sketch leads its reader as if conducting him on a tour: you will see this . . . and this. . . .) Countering the tightly controlled lines of sight in "Sevastopol in December" (the directed movement, the forced gaze) is the soundscape, in which sounds converge on the subject from near and far, from sources both visible and invisible.[25]

To focus on the *Stories'* sounds takes us beyond the critical trope of Tolstoyan didacticism—hinging on a viewer who sees in order to be instructed—and allows us to access something much more experimental, even proto-modernist. The *Stories'* sounds shift the emphasis from the self who is to be enlightened with discrete content received from without, to the concept of selfhood that is always in formation, engaged in processes of perception and cognition through which the self is simultaneously making sense of both itself and the world around it. While we might hazard that all sense organs can (at least potentially) function in this way—operating cognitive loops that shuttle awareness between proprioception and exteroception—it is worth paying particular attention to the way in which Tolstoy imbues seeing and hearing with antithetical qualities, resulting in a pervasive audiovisual split. Sights (as we encounter them in the touristic stroll) such as the horrifically injured people in the field hospital, are defined, graspable entities: vision offers a perspective on experiences or facts that exist outside of the self. Sounds, meanwhile, exist separately from their visible source and must be taken into the self for sense to be made: hearing is immersive and places the self inside an event.[26]

Before I proceed with illustration from the *Sevastopol Stories,* a cautionary word heeded from the discipline of sound studies, and a nod to Tolstoy's musical imagination: these distinctions between hearing and vision (hearing as immersive; vision as perspectival) are among those enumerated by the prominent sound studies scholar Jonathan Sterne in the "audiovisual litany," a list of binary distinctions between the two modes of sensory perception that Sterne faults for promoting a transhistorical or universalist phenomenological truth about

[24] For example, Gary Saul Morson, "The Reader as Voyeur: Tolstoi and the Poetics of Didactic Fiction," *Canadian-American Slavic Studies* 12/4 (Winter 1978), 465–80.

[25] The term "soundscape" was coined by R. Murray Schafer, who describes the soundscape as "any acoustic field of study" and consisting of "events *heard* not objects *seen*." Though recording technology may be an aid in capturing and studying soundscapes, "for the foundation of historical perspective," Schafer writes, "we will have to turn to earwitness accounts from literature and mythology, as well as to anthropological and historical records." Indeed, Tolstoy (along with Thomas Hardy and Thomas Mann) is singled out by Schafer as among literature's great earwitnesses; see his *The Soundscape: Our Sonic Environment and the Tuning of the World* (New York: Knopf, 1977), 8–9.

[26] Jonathan Sterne, "Sonic Imaginations," *The Sound Studies Reader*, ed. Jonathan Sterne (Abingdon, UK: Routledge, 2012), 9–10.

sound, blind to its own dependency on the "spirit/letter distinction in Christian spiritualism." These tenets of the "audiovisual litany," according to Sterne, "idealize hearing . . . as manifesting a kind of pure interiority," replicating a kind of thinking that is also at work in Romantic conceptualizations of music.[27] In Tolstoy's case, at least in his writings of the 1850s, these underlying assumptions now queried by Sterne form the ground on which a historicized understanding of sound in the *Sevastopol Stories* might be built. It is also worth noting, at this point, that the most famous representation of music in Tolstoy's work—the dangerously contagious passion transmitted by music in his late story "The Kreutzer Sonata" ("Kreitserova sonata," 1889)—upholds another essentializing distinction between the phenomenology of listening and of seeing. That is to say, to borrow Jean-Luc Nancy's programmatic diagnosis of this more general nineteenth-century condition, "The visual is tendentially mimetic, and the sonorous tendentially methexic (that is, having to do with participation, sharing contagion)."[28] In other words, seeing preserves the distance from and self-sufficiency of its object, while listening involves immersion and inter-animation, opening a circuit of sensational connection, both physical and emotional, which, for Tolstoy in the "The Kreutzer Sonata," is potently erotically charged.

In the *Sevastopol Stories*, these binary distinctions between sight and sound—perspectival/immersive, exteriority/interiority, mimetic/mexethetic—are played out on various levels of formal articulation. Conspicuously, the energy of these oppositions yields a narrative momentum to surge through a break between paragraphs:

> Take a good look at the faces, the bearing and movements of these men: in every crease of these bronzed, high-cheekboned countenances, in every muscle, in the breadth of these shoulders, in the thickness of these legs clad in their massive boots, in every calm, assured, unhurried movement may be seen those central characteristics that go to make up the Russian's strength—his stubbornness and straightforwardness. As you study these faces you will perceive that the danger, savagery and sufferings of war have added to those central distinguishing features the marks of a conscious sense of dignity and the traces of lofty feelings and thoughts.
>
> All of a sudden the noise of a most fearful explosion startles you out of your wits, delivering a jolt not only to your ears but to the whole

[27] Jonathan Sterne, *The Audible Past: Cultural Origins of Sound Reproduction* (Durham, NC: Duke University Press, 2003), 15.

[28] Jean-Luc Nancy, *Listening*, trans. Charlotte Mandell (New York: Fordham University Press, 2007), 10.

> of your being (*potriasaiushchii ne odni ushnye organy, no vse sushchestvo vashe*), making you tremble in every limb. Immediately afterwards you hear the fading whistle of the projectile, and a thick pall of powder smoke enshrouds you, likewise enveloping the platform and the black figures of the sailors moving to and fro on it. (54)

The reader's gaze here is first engaged in the activity of studying, and this mode of vision is implicitly fused with interpretation: the evidence yielded by this perspective of close, directed attention to external features corroborates the image of heroism (an image that will, however, be importantly modified in the final chapter of "Sevastopol in May"). Casting these soldiers in the image of bravery, this mode of vision, we might say, travels to, acts on, and remakes its object. Yet with the sound of the explosion, the nature of perception is transformed. The sound travels to and assaults the perceiving subject; it is pure sensation, not immediately spliced with interpretation. The explosive force is not just audible, but tangible (the physicality of perception is stronger still in the Russian, in which the explosion, more literally, "shakes not just the ear organs, but your whole being"). The contrast between the paragraphs is one of both sensory and temporal modality: the account of the gaze has the narrative leisure to fuse its visual description with the implicit valorization of bravery. Time here is determined by the act of description; it is created by and belongs exclusively to the narrative itself. On the other hand, the passage that begins with the sonic assault initiates a sequence of sensory perceptions that correspond to events unfolding in time, but that have not yet cohered into meaning. (The visual is still reported here, but now these images are obscured: the thick pall of smoke, the black figures.) The temporal markers "all of a sudden" and "immediately" register a sequence of direct sensory experience, into which, unlike the previous paragraph, prescriptive interpretation is not spliced.[29] Now, time belongs to events, sensations, experience themselves. Juxtaposed with the paragraph of interpreted-seeing, this paragraph of immediate hearing appears as if a direct transcription from reality: what happens happens in the external world and in the perceiving consciousness, not in the interpretative activity of representation. It is the shift from the visual to the audible that erases this layer of mediation, coming closer to capturing—and here placing the reader inside—the Tolstoyan narrative utopia of pure, unmediated present, a proto-stream-of-consciousness.

The paragraph continues to recount the bombardment: the succession of sounds (shrieking, slapping, palpable ringing, whistling, whining, rustling) are

[29] This distinction is also that of Lukács's famous opposition: description versus narration (in which narration is the ideologically favored term). Georg Lukács, "Narrate or Describe?" [1936], *Writer and Critic: and Other Essays*, trans. Arthur Kahn (London: Merlin Press, 1978), 110–48.

punctuated by the sentry's shouts of "Ca-a-nnon!," "Mortar!," and accompanied by fluctuating and at times incongruous feelings:

> You will hear the even whistle of a mortar shell, a sound that is quite pleasant and not at all easy to associate with anything dreadful; you will hear this whistling sound come nearer and nearer in an accelerating crescendo, and then you will see a black sphere and witness the shell's impact against the earth, its palpable ringing explosion. The shell-splinters will fly whistling and whining in all direction, stones will rustle through the air, and you will be spattered with mud. You will experience a sensation that is a strange blend of fear and enjoyment. At the moment you know the shell is heading in your direction, you are bound to think it is going to kill you; but a feeling of self-respect will sustain you, and no one will observe the knife that is lacerating your heart. When, however, the shell sails past, leaving you unscathed, you will recover your spirits and be seized, if only for a moment, by a sense of relief that is unutterably pleasant. (54–55)

One of the accomplishments of Tolstoy's early narrative art was the representation of states of consciousness in flux, what the contemporary writer and critic Nikolai Chernyshevsky described in his influential review as "the dialectic of the soul."[30] In the semiautobiographical account of *Childhood*, Tolstoy's protagonist, Nikolenka, struggles with the flux of feeling, and the troubling possibility flux may bring of experiencing two contradictory feelings at once. For Tolstoy—and for Nikolenka—the dynamic fluidity of self, conditioned by its interaction with the social and material world, constantly reposes questions of judgment, of others and of oneself, on the basis of moments of sensory and emotional experience.

With a shift from the domestic setting to the theater of war, the *Sevastopol Stories* continue this project: they dramatize the processes of cognition and self-cognition that depend upon the coordination of disparate and disjunctive elements of sensory experience—including, prominently, sounds. The first story, "Sevastopol in December," catalogues the ways in which the sounds of war may change their meanings. These modifications of interpretation depend in large part on the coordination and disjuncture between experiences of distance and proximity. As the war was fought around Sevastopol, the civilian zone of relative normality in the town existed in close proximity to the zone of conflict. At the beginning of the story "the majestic resonance of the firing in Sevastopol"

[30] N. G. Chernyshevsky, "L. N. Tolstoy's *Childhood* and *Boyhood* and *Military Tales*," *Belinsky, Chernyshevsky and Dobrolyubov: Selected Criticism*, ed. Ralph Matlaw (New York: E. P. Dutton, 1962), 97.

is heard in the distance, while the oars of the boat sound their "rhythmic beat" in the foreground (42). It is as if these two sounds worked together to uphold one particular image: the regular beat imposes an ennobling, martial order on the chaotic shelling in Sevastopol. The soundscape encompassing near and far is totalized and unified, with something of an epic, mythologized air—not unlike the "hammer strokes of Fate" that opened Woolf's essay.

Without the firsthand experience of conflict on the bastions, the sounds remain abstractions, a decorative or atmospheric backdrop to the theater of war—rousing typically romantic expectations of this place as an arena for deeds of heroism and patriotism. Here, even in sight of Sevastopol, the distance from war is, in effect, as great as it was from the drawing rooms of Petersburg, where the officer enlisted, moved by the pursuit of glory as a decorated hero. From a distance, it is possible to deny that the sounds of the artillery bespeak events linked through cause and effect—effects that extend not only in time but also in space, intruding into the zone away from the bastions, into the town, where the injured lie suffering and untreated. On returning to the town from the field hospital, the narrator observes,

> The funeral cortege will seem to you a thoroughly appealing martial spectacle, the sounds of the gunfire thoroughly appealing martial sounds, and with neither will you associate that clear and personally experienced awareness of suffering and death which you had at the dressing station. (48)

The dismantling of this abstracted totalized perspective comes, by degrees, with direct experience at closer quarters.

> The whistle, close at hand, of a shell or a cannonball, just at the very moment you start to climb the hill, gives you a nasty sensation. Suddenly you realize, in an entirely new way, the true significance of those sounds of gunfire you heard from the town. Some quiet, happy memory suddenly flickers to life in your brain; you start thinking more about yourself and less about what you observe around you, and are suddenly gripped by an unpleasant sense of indecision. (51)

Now the sounds are connected to their source and cause. Only as this connection is grasped does the profound disorder and unfathomability of the conflict zone become apparent. For we now realize that there is no mapped path of safe passage, no clear way to avoid with certainty the bullets and shells that fly around. But this precarious situation is never explicitly acknowledged: what is remarkable here is the way in which Tolstoy attempts to lay bare the inner movements of consciousness that comprise pre-articulate awareness—not the assessment of an external reality, but a turn inward. Is the unspecified memory of the past awakened by the sudden awareness of mortality? Or are we experiencing something prior to conscious awareness: sensations that are now so close and so intense that they sever ties with external reality? What Tolstoy shows us here might resemble that which has come to be understood, by theorists such as Brian Massumi, as "affect," an experience of intensity that is not conscious and thus evades linguistic representation.[31] There is a tension in this passage between words denoting cognition ("realize," "thinking," "indecision") and those denoting intuition ("the flickering memory," "an unpleasant sense"). Tolstoy is able to intimate the interaction of the two processes, and it is the experience of sound that shows the way to prelinguistic processes that evade representation—the apprehension of mortal danger and an experience of fear.[32] In the presence of this acoustic assault, subjectivity is altered: there is a rupturing of the reciprocity between externally received sounds and inwardly made sense that allows the individual to locate himself in the world. That "unpleasant sense of indecision" is not a conscious appraisal of the peril of battle but the loss of the means of understanding place, the loss of the agency to navigate and inhabit space. Though the body remains unharmed by the impact of shells or bullets, it nonetheless registers a form of sonic violence—the vibratory force felt through its bones and on its skin—that is psychological and emotional in its effect.[33]

As the protagonist of "Sevastopol in December" moves through the battery, the monolithic "majestic resonance" that he had heard from afar is broken down into its component parts, and the new language of sounds and their speakers is

[31] Brian Massumi, *Parables for the Virtual: Movement, Affect, Sensation* (Durham, NC, and London: Duke University Press, 2002), 24–33. Massumi (with reference to Bergson) also identifies a retrospective sense making of sensation that, to some extent, resembles the falsification of immediate experience that Tolstoy's narrative project seeks to expose and eliminate. Massumi finds that when the body is "absorbing impulses quicker than they can be perceived," this "anomaly is smoothed over retrospectively to fit conscious requirements of continuity and linear causality" (29). However, Massumi's more particular claim for the half-second "delay" between affect and consciousness has been subsequently called into question; see, for example, Ruth Leys, "The Turn to Affect: A Critique," *Critical Inquiry* 37 (2011), 434–72.

[32] In Tolstoy's first published work, *Childhood*, music functions as a privileged device for representing the child's amorphous consciousness: the protagonist recalls the feeling aroused in him when his mother played Beethoven at the piano: "[This feeling] was like a recollection, but a recollection of what? You seem to be recalling something that had never existed." (Tolstoy, *Childhood, Boyhood, Youth*, trans. Judson Rosengrant [London: Penguin, 2012], 38).

[33] The violence of acoustical assault in the context of our own twenty-first-century wars has been analyzed by Suzanne Cusick, "Towards an Acoustemology of Detention in the 'Global War on Terror,'" *Music, Sound and Space: Transformations of Public and Private Experience*, ed. Georgina Born (Cambridge: Cambridge University Press, 2012), 275–91.

learned: the bullets "that hum like bees or the ones that whistle rapidly by or twang with a noise like a plucked string" (52); the shriek and slap of cannon; the whistling and whining of mortar (54). The separation of sounds by their individual qualities effects the move from the mythic and majestic. The learned differentiation between the sound qualities of the artillery suggests the way toward a new intimate familiarity. Later, in the subsequent stories, we see how the ability to interpret them allows the sounds of battle to be decoded as news from the front: "they must have started hand-to-hand combat now because the firing's stopped" (77); "[e]veryone was saying—and indeed it was plain enough to hear—that there was a terrible bombardment underway" (112). Here, though, the sounds are still new, and not yet decoded as war reports. Rather, the protagonist seeks to fix them in his own experience and the artillery sounds are either named by sound words or described through analogy, with reference to familiar sounds—from nature, from musical instruments.

In this way, sounds are parsed into the discrete ideological domains of nature, culture (to wit, music), and war. The coexistence of ideas pulled from disparate realms is a given for the functioning of metaphor and analogy, but throughout the descriptions of Sevastopol, sounds from these three domains—nature, culture, and war—interact and overlap, often with an unease that signals the struggle of sense-making in war. The concluding paragraph of "Sevastopol in December" sets these three elements in contact with one another in a scene that is a kind of long-shot tableau.

> Already the day is drawing to a close. Just before it sets, the sun comes out from behind the grey storm-clouds that obscure the sky, and suddenly shines with a crimson light on the purple clouds, on the greenish sea bedizened with ships and sailboats and rocked by a broad even swell, on the white structures of the town, and on the people moving about the streets. The strains of an old waltz that is being played by the regimental band on the Boulevard come floating across the water, together with the booming of the guns from the bastions, which seems strangely to echo them. (57)

The sunset offers natural balance and closure to the dawn that had opened the story, where sounds of firing and the ships' bells all combined against the natural backdrop of the Bay. But that image felt much more stable than this one. Schooled in a new way of reading by the story that has gone before, we now detect the parts and tensions that this scene comprises. Crimea and the Black Sea as a "natural" setting was richly appropriated by Russian Romantic poets; it is a site that had long inspired elegiac meditation and contemplation of the self apart from society. This final paragraph's description of nature activates these associations, but extends its gaze to the signs and structures of human activity. The regimental music is at odds with the cosmic significance most often found in the sounds of the seascape of the Romantic poem. And added finally to the scene are the sounds of the guns from the bastions, as if echoing the sounds of the waltz, and left to hang in the air of the concluded story. This ethereal mixture is highly suggestive. Having exposed the disordered chaos of battle, Tolstoy points to an unlikely aesthetic pattern discernible in the scene—the as-if-coordinated sounds. At the same time, though, the illusion that the gunfire echoes or imitates the regimental music displaces the natural order of things. Nature is neither the force against which man struggles nor the one in which he finds solace. What remains is a man-made war. The natural order (be as it may nature mediated by culture) is a frame of reference that is ceding its place as the dominant, and war assumes ascendancy, shifting sense-making paradigms: the stars seen back in Russia will forever be taken as shells (76); the shadows of clouds moving across the land are mistaken for movements of the enemy by young, inexperienced soldiers.[34]

Writing of the sounds of Flanders heard from the Downs in 1916, Woolf, in her essay, addressed the same set of questions: the wager between nature, culture, and war in the remaking of interpretative—and literary—practices. Woolf's listeners attended to the distant sounds of the war according to the interpretative affordances at hand. Meanwhile, Tolstoy's hermeneutic mode can be elucidated through similarly local comparison: with a neighboring article in *The Contemporary,* the journal within which the *Stories* first appeared. In the "Miscellany" section of July 1855 edition (which aggregates several reports received from Sevastopol) an anonymous account describes the habituation of the residents of the town to a "concert of death that has been playing already six months in Sevastopol."[35] Added to this "concert of death," and blending with it in the night air, is the song of a signalman as he stands watch:

> The Russian soul was lost in sad reverie: leaning against the parapet, he let out (*zatianul)* at full voice, and span out (*tianul*) until dawn, one of those dear, native Russian songs that, in the words of one Russian writer calls and weeps and grips our heart . . ."
>
> "*These are not the white snows . . ."* sang the signalman, drawing out (*protiagivaia*) every syllable to eternity. "*That appear white in the wide field . . . caan-nonn! . . . that appear white. . . . mooor-tar!. . . bombshell! Appear white. . . ; ah, not white. . . mooor-tar! Coming by!. . .* And so on,

[34] *Sovremennik* 52 (July 1855), 7.

[35] Ibid., 128.

> all night long—but this song, with its insertions all sung in one and the same tone, amidst all these different sounds of firing, booming, and exploding shells, could only be conveyed, in all its beauty, by a talented musician.[36]

The writer of this anonymous account appeals to the famous question intoned by Nikolai Gogol in his novel *Dead Souls* (*Mertvye dushi*, 1842)—"what calls, and sobs, and clutches at the heart?"—asked by the novel's narrator as he contemplates the limitless and undifferentiated expanses of Russia, seemingly hearing them contained within the sound of native folk song.[37] Now, both Gogol's question and the signalman's song become absorbed into a contrapuntal wartime soundscape.

The sounds of war blend seamlessly into the plaintive, drawn-out melody of the Russian folksong, and the listener/writer in turn hears in it the song that accompanies Gogol's famous image of Russia as troika, flying down a never-ending road. The style of the coachman's song in *Dead Souls*—along with that of the signalman noted here—is of course the *protiazhnaia,* the plaintive, drawn-out folk song that, in the first half of the nineteenth century, "came to be seen as the essence of Russian creativity and the 'Russian soul' itself."[38] Here the song intones lyrical longing for Russia and stokes ardent yet melancholic speculation about national destiny in the war. The sounds of war themselves, assimilated into the distinct articulation of the genre of *protiazhnaia,* become bearers of this sentiment.

The contrast with Tolstoy's stories, appearing in adjacent volumes of the journal, is striking: the *Sevastopol Stories* also absorb the sounds of war, of nature, of musical and literary culture into a contrapuntal wartime soundscape. But for Tolstoy, these are not leveled by a traditional genre, which appropriates all these sounds for the articulation of romantic, mythic sentiment. Rather, Tolstoy makes a new genre of his own that does not level but estranges: a realist prose where reportage and fiction meet, and that transmits these sounds not for their sentiment, but in order to make sense, both psychological and moral.

[36] Ibid., 134.

[37] "But what then is the inapprehensible mysterious force that draws one to thee? Why is thy plaintive song, heard, why does it resound, unremitting, in the ears, as it carries through all thy length and breadth, from sea to sea? What is it, in this song? What calls, and sobs, and clutches at the heart? What sounds are these that painfully caress me and seek to plumb my soul and twine about my heart? Rus! What is it thou wantest from me?" Nikolai Gogol, *Dead Souls,* trans. Robert Maguire (London: Penguin, 2004), 251.

[38] Marina Frolova-Walker, *Russian Music and Nationalism: From Glinka to Stalin* (New Haven, CT, and London: Yale University Press, 2007) 30. Frolova-Walker identifies the *protiazhnaia* as the style of Gogol's coachman's song; see ibid., 39.

10

InConsequence

1853–56

HILLEL SCHWARTZ

> *Consequences:* a round game, in which a narrative of the meeting of a lady and gentleman, their conversation, and ensuing "consequences," is concocted by the contribution of a name or fact by each of the players, in ignorance of what has been contributed by the others. E.g., Jane Austen, *Sense & Sensibility* (1811) II. i. 11: "They met for . . . playing at cards, or consequences, or any other game that was sufficiently noisy."
>
> —*OED* 2nd edition, online

Begin with the Charge of the Light Brigade, not with the six hundred men but with dogs and horses. "Straight on the track of blood—their slavering jaws quivering with the lust of carnage—the dogs of war are slipped from their kennel and there is no man to whistle them back," wrote one of the fiercest partisans of the war, who went on to give Tennyson his refrain: "Under the inspiration of eternal equity, England has committed herself to the protection of a sacred cause and, until that cause be established free of blemish, her course is onward" (half a league onward).[1] At least 362 horses died under their riders during that single charge and retreat on 25 October 1854—more likely, 400 to 500, counting those later put down, "the blood pouring out in great profusion from many of them, owing to the bullet wounds they had received. They rushed up and down the valley even to the very mouth of the cannons, not knowing where to go or what to do."[2]

[1] *A Few Words Addressed to His Grace the Duke of Newcastle by an Englishman, on the Proximate Causes, the Principles, and the Conclusive Terms, of the Russian War* (London: Thomas Bosworth, 1855), 5. For the audible context of Tennyson's poem, listen to Stefanie Markovits, "Giving Voice to the Crimean War: Tennyson's 'Charge' and Maud's Battle-Song," *Victorian Poetry* 47/3 (2009), 481–503.

[2] Roy Dutton, *Forgotten Heroes: The Charge of the Light Brigade* (Oxton, UK: Infodial, 2007), 9, and 133 for account of private Robert Ashton.

Horses and men had been spurred into disaster following orders mis-taken in the smoke and din of Balaklava by Captain Louis Edward Nolan, a thirty-six-year-old veteran, expert horseman, and author of *The Training of Cavalry Remount Horses: A New System* (1852), and already earlier charged with purchasing remounts in Tunis, Turkey, and Syria. "Remount" was a bloodless metonym for a "horse suitable to replace one that is exhausted, injured, or has died." During World War I, that metonym itself was almost exhausted when the British Army's Remount Department became the world's most desperate horsetrader, as half a million of the Army's mounts died and had to be replaced. Of the six million horses deployed on all sides in the Great War, 50 to 60% were killed in battle or died of cold, exhaustion, disease, or mustard gas. By contrast, the mortality rate of horses and mules drafted for the Crimean War—for cavalry, for scouts or despatch riders, and to pull supply wagons, artillery, ambulance carts, portable telegraphy units—was 80%. Did 150,000 horses die? Did 200,000? In packets of letters, pages of diaries, remembrances and reports from Crimea, you must encounter dead or dying horses. They are everywhere. Come late to the peninsula, the photographer Roger Fenton was to write on 4 April 1855, "I never ride out without finding dead horses, even right away at the top of the hills."[3]

Horses had been dying in scores from the start, before reaching any front. Crowded into dank, cramped quarters below deck on ships leaving English ports, they sickened and died. Should they survive this brutal "middle passage," provisions and blankets for them on land were inadequate or inaccessible. Then, foretokened by a peculiar whistling of winds, the Great Storm of 14 November 1854 wiped out stocks for soldiers and mounts alike as it sank twenty allied vessels outside the port of Balaklava with "a harsh screaming sound, increasing in vehemence as it approached," so terrifying that "[n]early one-half of our cavalry horses broke loose," and in the ensuing sleet and snow, "[n]othing could be heard but the howling of the wind, the yelping of wild dogs driven into the enclosures, and the shrill neighing of terrified horses." Afterward, horses in the vicinity starved for forage and went crazy. "Before the end of November," wrote General Hamley,

> the neighbouring artillery camps were invaded by ravenous cavalry horses, galloping madly in at the sound of the feeding trumpet, and

[3] George E. Ryan, *Our Heroes of the Crimea: Being Biographical Sketches of Our Military Officers* (London: Geo. Routledge, 1855), 40–43; Michael Morpurgo, "War Horse," *The Guardian* (7 Jan. 2012); "Horses in World War One: Good-bye, Old Man," <http://www.scotlandswar.ed.ac.uk/sites/default/files/pdf_Horses.pdf> (this website, and all subsequent websites cited in this chapter, were accessible as of 13 August 2015); Kellow Chesney, *Crimean War Reader* (London: Severn House, [1960] 1975), 202, Fenton letter.

> snatching, undeterred by stick or stones, the hay and barley from the very muzzles of the right owners. Painful it was to see the frenzy of the creatures in their first pangs of hunger, more painful to see their quiet misery in the exhaustion that succeeded.

If *Times* correspondent W. H. Russell did write of the screaming of dying horses "waiting for the farriers to put them out of torment," their misery was more often passed over in silence, though near their end they must have been snorting, heaving, and (a usage new to the 1850s), "whistling" with broken wind, shrill, breathless.[4]

Whistling was elsewhere and otherwise deathly. The Battle of Inkerman on 5 November 1854 endorsed and inscribed those metallic whistles that had been carried by British light infantry officers since the 1700s, as larger artillery pieces took over battlefields and the volume of their noise overwhelmed that of drums or trumpets, and as the smoke of musket and cannon so billowed that in massed battles little could be seen or few commands heard. The Battle of Inkerman, wrote Russell, "admits of no description," which he meant literally:

> No one, however placed, could have witnessed even a small portion of the doings of this eventful day, for the vapours, fog, and drizzling mist obscured the ground to such an extent as to render it impossible to see what was going on at a distance of a few yards.[. . .] Our Generals could not see where to go. They could not tell where the enemy were—from what side they were coming, nor where they were coming to. In darkness, gloom, and rain they had to lead our line through thick, scrubby bushes and thorn brakes [and, through the smoke from muskets and cannon,] while every pace was marked by a corpse or man wounded by an enemy whose position was only indicated by the rattle of musketry and the rush of ball and shell.[5]

[4] Fannie (Mrs. Henry) Duberly, *Journal Kept during the Russian War, from the Departure of the Army from England in April 1854, to the Fall of Sebastopol*, 2nd ed. (London: Longman, Brown, Green, and Longmans, 1856), entries for 29 Apr., 8 May, 10 Nov, and 14 Nov. 1854; Guy Arnold, *Historical Dictionary of the Crimean War* (Lanham, MD: Scarecrow, 2002), 73 on the "Great Storm"; Sir John Adye, *A Review of the Crimean War to the Winter of 1854–5* (London: Hurst and Blackett, 1860), 150–54, the "gale"; Lady Alicia Blackwood, *A Narrative of Personal Experiences and Impressions during a Residence on the Bosphorus throughout the Crimean War* (London: Hatchard, Piccadilly, 1881), chap. 15, entry for Dec. 1855, on the whistling; General Sir Edward Hamley, *The War in the Crimea* (London: Seeley, 1900), chap. 7; Clive Ponting, *The Crimean War: The Truth behind the Myth* (New York: Random House, 2011), 163 for Russell.

[5] Michael Barthorp, *The British Army on Campaign (2): The Crimea 1854–56* (Oxford: Osprey, 1987), 47; <http://everything2.com/title/Inkerman+whistle>; MIS-041, silver plated metal lion

Russians and Turks were firing smoothbore muskets, awkward to load and useless to aim in any but a general direction, though both nations had begun to rearm with the Minié rifles, new to this war, whose bores had been spirally grooved for greater accuracy. In the hands of French and English soldiers and some Russians and Turks, the Minié rifles forced, or should have forced, a dramatic change in tactics, for they were effective at 300 yards, six times the range of smoothbores, and could be reloaded swiftly enough to make conventional frontal assaults and bayonet onslaughts futile. Since well-trained men with Miniés could pick off gunners manning nearby cannon, infantry could no longer count on artillery set up just beyond (outmoded) musket range to loft canisters or grapeshot that would open fields to hand-to-hand combat. Thus, around the Russian fortress complex at Sevastopol, the Crimean War gave rise to the first Eurasian instance of sustained trench warfare, in which soldiers knee deep in "yellow, liquid, foul-smelling mud," scratching at lice, scrabbling for footholds, scrambling for breath, listened to the war overhead.[6]

Of particular sonic note were the higher muzzle velocities of Minié rifles and their conical, grooved, soft-lead bullets (anachronistically called "balls"), which on their swift trajectories spun in the air to produce a distinctive whistling sound, a "sharp *twing-twinging.*" Bullets, cannonballs, and rockets had for centuries been described by soldiers as "whistling" (just as arrows could "whistle" through the air—and be designed by Chinese, Japanese, or Native American war artisans to whistle most shrilly), but the *twing-twinging* of the Minié bullets was no lazily conventional onomatopoeia. Indeed, twelve weeks into the Siege of Lucknow during the Indian Rebellion of 1857, with the British garrison under renewed attack, "a cry arose from the soldiers, 'The Minié! The Minié!'"—for through the musketry "they had detected the familiar whistle of the bullet of the Minié rifle, with which only European troops were armed," assuring them that relief

and whistle popular with the officers and NCOs of British Rifle and Light Infantry Corps throughout the 19th century, at <http://www.militaryheritage.com/images/whistle.jpg>; *Rules and Orders for the Discipline of the Light Infantry Companies in His Majesty's Army in Ireland. Given &c. the 15th Day of May 1772,* issued by [George Townsend, 1st Marquess Townsend], and printed in Appendix, 549–50, of Capt. R. H. Raymond Smithies, *Historical Records of the 40th (2nd Somersetshire) Regiment, Now 1st Battalion the Prince of Wales's Volunteers [...]* (Devonport, UK: A. H. Swiss, 1894). Cf. Massachusetts. 1st corps of cadets, *Massachusetts Volunteer Militia [. . .]: Standing Orders, 1890* (Boston: N. Sawyer & Son, 1890), 135–36 for descriptive list of whistle commands. For Russell on the battle: Chesney, *Crimean War Reader*, 123–24.

[6] Allan W. Howey, "The Widow-Makers," *Civil War Times Magazine* (Oct. 1999), for which see <http://www.historynet.com/mine-ball>; <http://www.historicalfirearms.info/post/46428681027/rifles-of-the-crimean-war-hostilities-between>; Robert L. O'Connell, *Of Arms and Men: A History of War, Weapons, and Aggression* (Oxford: Oxford University Press, 1990), 191; Wikipedia, "Minié ball/bullet" <https://en.wikipedia.org/wiki/Mini%C3%A9_ball>; Leo Tolstoy, *Sevastopol*, trans. Isabel F. Hapgood (New York: Thomas Y. Crowell, 1888), 24–25.

was at hand. Walking the trenches around Sevastopol with an English colonel, an auditor heard Minié bullets "mingled with the sound of badly cast or imperfect balls, which produced a sort of squeaking, melancholy sound (like *squae, squae, squae*), which amused the Colonel, who said he had heard some music like this in Spain, but preferred the *honest hiss* and *whiz* of the old musket-ball." *Hiss* and *whiz:* not whistle, not *twing-twinging.*[7]

So the sharper whistling in the air was further evidence of the modernity of this war, what with its "screw" or steam warships, its rail lines, the dit-dot of its telegraphy, the click of its war photography, and the chloroformed sleep of its amputees. About that last: even at three soccer fields' distance, Minié rifles were more instantly fatal or fully disabling than smoothbores closer in, since their heavier, faster, conical bullets cut a straight path through the body and, unlike round smoothbore balls, shattered bone. To hear such whistling in Crimea, then, whether from Minié rifles or the new "Whistling Dick" (grooved?) cannon, was to attend to no pretty concert like that given in London in 1851 by an Infant Whistler from Macon, Georgia, "not yet three yrs old." Guiding visitors across the Napoleonic battlefield of Waterloo just after the Crimean War, Sergeant Mundy told tourists,

> Gentlemen and ladies, I have often heard blustering young men, who have been in battles, say, that they gloried in being in the thickest of the fight, and were fond of the music of bullets whistling around them; [. . .] but gentlemen I am free to say when our regiment was ordered from that hill yonder [. . .] where the shot from the French batteries were fast thinning our ranks, to descend into that valley [. . .] where the bullets went over our heads, that was the most pleasant part of the day to *me.*[8]

[7] *Aldershottana: or, Chinks in My Hut: and Touch-and-Go Sketches from Court to Camp* (London: Savill and Edwards, 1856), 137; Charles Harcourt Ainslie Forbes-Lindsay, *India, Past and Present*, vol. 2 (Philadelphia: H. T. Coates, 1903), 227. For the acoustical physics of the bullet: Hermann Smith, *The Making of Sound in the Organ and in the Orchestra: An Analysis of the Work of the Air in the Speaking Organ Pipe of the Various Constant Types, and an Exposition of the Theory of the Air-Stream-Reed, Based upon the Discovery of the Tone of the Air, by Means of Displacement Rods* (New York: C. Scribner's Sons, 1911), 127–28. Whistling arrows have their own bibliography, unnecessary here.

[8] On devastations wrought by the Minié bullets, see n6, this chapter. For Whistling Dick: Sgt E. Gowing, in a letter entitled "Camp before Sebastopol," 29 Oct. 1854, cited by Eric Partridge, A *Dictionary of Slang and Unconventional English*, 8th ed. (London: Routledge, 2002), 1331. On Infant Whistler: Vera Brodsky Lawrence, *Strong on Music: Reverberations, 1850–1856*, vol. 2 (Chicago: University of Chicago, 1995), 194–95, preceded by an Infant Drummer. For Sergeant Mundy: John Ellis Edwards, *Random Sketches and Notes of European Travel in 1856* (New York: Harper & Brothers, 1857), 348–49.

The era of the romance of whistling shells and whistling bullets was dying out just as bullets and shells became whistling virtuosos. When Second Lieutenant Leo Tolstoy wrote about his months on the Russian fronts of Crimea, he seemed for paragraphs a captive to the whistling romance: "Again the sentry shouts [. . .] 'Mortar!' and you hear the monotonous, even rather pleasant whistle of the bomb, with which it is difficult to connect the thought of horror," but then "you hear this whistle approaching you, and increasing in swiftness, then you see the black sphere, the impact on the ground, the resounding explosion of the bomb which can be felt. With the whistle and shriek, splinters fly again, stones whiz through the air, and mud showers over you." Pages later the romance has entirely faded; any hint of whistling, once audible, is "fateful."[9]

Back home, *Svistok (The Whistle)* was soon to shrill from the masthead of a Russian journal whose title was apt because whistling (through closed if broken teeth or with fingers rudely in the mouth) was how audiences from Paris to St. Petersburg, Milan to Manchester, Berlin to Bankplassen made audible their impatience with the inept and their outrage at the arrogant or tiresome. Bankplassen? A district by the Akershus Fortress in central Christiania, Norway. Where sat the Christiana Theater, for which Bjornstjerne Bjornson in May 1856 organized *pipekonserter,* whistling concerts, to protest the performances of a Danish acting couple newly hired to permanent positions at Norway's only national theater. A patriot, Bjornson mustered his whistling claques in defense of Norwegian integrity; it was he, not I, who put whistlers on a military footing:

> A theater in the capital city is a nationality's most remote outpost against foreign countries. In the capital city the largest break between the foreign and that which is our own takes place, and the capital city influences most decisively by working inwards. It fights a great battle and has a great responsibility, and it requires troops and vigilant guards.

Across the Baltic, in St. Petersburg and well beyond, *The Whistle* would become known for a sardonic "laughter mixed with scandal" rather than Gogol's softer "laughter through tears." Fyodor Dostoevsky bristled in defense of that whistling after scandal, for "Voltaire whistled all his life, and not without sense, and not without results. And how furious they were at him, precisely for the whistle."[10]

[9] Tolstoy, *Sevastopol*, 30–31, 99. For an in-depth discussion of the sounds of battlefield see Alyson Tapp's contribution to this volume, 204–13

[10] *Svistok* was the satirical segment of the literary journal *Sovremennik,* appearing in nine issues between 1859 and 1863; Joseph Frank, *Dostoevsky: The Stir of Liberation, 1860–1865,* vol. 3 (Princeton, NJ: Princeton University Press, 1988), 76, 107; Ann Schmiesing, "The Christiania Theater and Norwegian Nationalism: Bjornson's Defense of the 1856 Whistle Concerts in 'Pibernes Program'," *Scandinavian Studies,* 76/3 (Fall 2004), 317–40, quotation on 317–18.

Whistling, as he well knew, was a dangerous affair in Czarist Russia, scarcely metaphorical, and dark if not fateful. Folk beliefs (as throughout Central Asia and Jewish and Arabic worlds) had it that even the most casual, blithe, uninsistent whistling invited ghosts, devils, ill fortune. That's why the postmaster's wife in Pushkin's story, "Dubrovsky" (1832), itches to send packing a foreign traveler who has been idly whistling for five hours straight while awaiting fresh horses at the post-house. "The Lord has sent us a whistler!" she tells her husband in a low voice, ". . . the accursed heathen!" How could she expect a visiting Frenchman to understand that no pious person whistles indoors? How would he know that it's witches and demons who whistle (around graveyards, as in Gogol's "St. John's Eve"), and murderous highwaymen (on empty roads, as in Turgenev's *Sportsman's Sketches),* and Nightingale the Robber (Solovei the Brigand, half-bird denizen of the Bryansk forest, who waylays and stuns with his terrible whistling all trespassers but hero Ilya Murometz, in a famous episode of East Slavic epic)? Russians, of course, true Russians did know what it meant to whistle, and when the new, radical editors of *The Contemporary* in 1857 issued *The Whistle* as a satirical supplement, they were not just blowing the whistle on liberals and conservatives alike; they were setting themselves in sharp acousti-political opposition to a more temperate reformist journal published by Alexander Herzen. But *The Whistle* (1859–63) was to Herzen's *The Bell* (*Kólokol,* 1857–67) and the later *Alarm Clock* (*Budil'nik,* 1866–1917) as toxic sarcasm was to solemn tocsin. Turgenev withdrew his support from *The Contemporary* as its *Whistle* blew into the wind.[11]

Herzen by then was in exile in London, where whistling had a more polyvalent public ambit. The English whistled *past* graveyards, as in Blair's enduringly popular "The Grave" of 1743, where

> Oft in the lone church-yard at night I've seen,
> By glimpse of moon-shine, chequering through the trees,
> The school-boy, with his satchel in his hand,
> Whistling aloud to bear his courage up

even as they told and retold the ghost story of the Seven Whistlers, spirits in the form of birds whose "[w]histling strangely, whistling sadly, whistling sweet

[11] Alexandr Sergeyevitch Pushkin, "Dubrovsky" [1832, published posthumously 1841], *The Complete Prose Tales,* trans. Gillon R. Aitken (New York: W. W. Norton, 1966), 236–37; Nikolai Gogol, "St. John's Eve: A Story Told by the Sacristan of the Dikanka Church," [serialized 1830, published 1831 in book *Evenings on a Farm near Dikanka*], trans. Isabel Hapgood, at <http://ebooks.adelaide.edu.au/g/gogol/nikolai/g61sj>; Albert H. Wratislaw, "XXXIV, Ilya of Murom and Nightingale the Robber," *Sixty Folk-Tales from Exclusively Slavonic Sources* (London: E. Stock, 1889), 169–70. On Turgenev and Herzen, see the comments by Vladimir Ilyich Lenin, *Remarks on Books: G. V. Plekhanov; N. G. Chernyshevsky* (St. Petersburg: Shipovnik, 1910), 58–59.

and clear" was the worst of omens.[12] Should authorities forbid the whistling of workers in mines (obscuring the whistling of canaries?), sailors at sea (inviting storm or the wraiths of the drowned), prisoners in jails (conniving?), or boys in school (unthinking), their most beloved English poets continued to exalt the carefree whistling of laborers ambling home from work, of ploughboys "compelled to frame their breath into a whistle" while guiding their workhorses in the fields *and* at play or rest. John Ruskin in 1871 was to make keen use of this idyll in a letter "to the Workmen and Labourers of Great Britain" on May Day, from Cumberland, where

> there was no dancing at all, and they could not even provide their own piping. They had their goblin to Pipe for them. They walked in procession after their steam plough, and their steam plough whistled to them occasionally in the most melodious manner it could. Which seemed to me, indeed, a return to more than Arcadian simplicity; for in old Arcadia, plough boys truly whistled as they went, for want of thought; whereas, here was verily a large company walking without thought, but not having any more even the capacity of doing their own Whistling.[13]

[12] Robert Blair, "The Grave" (1743), *The Book of Georgian Verse,* ed. William S. Braithwaite (London: G. Richards, 1909), 257; Alice E. Gillington, "The Seven Whistlers," *A Victorian Anthology, 1837–1895,* ed. Edmund C. Stedman (Boston: Houghton Mifflin, 1895), 608–9.

[13] On not whistling in mines: Frederick Engels, *Condition of the Working Class in England in 1844: With a Preface Written in 1892,* trans. Florence Kelley Wischnewetzky (London: George Allen & Unwin, 1892), 179 ("Every operative detected speaking to another, singing or whistling will be fined 6d."), also cited by John Lucas and Allan Chatburn, *A Brief History of Whistling* (Nottingham, UK: Five Leaves, 2013), which see for many other instances. In prison: Henry Mayhew and John Barry, *The Criminal Prisons of London and Scenes of Prison Life* (London: Griffin, Bohn, 1862), 139. In school: William M. Sewell, *A Year's Sermons to Boys, Preached in the Chapel of St. Peter's College, Radley,* vol. 2 (Oxford: A. Wassall, 1854), 343. At sea: E. Radford and M. A. Radford, *Encyclopedia of Superstitions,* ed. and rev. Christina Hole (New York: Barnes and Noble, [1948] 1996), 245, and 3 on sailors off season being hired to work the ropes for stage scenery, so prohibitions against idle or haphazard whistling were imported into the theater, where stage managers might signal scene changes with a low whistle, on which also see London Blue Bird, blog online at <http://thelondon-bluebird.wordpress.com/2012/08/16/witches-ghosts-whistling-and-cats-a-guide-to-theatrical-superstitions>. For ploughboys: John Case, *The Praise of Musicke* (Hildesheim, Germany: Olms, [1586], 1980) iii, 43; George Crabbe, "The Whistling Boy That Holds the Plough" (1810, published 1834 in his *Posthumous Poems*), in *Book of Georgian Verse,* ed. Braithwaite, 510; John Ruskin, *Fors Clavigera: Letters to the Workmen and Labourers of Great Britain* (Orpington, UK: George Allen, 1871), Letter V, May 1871, 10–22 excerpted at <http://www.pseudopodium.org/repress/ForsClavigera/05.html>. Contrast the attitude toward whistling in Walter Aimwell (i.e., William Simonds), *Whistler; or, The Manly Boy* (Lincoln, UK: Sheldon, Blakeman, 1858) with that in a children's book, *Whistling Horace* (Glasgow: J. S. Marr, 1878), whose anonymous moralizing author

Laments over the demise of incidentally happy whistling in the face of industrial-strength whistles, factory regimens, mercantile rectitude, bourgeois stuffiness, and urban clatter became an Anglo-American trope, with many a rearguard skirmish over such an innocent joy. Charles Dickens, though a "pure modernist" (opined Ruskin, who called him "a leader of the steam-whistle party par excellence"), was early at the barricades defending joyous human whistling rural and urban, but likely found himself more than a little dubious about what was being whistled on London streets during the Crimean War. In an 1854 contribution to Dickens's *Household Words*, in a long passage about "a half-noisy thoroughfare" that was so much "more excruciating than a wholly noisy one" because instead of just "one noise running through the day, you have two hundred noises at two hundred intervals in the day," Henry Morley detailed the London soundworld after dinner, the baby off to sleep . . . when

> came punctually a series of special nuisances that had their regular days for disturbing her; and we came to know their times. On Monday evenings there was a horn; after which (separate concern) a German band; organs; boys whistling "Pop goes the Weasel." Tuesday, Ethiopian serenaders; organs; boys whistling "Pop goes the Weasel." Wednesday, a detached performer on the bones; a brain-crushing machine drawn by a donkey—a man on a platform grinding all our heads in it; other organs; band of Scotch fiddlers scraping and scratching hideous strathspeys with unrosined horse-hair; boys whistling "Pop goes the Weasel." Thursday, ophicleides, cornopeans, and trombones; Indian beating tom-tom; acrobats and two drums; organs; boys whistling "Pop goes the Weasel." Friday, Ethiopian serenaders; psalm-singing by an old man playing the violincello, with two girls in white tuckers . . . ; organs; boys whistling "Pop goes the Weasel." Saturday, street fights and shouts; extra carts (butchers' carts very aggravating); German band; Ethiopians; hurdy-gurdy; harps and accordions; brain-crushing machine; knife-grinder (most excruciating); Finnan haddocks; hearthstones; and "Pop goes the Weasel" until eleven o'clock at night.[14]

twenty years later hears a plowboy's whistling as prime symbol of his contrary, careless, risky, and perhaps irredeemable nature.

[14] Charles Dickens, "Whistlers and Whistling" *All the Year Round* n.s. 9 (4 Jan. 1873), 182–85; Henry Morley, "Adeliza Castle," *Household Words* 22 (1854), 219–20. For Ruskin on Dickens: *Athenaeum* (30 Sept. 1905), 428. For rearguard defenses of whistling, see, e.g., Thomas F. Thiselton Dyer, "Whistling," *The Gentleman's Magazine* 254 (1883), 392–400. I will write at greater length elsewhere on these whistling apologias.

"Pop Goes the Weasel" was all the rage, and it was no lullaby. What exactly its refrain and early stanzas meant defied contemporaries no less than historians. The tune was simple: a jig dating back several centuries, familiar to country and genteel dancers. Queen Victoria and Prince Albert apparently delighted in the steps as set by Joseph Lowe, their tutor in dance at Balmoral. Danced, whistled, hummed, piped, barrel-organized, telegraphed, thrummed, music-boxed, parroted, and pianized, the music was everywhere, but at each streetcorner, in each tavern or parlor, on each of the "music hall" stages springing up in the 1850s, "Pop Goes the Weasel" was sung to a muddle of words. As I have been listening to the war in Crimea through the sonic filter of dying horses and then through the filter of whistling bullets and whistling winds, so now I mean to listen through the filter of a song still with us in the guise of a nonsense nursery rhyme but resisting, like any self-respecting weasel, its comeuppance.[15]

Here's one version of the more enigmatic stanzas:

Up and down the City Road
In and out of the Eagle,
That's the way the money goes,
Pop goes the weasel!

Half a pound of tuppenny rice,
Half a pound of treacle,
Mix it up and make it nice,
Pop goes the weasel!

Every night when I go out
The monkey's on the table;
Take a stick and knock it off,
Pop goes the weasel!

[15] David Joyce, quoted at <http://www.kickery.com/2008/08/pop-goes-the-we.html> on the jig; Joseph Lowe, *A New Most Excellent Dancing Master: The Journal of Joseph Lowe's Visits to Balmoral and Windsor (1852–1860) to Teach Dance to the Family of Queen Victoria*, ed. Allan Thomas (Stuyvesant, New York: Pendragon, 1992), 1, 11, 55, 65, 87; Charles MacKay, *Through the Long Day: Or, Memorials of a Literary Life during Half a Century*, vol. 1 (London: W. H. Allen, 1887), 137, barrel organs and much else; William Lynd, ed. "Ocean Telegraphy," *The Telegraphist* 10 (1 Sept. 1884), 120 ("from the needles striking the little ivory pins, and upon listening attentively the sounds were found to correspond with the tune of 'Pop goes the Weasel,' clicked out most distinctly by some brother wire-puller for his own amusement"); music box at <http://www.youtube.com/watch?v=kdSnOQ-rMdA>; jack-in-the-box version at <http://www.youtube.com/watch?v=Jp_0zhVFQuk>; T. S. Stevens, "Notes on an Intelligent Parrot," *Journal of the Trenton Natural History Society* 1 (1886–88), 347–55, at 349–50—"The longest lesson that I have attempted to teach her, is two verses of 'Pop goes the weasel,' containing forty-two words, which she can sing correctly. . . .[O]f course, no one will suppose that she knows the monkey from the weasel, or the priest from the cobbler's wife."

A penny for a ball of thread
Another for a needle,
That's the way the money goes,
Pop goes the weasel!

and equally enigmatic but added sometime later:

All around the cobblers bench
The monkey chased the people;
The donkey thought 'twas all in fun,
Pop goes the weasel![16]

Challenged by these riddling stanzas, folklorists, antiquarians, lexicographers, and various contributors to *Notes and Queries* have solved them by cobbling together seamen's and Cockney slang, underworld cant, urban history, Huguenot weaver traditions, snatches from theatrical history, and the flux of English orthography. I have further cobbled together a single somewhat plausible narrative: the Eagle—a tavern in London that became one of the first music halls, "The Grecian Saloon," where the song may have been sung—puts its talons into a workingman, quickly relieving him of his week's wages as a cobbler, tailor, or hatmaker wielding a weasel, a cloth-measuring device in the form of a spoked wheel that makes a popping sound when it arrives at the desired length. Penniless, the inebriant struggles to buy food for his family, though he is tempted to knock back the liquor in the glazed "monkey" jug on the table. His spiral is downward: unable to afford the necessities of his trade, he must pawn, or "pop," his weasel. Whence the donkey no one rightly knows, though "riding the donkey" was cant for cheating while weighing out fenced (or pawned?) goods. To be "on the monkery" was to be a vagrant on the country roads, the drinker's sad fate?[17]

[16] Mark Ford, *London: A History in Verse* (Cambridge, MA: Belknap Press of Harvard University Press, 2012), 294; W. W. Denslow, ed. and illus., *Denslow's Mother Goose: Being the Old Familiar Rhymes and Jingles of Mother Goose* (New York: McClure, Phillips, 1901); in the context of dance, <http://www.kickery.com/2008/08/pop-goes-the-we.html>; with other variations, <http://www.folklorist.org/song/Pop_Goes_the_Weasel>, as also <http://www.phrases.org.uk/meanings/pop-goes-the-weasel.html>.

[17] Albert Jack, illus. Lara Carlini. *Pop Goes the Weasel: The Secret Meanings of Nursery Rhymes* (New York: Perigee, 2009)*sive* "Pop Goes the Weasel"; Oxford English Dictionary, 2d ed. online, *sive* "Pop" and "Weasel"; and, in addition to those previously cited, these websites: <http://www.vam.ac.uk/content/articles/t/the-story-of-music-halls>; <http://treasuryislands.wordpress.com/2011/04/06/origins-pop-goes-the-weasel>; <http://en.wikipedia.org/wiki/Pop_Goes_the_Weasel>. S. J. Adair Fitz-Gerald, *Notes and Queries* (1905), ser. 10/4, 209–11; ser 10/3, 430; and ser. 10/4. 54–55; J. Foster Palmer (1905), ser. 10/3: 491; T. W. A. Lingard, *Notes & Queries* 171

Another narrative, less taken with Temperance, maintains that the pawning of the weasel itself enables, perhaps ennobles, the drinking spree. And another, more graceful narrative takes "Pop Goes the Weasel" seriously as the title of "an old English dance lately revived" whose sheet music was first published in 1850 without lyrics. The title refers to a dance figure in which a couple moves down one place by "popping" the second couple to the place above. . . . Or in which a couple bobs and weaves/weasels its way under an arch formed by other couples before "popping" into the open and shouting, e.g., "Active couple takes the next lady and the three circle around to the right then back to the left, ending with the lady going under the active couple's raised arms while all sing 'pop goes the weasel!' " I prefer this latter choreography, following which sense one or two mid-century racehorses were named "Pop Goes the Weasel"—in hopes (I surmise) that at the stretch they would pop into the lead (though in 1855 at the Wong-Nei Chong Stakes in Hong Kong, "Pop goes the Weasel bolted as usual"). Wasn't this the sense, too, in which the tune would accompany jack-in-the-box surprises and set the pace for the game of musical chairs (or "Snap-Tongs"), where with each Pop! another player loses her/his place? This sense was still operative in the later sound-surround of Scottish flyweight contender Benny Lynch's first matches during the 1930s, at which the crowd spontaneously whistled "Pop Goes the Weasel" in unison "at the bobbing dance of the young slender boxer in the ring."[18]

None of these narratives, however, does much to account for the timing of the rather abrupt revival and popularity of the tune, which was at the lips and hips of young and old by 1853 and for decades thereafter. One set of lyrics, in fact, was dedicated to extolling its extraordinary ubiquity:

> Now all the girls are going mad
> for—Pop goes the Weasel!
> And the finest tune we ever had
> is—Pop goes the Weasel!

(5 Sept., 1936) 176. On cant: Eric Partridge, *A Dictionary of the Underworld British and American* (New York: Bonanza, [1949], 1961), 197, 446.

[18] John T. Page, "Pop Goes the Weasel," *Notes and Queries* (1905), ser. 10/3, 491, with more on dance by J. Holden McMichael, 492, citing *The Home Circle*, 8/193, 183; <http://www.kickery.com/2008/08/pop-goes-the-we.html>. On the horses: William Levey, comp and ed., *The Victorian Ruff, Or, Pocket-Racing Companion for 1862* (Melbourne, Australia W. Levey, 1862), 62; *The General Stud Book, Containing Pedigrees of Each Horse from the Earliest Accounts to the Year 1892 Inclusive*, Vol. 7 3rd ed. (London: J. E., J. P., & C. T. Weatherby, 1894), 162; a Correspondent, "The Turf in Hong Kong," *Bell's Life in London and Sporting Chronicle* [Town Edition, 5 Oct. 1857]. For the boxer: John Burrowes, *Benny: The Life and Times of a Fighting Legend* (Edinburgh: Mainstream, 2011), viewed as an unpaginated Google book at <http://www.books.google.com/books?isbn=1780573472>.

It is danced by Albert and the Queen,
    Chummies done it round the green,
And many girls have ruined been,
    by—Pop goes the Weasel!

Organ boys grind in the street,
    Pop goes the Weasel!
The thing to make you feel your feet
    Is—Pop goes the Weasel!

It costs the young chaps such a lot
    To treat the girls to—you know what,
The Militiamen march and trot
    To Pop goes the Weasel! [. . .]

To hear it play'd some thousands hop,
    And last week a mad bull made a stop,
Then run into a music shop
    For Pop goes the Weasel! [. . .]

Now all the world, yes, even France,
    Like Pop goes the Weasel!
But we can make the Frenchmen dance
    Pop goes the Weasel!

Pop goes the Weasel gives delight
    And by your smiles I think I'm right.
If so, I'll try another night,
    Pop goes the Weasel![19]

Inevitably, the tune was pressed into service for the Crimean War, in a ballad called "Europe and Victory" as voiced by those about to go off a-soldiering:

Oh that we had a Wellington
    To give Old Nick the measles.
It would him please to stand at ease
    And pop the Russian weasel.

I wish we had a Nelson too
    To keep the tyrant under.
His wooden walls and cannon balls
    He would make to roar like thunder [. . .]

[19] *The Fashionable; Or London and Country Songster*, no. 1 (London, T. Goode, 1854), 4–6.

Maidens keep your spirits up
And never mind the measels.
We will return with victory
And then pop goes the weasel[!][20]

With each refrain of each variant, lyricists availed themselves of—or invented—another connotation of "weasel" and "pop": sexual (seduction, penetration, ejaculation); military (infiltration, detonation, concussion); economic (frivolity, generosity, bankruptcy); festive (exuberance and holiday); criminal (counterfeiting, stealing, knocking over the head). The very polyvalence of the song and its refrain became the subject of verses composed by W. R. Mandale and sung by the comic W. Lambert Edmonds at Cremorne Gardens in 1853:

I called upon a friend last week
To seek some explanation
Of this strange phrase, that now-adays
So charms the pop-ulation.
Quoth he, "I do apply it thus:—
My wife, sweet soul, (hem!) she's ill
Should she pop off, what then? Of course,
Why, pop goes my weasel!

Not feeling satisfied with this
Queer piece of information,
Unto Cremorne I went that night
For a little recreation;
While sauntering there a lady said,
"My frent, sare, if you please veel
Come join a partnare vis me,
In pop him go se veasel?"

"What is the meaning of this slang?
Cried I in desperation,
When a pop bottle came bang
On my bump of consternation;
Direct it struck me o'er the eye,
When a waiter cried, "Sir, be still,
I didn't mean to let it fly,
But pop went the weasel!" . . .

[20] *Europe and Victory* (London: E. [M. A.] Hodges, c. 1855), online as <http://ballads.bodleian.ox.ac.uk/static/images/sheets/25000/20169.gif>

Since then I've asked what it doth mean,
Of folks in every station;
Some grin and laugh, some cheer and chaff,
All's bother and vexation.
For I'm still as wise as e'er I was,
As full's an empty peashell,
As far as the true history goes,
Of pop goes the weasel.

Yet *popping* here and *popping* there,
And *popping* all about, sirs,
'Mongst *pop*lar trees in *pop*'lar airs,
It still keeps *popping* out, sirs;
Pop north and south, *pop* east and west,
Pop right or left, you'll see still,
Pop up and down, all o'er the town,
It's "*Pop* goes the weasel!"[21]

Clearly, no one set of lyrics or underlying theme explains the popularity of the song, which was as odd to contemporaries as to historians. Was its popularity rather a matter of its infectious rhythm? This is what the next variant suggested:

Queen Victoria's very sick,
Napoleon's got the measles,
Sebastopol is won at last—
Pop goes the weasel!

All around the cobbler's house,
The monkey chased the people,
And after them, in double haste,
Pop goes the weasel! [. . .]

Of all the dance that ever was planned,
To galvanize the heel and the hand,
There's none that moves so gay and grand
As Pop goes the weasel.[22]

[21] W. Lambert Edmonds, ed., *The Cremorne Comic Song-Book* (London: Thomas Allman, 1853), 9–11. Mandale had composed a number of other ballads published at the time, including "Soothing Words" (1852) and "Contentment" (1853).

[22] W. E. Tunis, *The Shilling Song Book: A Collection of 175 of the Most Favorite National, Patriotic, Sentimental, and Comic Ballads of the Day* (Niagara Falls, Canada: W. E. Tunis, 1860), 16.

Yea, that rhythm, with its blunt rhythmic shift at the end of each stanza, proved so distinctive that it could be deciphered (and deplored) even when its usually sprightly words were so redressed by the mistress of an infant school as to be implausibly didactic:

> Up and down life's broad highway,
> With its pleasures hollow,
> When wild comrades bid you stray,
> Stop! do not follow.
>
> There's a cup the drunkard sips
> Pleasant though you think it,
> Dash the poison from your lips:
> Stop! do not drink it.
>
> When your angry passions rise,
> Be the last to jangle.
> Temper wrath by soft replies.
> Stop! do not wrangle.
>
> Three great foes you vowed to fight,
> Oft when cannons rattle;
> Prudence is the soldier's might:
> Stop! wins the battle.

A certain Mr. Percival Prosser objected to such disguise; however stern the words, the underlying "street music" would still debase the ears of schoolchildren, most egregiously with the "unhallowed jingle" of that foolish tune, "Pop Goes the Weasel." The uplifting lyricist herself, E. E., responded in a tone of heated pragmatism that

> negro songs, and such "contagious and pestilent" [quoth Prosser] harmony *must* fall upon the ears of our school children; they cannot pass to and fro along the crowded streets and alleys without hearing them, nay, our rural villages and whistling cow-boys have caught the "unhallowed jingle" from hand-organs and travelling musicians.

Little ones after school hours *will* listen with "evident delight to 'Nelly Bly,' 'Lucy Long,' and even to 'Pop goes the Weasel.' " Exceedingly better was it "to associate these popular airs with songs of a moral and Christian character," rather than with "every species" of that "ribaldry and low wit" to which it had been Mr. Prosser's misfortune to listen.[23]

Ahh, but the rhythm of "Pop Goes the Weasel" was itself unstable. It appeared in short order at the head of a quadrille, a march, a polka, and a fantasia, and as the label for a change-ringing sequence for churchbells. Tell me, what was the time signature when "Pop Goes the Weasel" was whistled, as it was, by Tataiesi Owasjero, youngest of interpreters for the Grand Embassy from Japan to the United States in 1860? Or when taken up about that time by a young Maori gent "dressed in the black coat, trousers, and hat of Tottenham Court Road, jingling his money in his pocket, and singing 'Pop Goes the Weasel' "?[24]

Popular it was, no doubt at all about that, as a dance, a tune, a song, a satire, a farce, a burletta, an element of an allegorical extravanganza.[25] So enduringly popular that it could be one of the few airs that persons of lower station or duller ear might readily identify—as in this passage from "A Rivermouth Romance," which I excerpt exactly as written:

> As this letter has never been printed, and it is the only specimen extant of Mr. O'Rourke's epistolary manner, we lay it before the reader *verbatim et literatim*:
>
> febuary 1864
>
> mi bilovid wife
> fur the love of God sind mee pop gose the wezel. Yours till deth.
> larry O rourke

[23] E. E., "School Song for Boys," *National Society's Monthly Paper for 1855* 109 (1855), 253; "Street Music," letter to the editor of *National Society's Monthly Paper for 1856* 110 (1856), 16.

[24] Mervyn Slatter, "Musical Bouquet: A Study of a Music Publisher 1845–1917," online at <http://www.musicalbouquet.co.uk/foreword>; Michael Anthony Williams, "Call Changes," <http://www.campaniles.co.uk/maw/callchanges.html>; R. M. Devens, *American Progress: Or, The Great Events of the Greatest Century* (Chicago: Hugh Heron, 1883), 491; on Japanese translator; "Old and Young New Zealand," *London Review* (29 Aug. 1863), 235, reviewing two books, the second of which, *Old New Zealand; Being Incidents of the Native Customs and Character in the Old Time* (London: Smith, Elder, 1863), has as its author "A Pakeha Maori."

[25] William Bodham Donne, *Pop Goes the Weasel, or, the Adventures of the Weasel Family: A Farce in 1 Act* (Plays submitted to the Lord Chamberlain, received for licensing 28 Apr. 1853, intended for representation at the Standard Theatre; original in the British Library [MS 52939 S]); Pavilion Theatre, London playbill from 1854, in East London Theatre Archive, <http://www.elta-project.org/browse.html?type=person&id=3027>; George Dibdin Pitt, *Pop Goes the Weasel, or, The Devil's Dance: A Burletta in One Act* (Play submitted to the Lord Chamberlain, 1853; original in the British Library [MS 52939 V]); James Planché, "The New Haymarket Spring Meeting," *The Extravaganzas of J. R. Planché, Esq., 1825–1871*, vol. 5, ed. T. F. Dillon Croker and Stephen Tucker (London: Samuel French, 1879).

Thomas Bailey Aldrich's "Romance" was published in 1885, about the time that Sabine Baring-Gould, vicar of Mersea Island, Essex, and author of "Onward Christian Soldiers," was interviewing Roger Luxton in North Devon. Once illustrious as a song man, at the age of 76 he had forgotten his English ballads but could recall the downward trajectory of his career. He used to be welcome at all farmhouses, but now the farmers had pianos "and zing nort but twitery sort of pieces that have nother music nor sense in them; and they don't care to hear us, and any decent sort of music [. . .]. And now I reckon the labouring-folk be so tree-mendious edicated that they don't care to hear our old songs nother. 'Tis all *Pop goes the Weasel* and *Ehren on the Rhine* now. I reckon folks now have got different ears from what they used to have, and different hearts too. More's the pity."[26]

Certainly, the popularity of "Pop Goes the Weasel" and other twittery pieces was enabled by the rise of music halls and the penny press, and further assured by technical and commercial advances in the printing, advertisement, and distribution of sheet music as well as mass-produced music boxes, jack-in-the-boxes, and boxed sets of nursery rhymes.[27] And we are just beginning to explore how all this may have made for "different ears" or "different hearts."[28] Yet none of this explains the immediate and enduring currency of the tune and refrain of "Pop Goes the Whistle" above all others, overcoming such fierce competition in 1853 as "The Ratcatcher's Daughter" and "Villikins and his Dinah," now all but vorgotten.[29]

[26] Thomas Bailey Aldrich, "A Rivermouth Romance," *Marjorie Daw: And Other Stories* (Boston: Houghton, Mifflin, 1885), 270; Sabine Baring-Gould, *Old Country Life* (London: Methuen, 1890), 275–76. Cf. also Charles Dickens [?], "Quite Inexcusable! A Complete Story," *All the Year Round* (3 Aug 1891), 138.

[27] Derek B. Scott, *Sounds of the Metropolis: The 19th-Century Popular Music Revolution in London, New York, Paris and Vienna* (Oxford: Oxford University Press, 2008), 15–57.

[28] E.g., John M. Picker, *Victorian Soundscapes* (Oxford: Oxford University Press, 2003); Martin Hewitt and Rachell Cowgill, eds., *Victorian Soundscapes Revisited* (Horsforth, UK: Trinity and All Saints/Leeds Centre for Victorian Studies, 2007); Shelley Trower, *Senses of Vibration: A History of the Pleasure and Pain of Sound* (New York: Continuum, 2012); Hillel Schwartz, *Making Noise: From Babel to the Big Bang and Beyond* (Brooklyn, NY: Zone, 2011), Round Two

[29] MacKay, *Through the Long Day*, 137; G. K. Chesterton, "Censoring the Newspapers," *The Collected Works of G. K. Chesterton. Vol. 28: The Illustrated London News, 1908–1910* (San Francisco: Ignatius, 1987), 282, essay of 6 Mar. 1909; "A Word or Two about Scotch and English Song," *Sydney Morning Herald* (26 Apr. 1855), 3, in response to an article by Gerald Massey in *Hogg's Instructor* for December 1854: "We can well understand the disgust which leads young Mr. Massey to say [sarcastically] of the songs of today . . . 'Have we not also that delicious song, so full of lofty imagination, taste, and tenderness, "Villikins and His Dinah," which has been so popular and enchanted the people of England in this year of grace, 1854. This, and "Pop Goes the Weasel," have, most assuredly, been sung oftener in the last year than any other song in the world [. . .] Is it not enough to make a dog howl to hear such things?' "

I would enjoy nothing more than to attribute the popularity of the refrain, if not the tune, to one notorious figure during the Crimean War. I am thinking of James Brudenell, Seventh Earl of Cardigan, jealous of the dashing Captain Nolan (the Balaklava-disaster-enactor mentioned earlier), who upon presenting and defending the mis-taken orders, appeared to presume to take from him the glory of leading *his* Light Brigade. Riding with the 17th Lancers as they advanced at an all too conventional "walk" signaled by Cardigan's trumpeter (though trumpeter Harry Powell "never heard a sound of any sort, and many who were nearer to his Lordship, say the same as myself"), Nolan of a sudden galloped twenty yards ahead, waved his sword, shouted something unintelligible in "the thickest shower of shell, shot, grape, cannister, and Minié, from front and flanks [. . .] whistling and crackling of shells [. . .] beyond all description," and became the first casualty of the charge, his chest shredded by shrapnel from an exploding shell, his horse "giving out an almighty scream." Cardigan then took his horse at a canter to the front of the troops and drove onward, half a league, half a league onward, never looking back to appreciate the slaughter. Reaching the Russian lines with his life and mount intact, Cardigan found himself surrounded by enemy cavalry and about to be captured, so he wheeled around and back to safety through the still charging brigades. Whereupon he retired for the night to elegant quarters on his steam yacht in the harbor and popped the cork on a champagne supper. Was not "Pop Goes the Weasel" a spoof of that weasel of a man who twice in his career had slipped the knots of court martial (for bullying, and for cheating in a duel) and who now had proved himself as inhuman (quoth: it was "no part of a general's duty to fight the enemy among private soldiers") as he was incompetent?[30]

Alas, both the timing and the affect are off. The refrain had been in vogue for a year before Balaklava, and I find no reference to him in those stanzas—usually from across the Atlantic—that mocked the British and their war:

> Johnny Bull, he makes his brag,
> He can whip the whole creation,
> Why don't he take Sebastopol,
> By Pop goes the weasel.

[30] Dutton, *Forgotten Heroes*, 9, 14 (obituary of Nolan in *Illustrated London News* [25 Nov. 1854], 121; Private Anthony Sheridan, account in ibid. 120–1 [30 Oct. 1875]); Harry Powell, *Reflections of a Young Soldier during the Crimean War* (Oxford: Upston and Doe, 1876), 312, reproduced in ibid. 258–59; Capt. Godfrey Charles Morgan, letter to his father (30 Oct. 1854), ibid. 312. Cecil Woodham Smith, *The Reason Why: The Story of the Fatal Charge of the Light Brigade* (London: Constable, 1953), 262. For more on Nolan: John Sweetman, "Captain Louis Edward Nolan," *The Crimean War 1854–1856* (Osprey, 2001) 76–78; David Buttery, *Messenger of Death: Captain Nolan and the Charge of the Light Brigade* (Pen & Sword, 2008).

Worse, Cardigan was actually received as a hero on his return to the British Isles, praised for his derring-do at having led the charge, under orders and regardless of consequences. He was appointed Inspector General of the Cavalry, a post he held for five years until he retired to his estate and a life of horseracing, hunting, yachting, and dalliance.[31]

So I perforce put forward a more (sym)phonic explanation for the popularity of "Pop Goes the Weasel." I will argue by halves that it was *both the senselessness and the sound* of the refrain that resonated exactly at mid-century. After all, if (as Jean Harrowven claims) a slew of words had first been put to the jig during the 1830s by the Anglo-Jewish singer Charles Sloman, comedian and all-around "improvisatore," his song when performed "in such places of ill-repute as the Cyder Cellars and the Coal Hole" did not then catch on, self-referential as it was with regard to the flightiness of fads and fashions of catching on:

> Something new starts every day,
> Pop goes the Weasel,
> Fashion ever changes sway,
> Pop goes the Weasel.
>
> As one comes in another goes out,
> Pop goes the Weasel.
> The newest one, there is no doubt,
> Is Pop goes the Weasel.[32]

I will argue that the refrain held a cultural "potential energy" and that the shockwave of [p][ɒ][p] [g][ʊʊ][z] [o][ə] [w][I:][z][əl] bore direct, blatant analogy to what was happening not only in Crimea but in physics, engineering, and the countryside of the United Kingdom precisely at mid-century.

Make the acquaintance of the Scottish engineer, physicist, rifleman, pianist, cellist, and songster William John Macquorn Rankine (1820–1872). Son of

[31] "Pop Goes the Weasel," *Beadle's Dime Song Book: A Collection of New and Popular Comic and Sentimental Songs* (New York: Irwin P. Beadle, 1857), as reproduced and discussed in *Grosvenor Library Bulletin (Buffalo)* 1, 3 (Mar. 1919), 17–20; Jon W. Finson, *The Voices That Are Gone: Themes in Nineteenth-Century American Popular Song* (Oxford: Oxford University Press, 1997), 193; Florence Howe Hall, *Memories Grave and Gay* (New York: Harper & Brothers, 1918), 30, the Crimean War regarded from America; Dutton, *Forgotten Heroes*, 9; Saul David, *The Homicidal Earl: The Life of Lord Cardigan* (Boston: Little, Brown, 1997).

[32] Jean Harrowven, *Origins of Rhymes, Songs and Sayings* (London: Kaye & Ward, 1977), 273; William Tinsley, *Random Recollections of an Old Publisher*, vol. 2 (London: Simpkin, Marshall, Hamilton, Kent, 1900), 50–51 on Sloman, as also David Conway, *Jewry in Music: Entry to the Profession from the Enlightenment to Richard Wagner* (Cambridge: Cambridge University Press, 2012), 104.

David Rankine, a civil engineer and lieutenant in the rifle corps, he had studied the theory of music, the works of Aristotle, Newton's *Principia*, and the methods and principles of railway and locomotive construction, on which he worked in Scotland (with his father) and Ireland (with John Benjamin MacNeill). In 1848, that year of nearly achieved and aborted revolutions across continental Europe, he began to write and publish a series of historically significant papers on molecular physics in which he developed those first and second laws of thermodynamics almost simultaneously being refined and mathematically fixed by the slightly younger William Thomson (First Baron Kelvin). Thomson, like James Prescott Joule, came to thermodynamics through experiments with electricity, Rankine through experience with metallurgy and steam engines, so it was Rankine, not Thomson or Joule (or Rankine's more famous compatriot James Clerk Maxwell), who would formally introduced to science the notions of "stress," "strain," "adiabatic processes," and "potential energy." That last term, which in 1853 Rankine set in binary opposition to "actual" or "sensible" (later, in 1867, "kinetic") energy, he drew rather from his readings in the Greek of Aristotle than from prior expositions in the French of Carnot and Poncelet or the German of von Mayer and Humboldt. This theory meant that, at the start, "potential energy" had immediate philosophical implications with regard to how anything is realized—i.e., made real, achieved, accomplished, having impact or vigorous import.[33]

Rankine's definition of potential energy was part of a larger effort to clarify the idea of force, which had drawn to itself over a century (as tracked by the historian John Roche) such a confusion of expressions as Humboldt's *Kraft* and, in English, " 'living force,' 'ascending force,' 'fall-force,' 'duty,' 'hidden *vis viva*,' 'latent *vis viva*,' 'moment of activity,' 'source of work,' 'mechanical equivalent,' 'mechanical value,' or 'mechanical power.' " Back in 1807, Thomas Young had promoted the substitution of the word "energy" (ευεργεια, "activity, operation," as used in Aristotle's *Metaphysics*), making it "proportional to the labour expended in producing motion," but not until Thomson resumed the usage in 1849 did "energy" move to the fore. Again, though, it was Rankine, not Thomson, who in 1855 first defined energy "explicitly and unambiguously," and in print, as comprehending "every state of a substance which constitutes the capacity for performing

[33] P. G. Tait, "Memoir of Rankine," *Miscellaneous Scientific Papers by W. J. M. Rankine: From the Transactions and Proceedings of the Royal and Other Scientific and Philosophical Societies. . .*, ed. William J. Millar (London: C. Griffin, 1881); Ben Marsden, "Rankine, (William John) Macquorn (1820–1872)," *Oxford Dictionary of National Biography* (Oxford: Oxford University Press, 2004; online ed., 2011, at <http://www.oxforddnb.com/view/article/23133>; David F. Channell, "The Harmony of Theory and Practice: The Engineering Science of W. J. M. Rankine," *Technology and Culture* 23/1 (1982), 39–52; W. J. M. Rankine, "On the Phrase 'Potential Energy,' and On the Definitions of Physical Quantities," *The London, Edinburgh, and Dublin Philosophical Magazine and Journal of Science* 4/33 (1867), 88–92.

work." That was the nub: work, which meant putting things in motion. In 1867, Rankine further clarified: potential energy was the "energy of configuration," actual energy the "energy of activity." Potential energy described a latent state of work, a configuration such as that in which a cork stoppers an unopened bottle of champagne or a spring lies compressed inside a jack-in-the-box; kinetic energy was work in progress, the participial popping of the cork.[34]

Thanks to new investigations by oenological physicists, we know now that precisely 5% of the potential energy in champagne's confined bubbles of carbon dioxide goes toward the work of cork launch and lift off—at a temperature-dependent maximum speed of 31 mph; the rest is spent on the exclamatory shockwave, the "pop!" Rankine, who wrote in 1851 on the theory of the propagation of waves in his essay "On Laplace's Theory of Sound," would no doubt have appreciated this subsequently established fact, both as a physicist and as a convivial sort of Scotsman given to bursting into song, but I have led us here because he had a long-term, obstinate investment in the energy of molecular vortices that in-forms the "pop." Indeed, his understanding of the nature of heat, central to his influential manual on steam engines (1859), rested on a model of invisible whirling motions through whose vortical forces a practical theorist might unify the study of heat, elasticity, and light. The energy of these vortices, if calculated aright, might also reconcentrate energy in the universe and relieve physics of Thomson's alarming hypothesis of entropy, the "universal dissipation of the energy available to humanity."[35]

Which brings us spinning back to whistles. Although only fairly recently have scientists explained to their own satisfaction how spinning steam creates a vortex through which teakettles whistle, it has been understood more generally for much longer that the piercing sound of locomotive steam whistles probably had something to do with molecular vortices, and Rankine, like his "Mathematician in Love," had himself most likely

> studied (since music has charms for the fair)
> The theory of fiddles and whistles,

—even if he had not

[34] John Roche, "What Is Potential Energy?," *European Journal of Physics* 24/2 (2003), 185–96; Rankine, "On the Phrase 'Potential Energy'"; Keith Hutchison, "W. J. M. Rankine and the Rise of Thermodynamics," *British Journal for the History of Science* 14/1 (1981), 1–26.

[35] Gérard Liger-Belaira et al., "Champagne Cork Popping Revisited through High-Speed Infrared Imaging: The Role of Temperature," *Journal of Food Engineering* 116/1 (May 2013), 78–85; W. J. M. Rankine, "On Laplace's Theory of Sound," *The London, Edinburgh, and Dublin Philosophical Magazine and Journal of Science* 4/1 (1851), 225–27.

Then composed, by acoustic equations, an air,
Which, when 'twas performed, made the lady's long hair
Stand on end, like a porcupine's bristles.

Others of his time, in language vaguer and less metrical, seemed also to hear whistling—human whistling—as approximately vortical. We read, in a collaborative work of popular science on *The Principles of Physiology* in 1854,

> [b]efore leaving the subject of the human voice, whistling deserves a few words. The sound in whistling does not arise from the vibrations of the lips. [. . .] It has been supposed [. . .] that the air is thrown into sonorous vibration by friction against the borders of the opening [. . . and this vibration] throws the whole column of air in the mouth into vibrations, and the vibrations of this column of air, by a reciprocal influence, determine the rapidity of the vibrations of the air at the orifice. The only difference [. . .] between whistling and the sounds of a pipe is, that in whistling the whole column of air is in constant progressive motion through the tube and orifice, while in a pipe the air in the tube merely vibrates, and does not move as a current.

A column of air in "constant progressive motion" need only swirl around the mouth and through the orifice to initiate, in effect, the vortex and the whistle.[36]

But, really, it is neither here nor there whether his contemporaries supposed an invisible vortex to each whistling of their world with the same exactitude or passion as Rankine supposed a whirl of invisible vortices to each source of heat that drove it. For the whistling—especially of those whistles built into the increasingly powerful steam engines on the railways for which Rankine was very much responsible in theory and in engineering practice—was more prevalent, piercing, and premonitory than it had ever been before. Consider the first and last verses of the poem he wrote (a song, too, with music) in 1859 for the Iron Horse:

Put forth your force, my iron horse, with limbs that never tire!
The best of oil shall feed your joints, and the best of coal your fire;

[36] R. H. Henrywood and A. Agarwal, "The Aeroacoustics of a Steam Kettle," *Physics of Fluids* 25 (2013), published online: <https://doi.org/10.1063/1.4821782>; William J. Macquorn Rankine, "Mathematician in Love," *Songs and Fables* (Glasgow: James Maclehose; London: Macmillan, 1874), 4, stanza 3; William Somerville Orr, Richard Owen, and Robert Gordon Latham, *The Principles of Physiology* (London: Orr, 1854), 134. Cf. Robert C. Chanaud, "Aerodynamic Whistles," *Scientific American* 222/1 (1970), 40–47.

So off we tear from Euston Square, to beat the swift south wind,
As we rattle along the North-West rail, with the express train behind:—
Dash along, crash along, sixty miles an hour!
Right through old England flee!
For I am bound to see my love
Far away in the North Countrie. [. . .]

Now Thames and Trent are far behind, and evening's shades are come;
Before my eyes the brown hills rise that guard my true love's home:
Even now she stands, my own dear lass! beside the cottage door,
And she listens for the whistle shrill, and the blast-pipe's rattling roar:—
Roll along, bowl along, sixty miles an hour!
Right through old England flee!
For I am bound to see my love.
At home in the North Countrie.[37]

If street whistlers were having a tough time earning a living in London during the 1850s, Joshua Stoddard of Worcester, Massachusetts, in 1855 was patenting the steam calliope, which could out-whistle any strolling musician.[38] And if newspapers of the 1850s were reporting accident after fatal accident of pedestrians and lineworkers ignoring or befuddled by whistles from oncoming locomotives, rail companies and local municipalities soon put in place systems of "whistle-posts" that standardized the approach time of rail steam whistles used to alert all those at protected crossings.[39]

Granted a cultural history of energy and thermodynamics;[40] granted a study that makes Tennyson's *In Memoriam* (1851) a prooftext in early theorizing about entropy;[41] granted a literary history of the senses of vibration during the Victorian era;[42] granted such scholarship, I have some company in maneuvering the like of

[37] Rankine, "The Engine-Driver to His Engine," *Songs and Fables*, 12–14.

[38] Henry Mayhew, *London Labour and the London Poor: A Cyclopedia of the Conditions and Earnings of Those That Will Work, Those That Cannot Work, And Those That Will Not Work*, vol. 3 (London: Griffin, Bohn, 1851), chap. 4, "Street Vocalists," #5, The Whistling Man; J. C. Stoddard, "Apparatus for Producing Music by Steam or Compressed Air," U.S. Patent No. 13,668 (9 Oct. 1855).

[39] J. B. Owen, *Rules and Regulations, September 12, 1854* (Stratford: Eastern Countries Railway Company, 1857).

[40] Crosbie Smith, *The Science of Energy—A Cultural History of Energy Physics in Victorian Britain* (Chicago: University of Chicago Press, 1998).

[41] Barri J. Gold, "The Consolation of Physics: Tennyson's Thermodynamic Solution," *PMLA* 117/3 (2002), 449–64, elaborated in his *Thermopoetics: Energy in Victorian Literature and Science* (Cambridge, MA: MIT Press, 2010). Cf. Gillian Beer, "Helmholtz, Tyndall, Gerard Manley Hopkins: Leaps of the Prepared Imagination," *Open Fields: Science in Cultural Encounter* (Oxford: Oxford University Press, 1996), 242–71.

[42] Trower, *Senses of Vibration*.

war and whistles, physics and engineering, to dock with the senselessness and sound of the refrain, "Pop Goes the Weasel." First, the senselessness. Or rather, the elemental vacuousness of words that—once divorced from dance patterns—could be used to mean almost anything, including mere effervescence. The phrase had what I am willing to call lexical potential energy, an energy that could be easily, quickly enlisted to do highly kinetic, cultural work in dozens of contexts. It had strong potential at mid-century, and not earlier, because "popping" itself had by then accumulated a full head of steam and many atmospheres of pressure: the spluttering of increasingly powerful and plentiful locomotives; the expansion of steam boilers with their safety, or "pop," valves; the circulation of cheaper pistols, or "poppers," as well as a profusion of toy popguns; the appearance in 1847 of vulcanized-rubber toy balloons, designed in London by J. G. Ingram; the proliferation of carbonated and sparkling beverages, especially champagne. Especially this last. Benefiting from improved bottling operations, savvy use of liqueurs as additives to reduce explosive breakage, and the design of better corks with fitted wire muzzles during the 1840s, French production of champagne expanded from 300,000 bottles a year in 1800 to 20,000,000 bottles in 1850.[43]

Rankine's papers on a theory of potential energy were hardly in the minds of every comedian or soldier or busker singing "Pop Goes the Weasel," or every boy whistling it, or every organ grinder grinding it, or families toppling over musical chairs to the awful folly of its refrain. I would bet a sovereign that Rankine knew and sang (or whistled) the tune, but I am arguing now from cultural lexicography, not scientific biography. I am proposing that the notion of potential energy, of things "about to pop," was at full play across all spectra of experience, and it was this sense of things "about to pop" that contributed substantially to the initial enthusiasm for the war in Crimea as well as to the nervously exuberant unpredictability that went along with its (rash, shoddy, poorly conceived) preparations.

> Now do you think in battle Old Napier will ever flinch ?
> No I think he is a bull dog, and a sailor every inch
> And if he catches Nick the Bear he will stick him on a pole,
> *He will pop him in the weasel* and stop up his gully hole...[44]

[43] On "poppers" and "pop" valves: Partridge, *Dictionary of the Underworld*, 626. On popguns: *Every Little Boy's Book: A Complete Cyclopædia of in and Outdoor Games with and without Toys, Domestic Pets, Conjuring, Shows, Riddles, Etc.* (New York: Routledge, Warne, and Routledge, 1864), 17. On champagne: Rod Phillips, *A Short History of Wine* (New York: HarperCollins, 2000), 241; Stefan K. Estreicher, *Wine from Neolithic Times to the 21st Century* (New York: Algora, 2006), 91–92; James Simpson, *Creating Wine: The Emergence of a World Industry, 1840–1914* (Princeton, NJ: Princeton University Press, 2011), 132–48.

[44] J. Marks and M. Hyams (printers—and authors?), *Conversation of Old England in 1854* (London, 1854), single leaf.

What of that *weasel*? Although no one yet spoke about "weasel words," in English slang and cant "weasel" already had a slippery set of associations: aside from that popping measuring device, it could be a silver plate or vessel, an ermine-collared scholar's gown, a sixpence, a heavy iron—and, in the stanza just preceding, the throat or gullet, <*weasand*. Allegorical and zoo-literary references were also of diverse skins: weasel as an assiduous hunter; weasel as minion of the devil; weasel as cunning; weasel as mischievous; weasel as the sole deadly enemy of basilisks; weasel as lecherous; weasel as emblem of insemination through the ear, or of virgin birth and the Virgin Mary, held and caressed by many early Renaissance madonnas. Stuffed, weasels went a-courting at the Crystal Palace and attacked an owl in the Paris Exhibition of 1855. Alive and breathing, the animal was thought sleek, fast, bold: in 1855 the English Navy launched "Weasel," a 60-hp screw-driven gunboat with three (hard-popping?) cannon.[45]

If "popping" bespoke potential, the weasel espoused predicament, consequences unintended, madcap chase, escape, disappearance. Together, pop and weasel entertained that change of momentum which is surprise, yes, but also loss . . . and loss of significance. Confounding contemporaries themselves, "pop" went with "weasel" only so far as sound-and-fury that might begin in comedy but end in tragedy or nothing much at all. "Pop goes the weasel" was (do I risk a "hence" here?) a refrain ideal for a war conducted in near total disregard of the four previous wars in the Crimea—1768–74 (when plague killed 150,000 Russian troops), 1787–92, 1806–1812 (plague again), 1828–29 (plague and dysentery or cholera)—all fought amid bad weather, miscalculation, obstinacy, and miscommunication. It became bitingly germane for a war whose concluding treaty achieved almost none of the British Empire's ambitions, so that within a generation the war seemed to have been as senseless as the refrain.[46] The eye-popping splendor of its potential (Lord Cardigan in particular was a stickler for gold braid and flashy uniforms) had ended in bedraggle.[47]

[45] Lord Ronald Sutherland Gower, *Records and Reminiscences* (London: J. Murray, 1903), 66; Canada, Executive Committee for the Paris Exhibition, *Canada at the Universal Exhibition of 1855* (Toronto: J. Lovell, 1856), 256; Maurizio Bettini, *Women and Weasels: Mythologies of Birth in Ancient Greece and Rome* (University of Chicago, 2013), esp. 120–22; P. Benyon, "Late 18th, 19th, and Early 20th Century Naval and Naval Social History," http://www.pbenyon.plus.com/18-1900/W/05203.html; "The Baltic Fleet," *Illustrated London News* (28 July 1855), 121.

[46] For the treaty's failure, Brian James, "Allies in Disarray: The Messy End of the Crimean War," *History Today* 58/3 (2008), 24–31.

[47] Stefanie Markovits, *The Crimean War in the British Imagination* (Cambridge: Cambridge University Press, 2009), esp. 210–18, her final chapter, on Elizabeth Thompson's 1874 painting, "Roll Call."

Even today, such historians as Orlando Figes who mean to fix a pivotal place for the war in the modern history of the Middle East, in modern relations between Russia and Europe or Islam and the Christian West, must battle against a gale force of 150 years of arm waving and such titles as A. J. Barker's *The Vainglorious War* or, more recently, Alexis Troubetzkoy's *History's Most Unnecessary Struggle*.[48] In the annals of war, the Crimean War was veritably a flash in the pan compared to the 20,000,000 casualties of the Taiping Rebellion (1850–64). And in the larger scheme of things, what events of 1854–56 were most consequential: The Crimean War? The cholera epidemic in London that killed 10,000 but led to a new epidemiology of the disease? The opening of Japan to Western and Russian commerce and armaments? The Kansas-Nebraska Act that made a war between the States almost inevitable? Smith and Wesson patenting metal bullet cartridges? Abraham Gesner patenting kerosene? Isaac Singer patenting the first sewing-machine motor? Bessemer patenting a new process for making steel? Elisha Otis completing work on the safety elevator? Pope Pius IX defining, ex cathedra, the dogma of the Immaculate Conception? The industrial promotion of wood-pulp paper? The drilling of the world's first oil wells? The Woman's Rights Petition to the New York Legislature ("Resolved, That women are human beings whose rights correspond with their duties . . .")? George Boole publishing his treatise on algebraic logic, whose operators would be crucial to computer programs? Walt Whitman publishing *Leaves of Grass*? Richard Wagner mounting his *Die Walküre*?

An opera? How foolish of me. In the Western scheme of things, neither poetry nor music has been creditable with greater historical heft or consequence. So it must be with unequivocal foolishness that I move to the second half of my argument: that it was the **sound** of the refrain that made for its popularity. Rankine as a merry musician may have appreciated its voiceless bilabial plosives [p] [p], the voiced velar plosive [g], the voiced sibilant fricative es=[z], and the whistle and whizz of the voiced bilabial-velar semi-vowel [w] with the voiced sibilant fricative s=[z], ending on the howl of the reduced vowel and voiced alveolar lateral [əl]. What more can I say? I can say that the sound sequence of plosives, fricatives, velar semivowel, and alveolar lateral of **pop→weasel** essentially reproduced the sound sequence of the following:

- The progression of illness as mucus or fluid accumulates in the lungs and throat. This sequence had become audibly familiar as Scottish physicians

[48] Orlando Figes, *The Crimean War: A History* (London: Penguin, 2010), xxi–xxii; A. J. Barker, *The Vainglorious War, 1854–56* (Weidenfeld and Nicolson, 1970); Alexis Troubetzkoy, *A Brief History of the Crimean War: History's Most Unnecessary Struggle* (New York: Carroll and Graf, 2006), and cf. Candan Badem, *The Ottoman Crimean War: 1853–1856* (Leiden: Brill, 2010), 2.

during the 1830s and English physicians during the 1840s adopted the use of the stethoscope, for then could be heard more clearly (and explained to onlisteners at the bedside) the rales or crackles like "the sound that you hear when burning wood crackles and **pop**s in a fireplace," signs of that pneumonia toward which many diseases or infirmities tend in later stages, through the moaning (**ohs**) of pain or disorientation, to the high-pitched **wheez**ing or stridor of the death rattle.[49]

- The death throes of war horses shot and falling, their joints **pop**ping as they fell wounded to the ground, **groan**ing, then **wheez**ing/whistling until they could no longer breathe.
- The acoustic experience of mid-century infantry under attack, first, from the distant **pop**ping of mortar or artillery, then from the **whistl**ing of shells overhead as well as of the new conical bullets from Minié rifles.
- Tinnitus, or "noises in the head," often brought on by proximity to an explosion or sudden loud percussion or **pop**ping, leading to a persistent, bothersome, sometimes tortuous inner soundworld of more or less intense roaring, buzzing, or piercing **whistl**ing, like that of a teakettle (said one woman) or locomotives. The incidence of tinnitus was increasing at mid-century due to the regular use of ototoxins like quinine or opium, to the consumption of cheaper sweets that made for dental problems and otolaryngological complications, to long hours of unprotected civilian work near loud steam-powered machinery (saws, looms, locomotives, steam ships), and to unprotected military training with more powerful cannon, guns, and ship engines.[50]
- Irritable heart, or what would later be called shell shock, and which was originally remarked during the Crimean War among soldiers who were "utterly unnerved and agitated violently by the merest trifles." Any **pop**ping within hearing (akin to artillery or musketry) reverberated in them as tremors, shakes, jolts, jars, jerks, shocks, succussions, trepidations, quivers, quavers, jactitations, quassations, twitters, flickers, flutters, trembles (< "315. [Irregular motion] Agitation," from Peter Roget's *Thesaurus*, first published in 1852). Two days after the Battle of Inkerman, Lieutenant Lleuellyn of the 46th Regiment arrived in Crimea and went into the trenches in search of his division, all of whose officers had been killed; he found a straggle of infantry "dazed and stupefied and unable to give us any idea of our position or chances." When the sickest among them were brought at last to Scutari, Florence Nightingale learned quickly to be sensitive by proxy to

[49] Schwartz, *Making Noise*, 202–21; for the sounds, East Tennessee State University, <http://faculty.etsu.edu/arnall/www/public_html/heartlung/breathsounds/contents.html>.

[50] Schwartz, *Making Noise*, 371–73.

> the smallest noises, especially the **whis**pering of orderlies and the ru**stle** of crinolines.[51]

That's all that I can say, and all I can say has been dys-lexic. That is, I have been keying sets of sung words and their recurrent refrain to a historical moment less by dictionary specifics, metaphorical reach, or musical structure than by a superfluidity of meanings and a plosive pattern of sound. Whether such dyslexia deserves to be heard out, time may never tell, and besides, it's of utterly no consequence.

After/words: On 28 September 1987, a century and a third after the Crimean War, a century and a third after Alfred Russell Wallace began making entries on natural selection in his Species Notebook while in far-off Borneo and New Guinea, "Encounter at Farpoint" aired as the series premiere of "Star Trek: The Next Generation."[52] Almost immediately, Captain Picard and his crew are accosted by Q, an omnipotent being who puts humanity on trial as an irredeemably savage species. Q does not cite, as well he could, the deaths of 750,000 soldiers and 200,000 horses in Crimea, or the uncounted hundreds of thousands of noncombatants who were starved, infected, or massacred during that "first 'total war,'" and never mind all else that ensues during the episode.[53] Except this: while humanity stands trial, the one android aboard the Enterprise-D starship, Commander Data, is working hard to perfect his own humanity—by trying to learn to whistle, fluently and tunefully, "Pop Goes the Weasel."

[51] Edgar Jones and Simon Wessely, *Shell Shock to PTSD: Military Psychiatry from 1900 to the Gulf War* (Hove and New York: Psychology Press, 2005), 3–4; Schwartz, *Making Noise*, 268–72.

[52] Alfred Russel Wallace, *On the Organic Law of Change: A Facsimile Edition and Annotated Transcription of Alfred Russel Wallace's Species Notebook of 1855–1859*, ed. James T. Costa (Cambridge, MA: Harvard University Press, 2013).

[53] Figes, *Crimea*, xix.

BIBLIOGRAPHY

Primary

I) NEWSPAPERS, JOURNALS, AND OTHER PRINTED TEXT

Cimze, Jānis. "Dseeşmu rohta jauneķļeem un wihreem (Preekşchrunnas weetâ)." *Mājas Viesis* (1 May 1871): 140–42.

de Staël, Germaine. "Sulla maniera e l'ultilità delle traduzioni." Trans. Pietro Giordani. *Biblioteca italiana* 1/1 (1816): 9–18.

Dickens, Charles [?]. "Quite Inexcusable! A Complete Story." *All the Year Round* (3 Aug. 1891), 138.

Dickens, Charles. "Whistlers and Whistling." *All the Year Round* n.s. 9 (4 Jan. 1873), 182–85.

Dyer, Thomas F. Thiselton. "Whistling." *The Gentleman's Magazine* 254 (1883), 392–400.

[E. E]. "School Song for Boys." *National Society's Monthly Paper for 1855* 109 (1855), 253.

[E. E.]. "Street Music." *National Society's Monthly Paper for 1856* 110 (1856), 16.

Europe and Victory. London: E. [M. A.] Hodges, c. 1855. <http://ballads.bodleian.ox.ac.uk/static/images/sheets/25000/20169.gif>.

Every Little Boy's Book: A Complete Cyclopaedia of in and Outdoor Games with and without Toys, Domestic Pets, Conjuring, Shows, Riddles, Etc. New York. Routledge, Warne, and Routledge, 1864

Fitz-Gerald, S. J. Adair. *Notes and Queries* (1905): ser. 10/4, 209–11; ser 10/3, 430 and ser. 10/4. 54–55.

Konchevskii, A. K. "Proshloe i nastoiashchee v pesniakh Kryma." *Obshchestvenno-nauchnyi zhurnal* 1 (1925): 31–33.

Kronvalds, Atis. "Isbraukums pa tehwa-semmi." *Baltijas Vēstnesis* (30 July 1870), 237.

Lingard, T. W. A. *Notes & Queries* 171 (5 Sept. 1936), 176.

List of Killed and Wounded at the Battle of the Alma on Wednesday, September 20, 1854, with the Official Sketch of the Battle Field, and Other Illustrations. 5th ed. London: Clarke & Beeton, 1854.

Lynd, William, ed. "Ocean Telegraphy." *The Telegraphist* 10 (1884): 120.

Marks, J., and M. Hyams (printers—and authors?). *Conversation of Old England in 1854*. London, 1854.

Marx, Karl. *New York Daily Tribune* (7 Mar. 1853), 4.

Mitchell, Stanley W. "Phantom Limbs." *Lippincott's Magazine of Popular Literature and Science* 8 (1871): 563–69.

Morley, Henry. "Adeliza Castle." *Household Words* 22 (1854): 219–20.

"Music Really for the Million." *Punch, or the London Charivari* 28 (1855), 251.

"Old and Young New Zealand." *London Review* (29 Aug. 1863), 235.

Page, John T. "Pop Goes the Weasel." *Notes and Queries* (1905): ser. 10/3, 491.

Palmer, J. Foster. *Notes and Queries* (1905): ser. 10/3: 491.

Rankine, W. J. M. "On Laplace's Theory of Sound." *The London, Edinburgh, and Dublin Philosophical Magazine and Journal of Science* 4/1 (1851): 225–27.

Rankine, W. J. M. "On the Phrase 'Potential Energy,' and On the Definitions of Physical Quantities." *The London, Edinburgh, and Dublin Philosophical Magazine and Journal of Science* 4/33 (1867): 88–92.

Ruskin, John. *Athenaeum* (30 Sept. 1905), 428.

Russell, William Howard. "Arrival of the Wounded in the Bosphorus." *The Times* (London, 9 Oct. 1854), 8.

Russell, William Howard. "From Our Special Correspondent." *The Times* (London, 10 Oct. 1854), 7–8.

Russell, William Howard. "The War; The British Expedition; The Siege of Sebastopol (From Our Special Correspondent)." *The Times* (London, 28 Oct. 1854), 7.

Schönberg, E. F. "Rekruşcha-dseeşmiņa." *Latviešu Avīzes* (18 Mar. 1854), 48.

Tolstoy, Leo. "A History of Yesterday" (1851). Trans. George L. Kline. *Russian Review* 8/2 (1949): 142–60.

"The Turf in Hong Kong." *Bell's Life in London and Sporting Chronicle* (town ed., 5 Oct. 1857), 5.

II) MATERIALS CONSULTED IN MANUSCRIPT

Bartelemy. "Le bombardement d'Odessa." June 1854, Soho Theatre. British Library MS 52947 V.

Borough, R. B. "The Overland Journey to Constantinople." Apr. 1854, Adelphi Theatre. British Library MS 52947 C.

Cooper, F. F. "The Soldier's Wife." Nov. 1854, Strand Theatre. British Library MS 52950 W.

Donne, William Bodham. Standard Theatre, Apr. 1853. *Pop Goes the Weasel, or, the Adventures of the Weasel Family: A Farce in 1 Act.* British Library MS 52939 S.

Gliński, Ludwik, collection of postcards and marches. National Library in Warsaw. BN Mus. II.17.834.

Korczak, Katarzyna. "Pieśń ujdzie cało." Joanna Kaczyńska, personal collection, Warsaw, Poland.

Pitt, George Dibdin. *Pop Goes the Weasel, or, The Devil's Dance: A Burletta in One Act.* British Library MS 52939 V.

Simpson, J. P. "Schamyl, the Warrior Prophet." Nov. 1854, Princess's Theatre. British Library MS 52950 D.

Sparling, Marcus. No. 21a in the Roger Fenton Crimean War photograph collection. Library of Congress Prints and Photographs Division. Washington, DC.

Stirling, E. "Sebastopol from Our Own Correspondent." Oct. 1854, Marylebone Theatre. British Library MS 52949 Y.

Stocqueler, J. H. [Joachim Hayward]. "The Battle of the Alma: Grand Military Spectacle in Two Acts." British Library Manuscripts Collection MS 52950 H.

"The War in Turkey, or The Struggle For Liberty." Apr. 1854, Britannia Saloon. British Library MS 52946 KK.

III) PRINTED MUSIC

Cimze, Jānis. *Dseeşmu rohta jaunekļeem un wihreem,* vol. 2, *Lauka puķķes: Latweeşchu tautas-dseeşmas* Leipzig: G. Kreysing, 1872.

Farmer, Henry, and Edward Farmer. "The Battle of the Alma." London: Leoni Lee, 1855.

The Fashionable; Or London and Country Songster. 12 nos. London: T. Goode, 1854–. https://books.google.com/books?id=1DBYAAAAcAAJ&printsec=frontcover&source=gbs_ge_summary_r&cad=0#v=onepage&q&f=false.

Glover, Stephen. "National Music Played at the Battle of Alma, The British Grenadiers, Partant pour la Syrie, La Marseillaise, God Save the Queen, and Rule Britannia." London: Charles Jeffreys, 1854.

Glover, Stephen, and Henry Abrahams. "A Voice Was Heard in England." London: D'Almaine, 1855.

Glyde, W. D., and G. P. R. Pulman. "The Ballad of Alma." London: Jewell & Letchford, 1854.

Green, Plunkett, and Samuel Liddle. "The Kerry Recruit (An Irish Recruiting Song of the Time of the Crimean War)." London: Stainer & Bell, 1938.
Ireson, William. "The Alma March." London: Campbell, Ransford, 1854.
Jefferys, Charles. "The Heroes of Alma." London: Charles Jefferys, 1854.
Lindahl, Albert. "Alma: A Battle Piece for Piano." London: Jullien, 1854.
Mayer, J. "Grand Military Funeral March of the Battle of the Alma." London: Jewell & Letchford, 1854.
Musard, Phillippe. "Schamyl." London: Campbell, Ransford, 1854.
Osborne, Charles. "On Duty: A Tale of the Crimea." London: R. Maynard, 1893.
Shrivall, Frederick R., and H. Montagu. "Alma, 'Tis Sweet for Our Country to Die," in *A Trip to the Crimea: A New Musical Entertainment*. London, Addison Hollier & Lucas, 1857.
Smith, Charles William. "The Alma Polka." Liverpool, UK: John Smith, 1854.
Smyth, James. "The Warrior's Return from the Crimea." London: Lafleur & Sons, 1874.
Weetman, Matthew Henry. "England's Bygone Days: Song Composed at the Siege of Sebastopol by British Soldiers." London: B. Williams.
Wilson, William. "The Battle of the Alma Fantasia." London: T. E. Purday, 1854.

IV) BOOKS

Adye, John. *A Review of the Crimean War to the Winter of 1854–5*. London: Hurst and Blackett, 1860.
Aimwell, Walter [William Simonds]. *Whistler; or, The Manly Boy*. Lincoln, UK: Sheldon, Blakeman, 1858.
Aldershottana: Or, Chinks in My Hut; And Touch-and-Go Sketches from Court to Camp. London: Savill and Edwards, 1856.
Aldrich, Thomas Bailey. *Marjorie Daw: And Other Stories*. Boston: Houghton, Mifflin, 1885.
Arrighi, Cletto. *La Scapigliatura e il 6 febbraio (un dramma in famiglia)*. Milan: Francesco Sanvito, 1862.
Baddeley, John Frederick. *The Russian Conquest of the Caucasus*. New York: Longmans, Green, 1908.
Baring-Gould, Sabine. *Old Country Life*. London: Methuen, 1890.
Bazancourt, Barone de. *Cinque mesi al campo di Sebastopoli*. Trans. Giuliano Landucci. Florence: Tipografia Giacomo Terni, 1855.
Biriukov, P. I. *Biografiia*. Moscow: Posrednik, 1911.
Biriukov, P. I. *Tolstoi*. 4 vols. Moscow: Posrednik, 1906–1923.
Blackwood, Lady Alicia. *A Narrative of Personal Experiences & Impressions during a Residence on the Bosphorus throughout the Crimean War*. London: Hatchard, Piccadilly, 1881.
Braithwaite, William S., ed. *The Book of Georgian Verse*. London: G. Richards, 1909.
Buzzard, Thomas. *With the Turkish Army in the Crimea and Asia Minor: A Personal Narrative*. London: John Murray, 1915.
Canada, Executive Committee for the Paris Exhibition. *Canada at the Universal Exhibition of 1855*. Toronto: J. Lovell, 1856.
Canale, Michele Giuseppe. *Della Crimea, del suo commercio, e dei suoi dominatori, dalle origini fino ai dì nostri*. Genoa, Italy: Tipografia del Regio Istituto dei Sordomuti, 1855.
Case, John. *The Praise of Musicke*. Hildesheim, Germany: Olms, (1586) 1980.
Cevdet Pasa, Ahmed. *Tezâkir*. 12 vols. Ed. Cavid Baysun. Ankara, Turkey: Türk Tarih Kurumu Basımevi, 1953.
Chesterton, G. K. "Censoring the Newspapers." *The Collected Works of G. K. Chesterton*. Vol. 28, *The Illustrated London News, 1908–1910*. San Francisco: Ignatius, 1987.
Claridge, R. T. *A Guide along the Danube, from Vienna to Constantinople, Smyrna, Athens, the Morea, the Ionian Islands, and Venice, from Notes of a Journey Made in the Year 1836*. London: R. C. Westley, 1837.
Clausewitz, Carl von. *On War*. Trans. Michael Howard and Peter Paret. Princeton, NJ: Princeton University Press, 1970.
The Crimea: Its Towns, Inhabitants, and Social Customs; By a Lady Resident near the Alma. London: Partridge, Oakey, 1855.

Czajkowski, Michał. *Moje wspomnienia o wojnie w roku 1864*. Ed. Józef Fijałek. Warsaw: Wydawnictwo Ministerstwa Obrony Narodowej, 1962.

D'Ayala, Mariano. *I Piemontesi in Crimea: Narrazione storica*. Florence: Tipografia Barbera, 1858.

Davis, Elizabeth, and Jane Williams. *The Autobiography of Elizabeth Davis: A Balaclava Nurse*. Ed. Deirdre Beddoe. Cardiff, UK: Honno, 1987.

De Amicis, Edmondo. *Constantinople*. Trans. Caroline Tilton. New York and London: Putnam's, (1878) 1888.

Denslow, W. W., ed. and illus. *Denslow's Mother Goose: Being the Old Familiar Rhymes and Jingles of Mother Goose*. New York: McClure, Phillips, 1901.

Devens, R. M. *American Progress; or, The Great Events of the Greatest Century*. Chicago: Hugh Heron, 1883.

Dodd, George. *Pictorial History of the Russian War 1854–5–6: With Maps, Plans, and Wood Engravings*. Edinburgh and London: W. & R. Chambers, 1856.

Duberly, Frances Isabella. *Mrs Duberly's War: Journal and Letters from the Crimea, 1854–6*. Intro. and ed. Christine Kelly. Oxford: Oxford University Press, 2007.

Earp, G. Butler. *The History of the Baltic Campaign of 1854: From Documents and Other Materials Furnished by Vice-Admiral Sir C. Napier, K.C.B.* London: Richard Tentley, 1857.

Eckardt, Julius. *Die baltischen Provinzen Rußlands: Politische und culturgeschichtliche Aufsätze*. Leipzig: Duncker & Humbolt, 1868.

Edmonds, W. Lambert, ed. *The Cremorne Comic Song-Book*. London: Thomas Allman, 1853.

Edwards, John Ellis. *Random Sketches and Notes of European Travel in 1856*. New York: Harper & Brothers, 1857.

Emerson, Ralph Waldo. *Essays: First Series*. Cambridge, MA: Harvard, 1841.

Engels, Frederick. *Condition of the Working Class in England in 1844 with a Preface Written in 1892*. Trans. Florence Kelley Wischnewetzky. London: George Allen & Unwin, 1892.

Farmer, Henry George. *Military Music*. London: Max Parrish, 1950.

A Few Words Addressed to His Grace the Duke of Newcastle by an Englishman, on the Proximate Causes, the Principles, and the Conclusive Terms, of the Russian War. London: Thomas Bosworth, 1855.

Forbes-Lindsay, Charles Harcourt Ainslie. *India, Past and Present*. Vol. 2. Philadelphia: H. 4T. Coates, 1903.

Galt, Edwin. *The Camp and the Cutter; or, A Cruise to the Crimea*. London: Thomas Hodgson, 1856.

The General Stud Book, Containing Pedigrees of Each Horse from the Earliest Accounts to the Year 1892 Inclusive. Vol. 7. 3rd ed. London: J. E., J. P., & C. T. Weatherby, 1894.

Gillington, Alice E. "The Seven Whistlers." *A Victorian Anthology, 1837–1895*, ed. Edmund C. Stedman. Boston: Houghton Mifflin, 1895.

Gogol, Nikolai. *Dead Souls*. Trans. Robert Maguire. London: Penguin, 2004.

Gower, Ronald Sutherland. *Records and Reminiscences*. London: J. Murray, 1903.

Guarnieri, Luigi. *La battaglia di Adua e il popolo italiano*. Turin, Italy: Roux Frassati, 1897.

Hall, Florence Howe. *Memories Grave and Gay*. New York: Harper & Brothers, 1918.

Hamley, Edward. *The War in the Crimea*. London: Seeley, 1900.

Hamlin, Cyrus. *Among the Turks*. New York: American Tract Society, 1877.

A Handbook for Travellers in the Ionian Islands, Greece, Turkey, Asia Minor, and Constantinople. London: John Murray, 1840.

A Handbook for Travellers in Turkey: Describing Constantinople, European Turkey, Asia Minor, Armenia, and Mesopotamia. 3rd ed. London: John Murray, 1854.

Harwood, John. *Stamboul and the Sea of Gems*. London: Richard Bentley, 1852.

Henry, William. *The Baltic, 1854. The Crimea, 1854–1855*. London: Hatchards, 1884.

Hodge, Edward Cooper. *"Little Hodge": Being Extracts from the Diaries and Letters of Colonel Edward Cooper Hodge Written during the Crimean War, 1854–1856*, ed. Marquess of Anglesey, London: Leo Cooper, 1971.

Hornby, Emelia Blithynia (Lady). *Constantinople during the Crimean War*. London: Richard Bentley, 1863.

Hughes, Robert Edgar. *Two Summer Cruises with the Baltic Fleet, in 1854–1855; Being the Log of the "Pet" Yacht, 8 Tons, R.T.Y.C.* London: Smith, Elder, 1855.

Ibn ʿArabī, Muhyīddīn. *Divine Sayings: 101 Hadīth Qudsī.* Trans. Stephen Hirtenstein and Martin Notcutt. Oxford: Anqa Publishing, 2004.

Kant, Immanuel. *Von den verschiedenen Racen der Menschen.* Königsberg: H. Hartung, 1755.

Kant, Immanuel. *Anthropology from a Pragmatic Point of View.* Trans. Robert B. Louden. Cambridge: Cambridge University Press, 2006.

Kant, Immanuel. *Political Writings.* Ed. Hans Reiss. Cambridge: Cambridge University Press, 1991.

Kastner, George. *Manuel général de musique militaire à l'usage des armées françaises.* Paris: Didot Frères, 1848.

Kaudzīte, Matīss. *Kreewu walsts attihstibas gahjums jeb Kreewijas wehstures pahrşkats* (Evolution of the Russian State, or Survey of the History of Russia). Riga: Schnakenburgs, n.d.; imperial censor cleared 19 Aug. 1887.

Kinglake, Alexander William. *The Invasion of the Crimea: Its Origin, and an Account of Its Progress Down to the Death of Lord Raglan.* 8 Vols. London: Blackwood and Sons, 1888.

Koczorowski, Stanisław P. *Adam Mickiewicz et la pensée française, 1830–1835.* Paris: Gebethner et Wolff, 1929.

Kronwald, Otto [Atis Kronvalds]. *Nationale Bestrebungen.* Dorpat [Tartu, Estonia]: C. Mattiesen, 1872.

Lake, Atwell. *Kars and Our Captivity in Russia.* London: Richard Bentley, 1856.

Lenin, Vladimir Ilyich. *Remarks on Books: G. V. Plekhanov; N. G. Chernyshevsky.* St. Petersburg: Shipovnik, 1910.

Levey, William, comp. and ed. *The Victorian Ruff, Or, Pocket-Racing Companion for 1862.* Melbourne, Australia: W. Levey, 1862.

Lowe, Joseph. *A New Most Excellent Dancing Master: The Journal of Joseph Lowe's Visits to Balmoral and Windsor (1852–1860) to Teach Dance to the Family of Queen Victoria.* Ed. Allan Thomas. Stuyvesant, New York: Pendragon Press, 1992.

MacKay, Charles. *Through the Long Day; or, Memorials of a Literary Life during Half a Century.* Vol. 1. London: W. H. Allen, 1887.

Marx, Karl. "Communism, Revolution, and a Free Poland." *Political Writings,* by Karl Marx. Vol. 1. New York: Penguin, 1973.

Marx, Karl, and Friedrich Engels. *Die deutsche Ideologie.* Vol. 3 of *Marx-Engels-Werke,* 43 vols. Berlin: Institute for Marxism and Leninism, 1958.

Massachusetts, 1st corps of cadets. *Massachusetts Volunteer Militia: Standing Orders, 1890.* Boston: N. Sawyer & Son, 1890.

Mayhew, Henry. *London Labour and the London Poor: A Cyclopedia of the Conditions and Earnings of Those That Will Work, Those That Cannot Work, and Those That Will Not Work.* Vol. 3. London: Griffin, Bohn, 1851.

Mayhew, Henry, and John Barry. *The Criminal Prisons of London and Scenes of Prison Life.* London: Griffin, Bohn, 1862.

Mazzini, Giuseppe. *Scritti letterari.* 2 vols. Milan: Bietti, 1933.

Mickiewicz, W. *Żywot.* Vol. 4. Poznań, Poland: Drukarnia Dziennika Poznańskiego, 1895.

Olesnitskii, A. A. *Pesni krymskikh turok.* Moscow: Tipografiia n-v Gattsuka, 1910.

Orr, William Somerville, Richard Owen, and Robert Gordon Latham. *The Principles of Physiology.* London: Orr, 1854.

Pagani, Carlo Osvaldo. *Alfonso La Marmora: Pagine nuove; Ricordi storici della campagna di Crimea.* Rome: Carlo Voghera, 1880.

Paget, George. *The Light Cavalry Brigade in the Crimea: Extracts from the Letters and Journal of the Late Gen. Lord George Paget, K.C.B., during the Crimean War.* London: John Murray, 1881.

Partridge, Eric. *A Dictionary of the Underworld: British and American.* New York: Bonanza, (1949) 1961.

Pictures from the Battle Fields: By "The Roving Englishman." [Murray, Eustace Clare Grenville]. 2nd ed. London: Routledge, 1855.

Planché, James. "The New Haymarket Spring Meeting." *The Extravaganzas of J. R. Planché, Esq., 1825–1871.* Ed. T. F. Dillon Croker and Stephen Tucker. Vol. 5. London: Samuel French, 1879.

Powell, Harry. *Reflections of a Young Soldier during the Crimean War.* Oxford: Upston and Doe, 1876.

Przyjemski, Juliusz Rawicz. *Sketches of the Polish Mind.* London: William and Frederick G. Cash, 1857.

Pushkin, A. S. "Dubrovsky" ([1832], published posthumously 1841). *The Complete Prose Tales.*Trans. Gillon R. Aitken. New York: W. W. Norton, 1966.

Pushkin, A. S. *Evgenii Onegin.* St. Petersburg: Smirdin, 1833.

Qarakhi, Muhammad Tahir al-. "The Shining of Daghestani Swords in Certain Campaigns of Shamil (Selected Passages)." Trans. Ernest Tucker and Thomas Sanders. *Russian-Muslim Confrontation in the Caucasus: Alternative Visions of the Conflict between Imam Shamil and the Russians, 1830–1859.* Ed. Thomas Sanders, Ernest Tucker, and Gary Hamburg. London: Routledge, 2004.

Rankine, William J. Macquorn. *Songs and Fables.* Glasgow: James Maclehose, 1874.

Ricci, Agostino. *In Crimea.* Turin, Italy: Roux, Frassati, 1896.

Ricordo pittorico militare della spedizione sarda in Oriente, Pubblicato d'ordine del ministero di guerra per conto dello stato maggiore. Turin, Italy: Ministero di guerra, 1857.

Rīgas Latviešu Biedrība sešos gadu desmitos 1868–1928. Riga: Rīgas Latviešu Biedrība, 1928.

Rink, Friedrich Theodor. *Immanuel Kants physische Geographie.* Mainz and Hamburg: Gottfried Vollmer, 1804.

Robinson, Frederick. *Diary of the Crimean War.* London: Richard Bentley, 1856.

Ruskin, John. *Fors Clavigera: Letters to the Workmen and Labourers of Great Britain.* Orpington, UK: George Allen, 1871.

Russell, William Howard. *The British Expedition to the Crimea.* London and New York: Routledge and Co., 1858.

Russell, William Howard. *The British Expedition to the Crimea*: New and Revised Edition. London and New York: Routledge and Sons, 1877.

Russell, William Howard. *Russell's Despatches from the Crimea, 1854–1856,* ed. Nicolas Bentley. London: Andre Deutsch, 1966.

Ryan, George E. *Our Heroes of the Crimea: Being Biographical Sketches of Our Military Officers.* London: Geo. Routledge, 1855.

Sewell, William M. *A Year's Sermons to Boys, Preached in the Chapel of St. Peter's College, Radley.* Vol. 2. Oxford: A. Wassall, 1854.

Shklovskii, V. B. *O teorii prozy.* Moscow: Krug, (1919) 1925.

Skujenieks, Marģeris. *Latvija: Zeme un eedzīvotaji.* 2nd ed. Riga: Valsts statistiskās pārvalde, 1922.

Smith, Adam. *Theory of Moral Sentiments.* Ed. D. D. Raphael and A. L. Macfie. Oxford: Clarendon, (1759) 1976.

Smith, Albert. *A Month at Constantinople.* 2nd ed. London: David Bogue, 1850.

Smith, Hermann. *The Making of Sound in the Organ and in the Orchestra: An Analysis of the Work of the Air in the Speaking Organ Pipe of the Various Constant Types, and an Exposition of the Theory of the Air-Stream-Reed, Based upon the Discovery of the Tone of the Air, by Means of Displacement Rods.* New York: C. Scribner's Sons, 1911.

Smithies, R. H. Raymond. *Historical Records of the 40th (2nd Somersetshire) Regiment, Now 1st Battalion the Prince of Wales's Volunteers [. . .].* Devonport, UK: A. H. Swiss, 1894.

Stendhal [Marie-Henri Beyle]. *The Charterhouse of Parma.* Trans. Margaret Mauldon. Oxford: Oxford University Press, 1997.

Stevens, T. S. "Notes on an Intelligent Parrot." *Journal of the Trenton Natural History Society* 1 (1886–88): 347–55.

Stocqueler, Joachim Hayward. *The British Soldier: An Anecdotal History of the British Army from Its Earliest Formation to the Present Time*. London: Orr, 1856.

Strachey, Lytton. *Eminent Victorians: Cardinal Manning, Florence Nightingale, Dr. Arnold, General Gordon*. London: Chatto & Windus, 1918.

Tait, P. G. "Memoir of Rankine." *Miscellaneous Scientific Papers by W. J. M. Rankine: From the Transactions and Proceedings of the Royal and Other Scientific and Philosophical Societies. . . .* Ed. William J. Millar. London: C. Griffin, 1881.

Tinsley, William. *Random Recollections of an Old Publisher*. Vol. 2. London: Simpkin, Marshall, Hamilton, Kent, 1900.

Tolstoi, L. N. *Biblioteka L'va Nikolaevicha Tolstogo v Iasnoi Poliane: Knigi na russkom yazyke*. Vol. 1. Moscow: Izdatel'stvo Sovetskaia Rossiia, 1958.

Tolstoy, L. N. *Childhood, Boyhood, Youth*. Trans. Judson Rosengrant. London: Penguin, 2012.

Tolstoy, L. N. "Hadji Murat." *The Cossack and Other Stories*. Trans. David McDuff and Paul Foote. London: Penguin, 2007.

Tolstoi, L. N. *Polnoe sobranie sochinenii*. 90 vols. Moscow: Gosudarstvennoe izdatel'stvo "Khudozhestvennaia literatura," 1928–58.

Tolstoy, L. N. *Sevastopol*. Trans. Frank D. Millet. New York: Harper & Brothers, 1887.

Tolstoy, L. N. *Sevastopol*. Trans. Isabel F. Hapgood. New York: Thomas Y. Crowell, 1888.

Tolstoy, L. N. *The Sevastopol Sketches*. Trans. David McDuff. London: Penguin, 1986.

Tunis, W. E. *The Shilling Song Book: A Collection of 175 of the Most Favorite National, Patriotic, Sentimental, and Comic Ballads of the Day*. Niagara Falls, Canada: W. E. Tunis, 1860.

Twain, Mark. *The Innocents Abroad; or, The New Pilgrims' Progress; Being Some Account of the Steamship Quaker City's Pleasure Excursion to Europe and the Holy Land with Descriptions of Countries, Nations, Incidents, and Adventures as They Appeared to the Author*. Hartford, CT: H. H. Bancroft, 1869.

Whistling Horace. Glasgow: J. S. Marr, 1878.

Williams, Jane. *The Autobiography of Elizabeth Davis: A Balaclava Nurse, Daughter of Dafydd Cadwaladyr*. 2 vols. London: Hurst and Blackett, 1857.

Wolski, Włodzimierz. "Marsz Sybirski." *Promyki: Nowe Poezje Liryczne*. Brussels: J.-H. Dehoue, 1869.

Wolski, Włodzimierz. *Śpiewy Powstańcze*. 2nd ed. Paris: Drukarnia Polska, 1863.

Woolf, Leonard. *Essays on Literature, History, Politics, Etc.* London: Hogarth Press, 1927.

Woolf, Virginia. "Heard on the Downs: The Genesis of Myth". *The Essays of Virginia Woolf*. Ed. Andrew McNeillie. Vol. 3. London: Hogarth Press, (1916) 1991.

Woolf, Virginia. *To the Lighthouse*. San Diego: Harcourt, 1927.

Wratislaw, Albert H. *Sixty Folk-Tales from Exclusively Slavonic Sources*. London: E. Stock, 1889.

Zeiferts, Teodors. *Brīvzemnieks: Tautas darbinieka mūžs un laikmeta aina*, Vol. 1. Riga: R. L. B. Brīvzemnieka rakstu komisija, 1929.

Secondary

Abu-Manneh, Butrus. "The Islamic Roots of the Gülhane Rescript." *Die Welt des Islams* 34 (1994): 173–203.

Akbulut, Uğur. "Suriye'ye İlk Telgraf Hatlarının Çekilmesi." *History Studies: International Journal of History* 2 (2010): 1–11.

Akyıldız, Ali. "Tanzimat." *Türkiye Diyanet Vakfı İslâm Ansiklopedisi* 40 (2011): 1–10.

Aliev, Fevzi. *Antologiia krymskoi narodnoi muzyki*. Ak'mesdzhit [Simferopol], Crimea: Krymuchpedgiz, 2001.

Alston, Charlotte. *Tolstoy and His Disciples: The History of a Radical International Movement.* London: I. B. Tauris, 2014.
Ana María Ochoa Gautier. *Aurality: Listening and Knowledge in Nineteenth-Century Columbia.* Durham, NC, and London: Duke University Press, 2014.
Ana María Ochoa Gautier. "Silence." *Keywords in Sound.* Ed. David Novak and Matt Sakakeeny. Durham, NC, and London: Duke University Press, 2015.
Anderson, Benedict. *Imagined Communities: Reflections on the Origins and Spread of Nationalism.* London: Verso, 1991.
Anderson, Benedict. *The Spectre of Comparisons: Nationalism, Southeast Asia and the World.* London: Verso, 1998.
Aracı, Emre. "A Levantine Life: Giuseppe Donizetti at the Ottoman Court." *The Musical Times* 1880 (2002): 49–56.
Aracı, Emre. *Naum tiyatrosu: 19. yüzyıl İstanbul'unun İtalyan operası.* Istanbul: Yapı Kredi Yayınları, 2010.
Aravamudan, Srinivas. "Introduction: Perpetual War." *Periodical of the Modern Language Association* 124/5 (2009): 1505–14.
Arcangeli, Massimo. *La Scapigliatura poetica milanese e la poesia italiana fra Otto e Novecento: Capitoli di lingua e di stile.* Rome: Aracne, 2003.
Arnold, Guy. *Historical Dictionary of the Crimean War.* Lanham, MD: Scarecrow, 2002.
Asad, Talal. "The Idea of an Anthropology of Islam." *Qui parle* 17/2 (2009): 1–30.
Attali, Jacques. *Noise: The Political Economy of Music.* Trans. Brian Massumi. Minneapolis, MN, and London, UK: University of Minnesota Press, 1985.
Austin, John Langshaw. *How to Do Things with Words.* 2nd ed. Ed. J. O. Urmson and Marina Sbisà. Cambridge, MA: Harvard University Press, 1975.
Badem, Candan. *The Ottoman Crimean War, 1853–1856.* Leiden, The Netherlands: Brill, 2010.
Badem, Candan. "Sadyk Pasha in the Light of the Ottoman Archives." *The Crimean War 1853–1856: Colonial Skirmish or Rehearsal for World War? Empires, Nations, and Individuals.* Ed. Jerzy W. Borejsza. Warsaw: Wydawnictwo Neriton, 2011.
Baha, Ahmed. *Telegrafçılıkta Ana Dilimiz ve Mustafa Efendi.* Istanbul: Güneş Matbaası, 1933.
Baltacan, Mehmet. "The Relationship between Turkish and Armenia Regarding the Ottoman Empire and Contributions of Armenian Artists to the Turkish Opera." *International Journal of Social, Business, Psychological, Human Science and Engineering* 8/5 (2014): 1229–35.
Barker, J. *The Vainglorious War, 1854–56.* London: Weidenfeld and Nicolson, 1970.
Barnett, Michael. *Empire of Humanity: A History of Humanitarianism.* Ithaca, NY: Cornell University Press, 2011.
Barthorp, Michael. *The British Army on Campaign (2): The Crimea, 1854–56.* Oxford: Osprey, 1987.
Baudrillard, Jean. *For a Critique of the Political Economy of the Sign.* Trans. Charles Levin. St. Louis, MO: Telos, (1972) 1981.
Baudrillard, Jean. *Simulacra and Simulation.* Trans. Sheila Faria Glaser. Ann Arbor: University of Michigan Press, 1994.
Baugh, Daniel. *The Global Seven Years War, 1754–1763: Britain and France in a Great Power Contest.* Harlow, UK: Longman, 2011.
Bazin, André. *Qu'est-ce que le cinéma? Cinema et sociologie.* Vol. 3. Paris: Cerf, 1961.
Beck, Ulrich. *World at Risk.* Cambridge: Polity, 2007.
Beer, Gillian. *Open Fields.* Oxford: Oxford University Press, 1996.
Bektaş, Yakup. "The Sultan's Messenger: Cultural Constructions of Ottoman Telegraphy, 1847–1880." *Technology and Culture* 41/4 (2000): 669–96.
Belitser, Natalya. "'Indigenous Status' for the Crimean Tatars in Ukraine: A History of a Political Debate." *"Fuzzy Statehood" and European Integration in Central and Eastern Europe.* Ed. Judy Batt. University of Birmingham, UK: Economic and Social Research Council, 2002.
Belknap, Robert L. "Survey of Russian Journals, 1840–1880." *Literary Journals in Imperial Russia.* Ed. Deborah A. Martinsen. Cambridge: Cambridge University Press, 1997.

Benyon, P. "Late 18th, 19th, and Early 20th Century Naval and Naval Social History." <http://www.pbenyon.plus.com/18-1900/W/05203.html>.

Bergeron, Katherine. *Voice Lessons: French Mélodie in the Belle Epoque*. New York: Oxford University Press, 2009.

Berkes, Niyazi. *The Development of Secularism in Turkey*. Montreal: McGill University Press, 1964.

Bettini, Maurizio. *Women and Weasels: Mythologies of Birth in Ancient Greece and Rome*. Chicago: University of Chicago, 2013.

Bhabha, Homi. *The Location of Culture*. London: Routledge, 1994.

Biehl, João, Byron Good, and Arthur Kleinman, eds. *Subjectivity*. Berkeley: University of California Press, 2007.

Biersack, John, and Shannon O'Lear. "The Geopolitics of Russia's Annexation of Crimea: Narratives, Identity, Silences, and Energy." *Eurasian Geography and Economics* 55/3 (2014): 247–69.

Bilenky, Serhiy. *Romantic Nationalism in Eastern Europe: Russian, Polish, and Ukrainian Political Imaginations*. Stanford, CA: Stanford University Press, 2012.

Bill, Stanley. "Seeking the Authentic: Polish Culture and the Nature of Postcolonial Theory." *Nonsite* 12 (2014): published online. <http://nonsite.org/article/seeking-the-authentic-polish-culture-and-the-nature-of-postcolonial-theory>.

Biskupski, M. B. "The Militarization of the Discourse of Polish Politics and the Legion Movement of the First World War." *Armies in Exile*. Ed. David Stefancic. New York: Columbia University Press and East European Monographs, 2005.

Bobe, Mendels. *Ebreji Latvijā*, trans. Rita Bogdanova, Eva Sinkēviča, and Ilmārs Zvirgzds. Riga: Shamir, 2006.

Bogucka, Maria. *The Lost World of the "Sarmatians": Custom as the Regulator of Polish Social Life in Early Modern Times*. Warsaw: Polish Academy of Sciences, 1996.

Born, Georgina. "On Musical Mediation: Ontology, Technology and Creativity." *Twentieth-Century Music* 2/1 (2005): 7–36.

Bostridge, Mark. *Florence Nightingale: The Woman and Her Legend*. London: Penguin, 2009.

Bourke, Richard. *Empire and Revolution: The Political Life of Edmund Burke*. Princeton, NJ: Princeton University Press, 2015.

Boym, Svetlana. *The Future of Nostalgia*. New York: Basic Books, 2001.

Bratton, Jacky. "Theatre of War: The Crimea on the London Stage." *Performance and Politics in Popular Drama*. Ed. David Bradby, Louis James, and Bernard Sharratt. Cambridge: Cambridge University Press, 1980.

Bruyn, Frans de and Shaun Regan. *The Culture of the Seven Years' War: Empire, Identity, and the Arts in the Eighteenth-Century Atlantic World*. Toronto: University of Toronto Press, 2014.

Burrowes, John. *Benny: The Life and Times of a Fighting Legend*. Edinburgh: Mainstream, 2011.

Burton, Annette. *Dwelling in the Archive: Women Writing House and Home in Late Colonial India*. Oxford: Oxford University Press, 2003.

Butler, Shane. *The Ancient Phonograph*. New York: Zone, 2015.

Buttery, David. *Messenger of Death: Captain Nolan and the Charge of the Light Brigade*. Philadelphia: Pen & Sword, 2008.

Caminiti, Alberto. *La guerra di Crimea, 1853–1856*. Genoa, Italy: Edizioni Liberodiscrivere, 2013.

Carli, Alberto. *Anatomie scapigliate: L'estetica della morte tra letteratura, arte e scienza*. Novara, Italy: Interlinea, 2004.

Certeau, Michel de. *The Writing of History*. Trans. Tom Conley (1988). New York: Columbia University Press, 1975.

Chanaud, Robert C. "Aerodynamic Whistles." *Scientific American* 222/1 (1970): 40–47.

Chandler, James. *An Archaeology of Sympathy: The Sentimental Mode in Literature and Cinema*. Chicago: University of Chicago Press, 2013.

Channell, David F. "The Harmony of Theory and Practice: The Engineering Science of W. J. M. Rankine." *Technology and Culture* 23/1 (1982): 39–52.

Chernyshevsky, N. G. "L. N. Tolstoy's *Childhood* and *Boyhood* and *Military Tales.*" *Belinsky, Chernyshevsky and Dobrolyubov: Selected Criticism.* Ed. Ralph Matlaw. New York: E. P. Dutton, 1962.

Chertkov, V. G., and N. N. Gusev, eds. *Tolstoi o Tolstom: Novye materialy*. 4 vols. Moscow: Tolstoy Museum, 1924–28.

Chesney, Kellow. *Crimean War Reader*. London: Severn House, 1975.

Chion, Michel. *Audio-Vision: Sound on Screen*. Trans. Claudia Gorbman. New York: Columbia University Press, 1994.

Clifford, James. "Diasporas." *Cultural Anthropology* 9/3 (1994): 302–38.

Cohen, Margaret, and Carolyn Diever, eds. *The Literary Channel: The Inter-National Invention of the Novel.* Princeton, NJ: Princeton University Press, 2002.

Colley, Linda. *Britons: Forging the Nation, 1707–1837.* New Haven, CT: Yale University Press, 2005.

Comment, Bernard. *The Panorama*. London: Reaktion Books, 1999.

Conboy, Martin. *Journalism: A Critical History*. London: Sage, 2004.

Constable, Marianne. *Just Silences: The Limits and Possibilities of Modern Law*. Princeton, NJ: Princeton University Press, 2005.

Conway, David. *Jewry in Music: Entry to the Profession from the Enlightenment to Richard Wagner.* Cambridge: Cambridge University Press, 2012.

Conway, Stephen. *A Short History of the American Revolutionary War.* London: I. B. Tauris, 2013.

Cooke, Brian. *The Grand Crimean Central Railway: The Story of the Railway Built by the British in the Crimea during the War of 1854–1856*. Knutsford: Cavalier House, 1997.

Corbin, Alain. *Village Bells: Sound and Meaning in the Nineteenth-Century French Countryside.* Trans. Martin Thom. London: Macmillan, (1995) 1998.

Cottrell, Stephen. *The Saxophone*. New Haven, CT: Yale University Press, 2012.

Coward, T. A. *The Birds of the British Isles and Their Eggs: Second Series Comprising the Families Anatidae to Phasianidae*. London: Frederick Warne, (1920) 1950.

Cusick, Suzanne G. "Music as Torture/Music as Weapon." *Trans: Revista Transcultural de Música* 10 (2006). <http://www.sibetrans.com/trans/articulo/152/music-as-torture-music-as-weapon>. Accessed 4 Apr. 2016.

Cusick, Suzanne G. "Musicology, Torture, Repair." *Radical Musicology* 3 (2008): published online. http://www.radical-musicology.org.uk.

Cusick, Suzanne. "Towards an Acoustemology of Detention in the 'Global War on Terror.'" *Music, Sound and Space: Transformations of Public and Private Experience.* Ed. Georgina Born. Cambridge: Cambridge University Press, 2012.

Dahlén, Ashk. *Islamic Law, Epistemology and Modernity: Legal Philosophy in Contemporary Iran.* London: Routledge, 2004.

Dainotto, Roberto M. *Europe (in Theory)*. Durham, NC: Duke University Press, 2007.

Dalgarno, Emily. "A British *War and Peace?* Virginia Woolf Reads Tolstoy." *Modern Fiction Studies* 50/1 (2004): 129–50.

Daughtry, J. Martin *Listening to War: Sound, Music, Trauma, and Survival in Wartime Iraq.* Oxford: Oxford University Press, 2015.

Daughtry, J. Martin "Thanatosonics: Ontologies of Acoustic Violence." *Social Text* 32/2 (2014): 25–51.

Davies, James Q. "A Musical Souvenir: London in 1829." PhD thesis, University of Cambridge, 2005.

Davies, James Q., and Ellen Lockhart, eds. *Sound Knowledge: Music and Science in London, 1789–1851*. Chicago: Chicago University Press, 2016.

Davison, Roderic. *Essays in Ottoman and Turkish History, 1774–1923: The Impact of the West.* Austin: University of Texas Press, 1990.

Davison, Roderic. *Reform in the Ottoman Empire, 1856–1876*. Princeton, NJ: Princeton University Press, 1963.

Del Principe, David. *Rebellion, Death and Aesthetics in Italy: The Demons of Scapigliatura.* Madison, WI: Fairleigh Dickinson University Press, 1996.

DeLanda, Manuel. *A New Philosophy of Society.* London: Continuum, 2006.

Deleuze, Gilles. *The Logic of Sense*. Trans. Constantin V. Boundas. London: Continuum, (1969) 2003.

Deleuze, Gilles, and Felix Guattari. *A Thousand Plateaus: Capitalism and Schizophrenia*. Trans. Brian Massumi. Minneapolis: University of Minneapolis Press, 1987.

Derrida, Jacques. *Archive Fever: A Freudian Impression*. Trans. Eric Prenowitz. Chicago: University of Chicago Press, 1996.

Derrida, Jacques. *De la grammatologie*. Paris: Les Éditions de Minuit, 1967.

Dixon, Thomas. *From Passions to Emotions: The Creation of a Secular Psychological Category*. Cambridge: Cambridge University Press, 2002.

Dolar, Mladen. *A Voice and Nothing More*. Cambridge, MA: MIT Press, 2006.

Dutton, Roy. *Forgotten Heroes: The Charge of the Light Brigade*. Oxton, UK: Infodial, 2007.

Dworkin, Ronald. *Law's Empire*. Cambridge, MA: Harvard University Press, 1986.

Edgerton, David. *The Shock of the Old: Technology and Global History since 1900*. London: Profile, 2006.

Ehrlich, Cyril. *The Piano: A History*. Oxford: Clarendon, 1990.

Eikhenbaum, Boris. *The Young Tolstoy*. Trans. Gary Kern. Ann Arbor, MI: Ardis Press, 1972.

Eile, Stanisław. *Literature and Nationalism in Partitioned Poland, 1795–1918*. New York: St. Martin's, 2000.

Eisenstein, Sergei. *Montazh*. Moscow: Muzei kino, 2000.

Elden, Stuart, and Eduardo Mendieta, eds. *Reading Kant's Geography*. New York: State University of New York Press, 2011.

Ellingson, Ter. *The Myth of the Noble Savage*. Berkeley and Los Angeles: University of California Press, 2001.

Eppinger, Monica. "Silencing and Backtalk: Scenes from the Crimean Occupation." *Anthropoliteia*, published online on 16 Mar. 2014. <http://anthropoliteia.net/>. Accessed 10 Feb. 2016.

Esch, Deborah. *In the Event: Reading Journalism, Reading Theory*. Stanford, CA: Stanford University Press, 1999.

Esposito, Roberto. *Communitas: The Origin and Destiny of Community*. Trans. Timothy Campbell. Stanford, CA: Stanford University Press, 2009.

Esse, Melina. "Encountering the *improvvisatrice* in Italian Opera." *Journal of the American Musicological Society* 66/3 (Fall 2013): 709–70.

Estreicher, Stefan K. *Wine from Neolithic Times to the 21st Century*. New York: Algora, 2006.

Fabre, Cécile. *Cosmopolitan War*. Oxford: Oxford University Press, 2012.

Fadel, Khaled Abou al-. "The Islamic Legal Tradition: A Comparative Law Perspective." *The Cambridge Companion to Comparative Law*. Ed. Mauro Bussani and Ugo Mattei. Cambridge: Cambridge University Press, 2012.

Farinelli, Giuseppe. *La Scapigliatura: Profilo storico, protagonisti, documenti*. Rome: Carrocci, 2003.

Favret, Mary A. *War at a Distance: Romanticism and the Making of Modern Wartime*. Princeton, NJ, and Oxford: Princeton University Press, 2010.

Feld, Steven, and Keith H. Basso, eds. *Senses of Place*. Santa Fe, NM: School of American Research Press, 1996.

Feldman, Martha. "The Interstitial Voice: An Opening." *Journal of the American Musicological Society* 68/3 (2015): 653–59.

Figes, Orlando. *The Crimean War: A History*. New York: Picador, 2010.

Findley, Carter Vaughn. *Bureaucratic Reform in the Ottoman Empire: The Sublime Porte, 1789–1922*. Princeton, NJ: Princeton University Press, 1980.

Finger, Stanley, and Meredith P. Hustwit. "Five Early Accounts of Phantom Limb in Context: Paré, Descartes, Lemos, Bell, and Mitchell." *Neurosurgery* 52/3 (2003): 675–86.

Finnin, Rory. "Captive Turks: Crimean Tatars in Pan-Turkist Literature." *Middle Eastern Studies* 50/2 (2014): 291–308.

Finson, Jon W. *The Voices That Are Gone: Themes in Nineteenth-Century American Popular Song*. Oxford: Oxford University Press, 1997.

Fisher, Alan. *The Crimean Tatars*. Stanford, CA: Hoover Institution Press, 1978.

Ford, Mark. *London: A History in Verse.* Cambridge, MA: Belknap Press of Harvard University Press, 2012.

Foster, John Burt, Jr. *Transnational Tolstoy: Between the West and the World.* London: Bloomsbury, 2013.

Foucault, Michel. *The Archaeology of Knowledge.* Trans. A. M. Sheridan Smith. New York: Tavistock, 1986.

Foucault, Michel. *Language, Counter-Memory, Practice: Selected Essays and Interviews.* Ed. Donald F. Bouchard. Trans. Donald F. Bouchard and Sherry Simon. Ithaca, NY: Cornell University Press, 1977.

Fox, Aaron. *Real Country: Music and Language in Working-Class Culture.* Durham, NC: Duke University Press, 2004.

Frank, Joseph. *Dostoevsky: The Stir of Liberation, 1860–1865.* Vol. 3. Princeton, NJ: Princeton University Press, 1988.

Franklin, Robert R. "Tsar Alexander II and President Abraham Lincoln: Unlikely Bedfellows?" *University of Hawai'i at Hilo: Hawaii Community College* 10 (2012): 74–84.

Frary, Lucien J., and Mara Kozelsky, eds. *Russian-Ottoman Borderlands: The Eastern Question Reconsidered.* Madison: University of Wisconsin Press, 2014.

Freeden, Michael. "Emotions, Ideology and Politics." *Journal of Political Ideologies* 18/1 (2013): 1–10.

Frolova-Walker, Marina. *Russian Music and Nationalism: From Glinka to Stalin.* New Haven, CT, and London: Yale University Press, 2007.

Furneaux, Holly. *Men of Feeling: Emotion, Touch and Masculinity in the Crimean War.* Oxford: Oxford University Press, 2016.

Fussell, Paul. *The Great War and Modern Memory.* Oxford: Oxford University Press, 1975.

Gammer, Moshe. *Muslim Resistance to the Tsar: Shamil and the Conquest of Chechnia and Daghestan.* London: Frank Cass, 1994.

Geczy, Adam. *Fashion and Orientalism: Dress, Textiles and Culture from the 17th to the 21st Century.* London and New York: Bloomsbury, 2013.

Ghidetti, Enrico. *Tarchetti e la Scapigliatura lombarda.* Naples, Italy: Libreria scientifica editrice, 1968.

Gitelman, Lisa. *Always Already New: Media, History and the Data of Culture.* Cambridge, MA: MIT Press, 2008.

Gitelman, Lisa. *Paper Knowledge: Towards a Media History of Documents.* Durham, NC: Duke University Press, 2014.

Gitelman, Lisa. *Scripts, Grooves and Writing Machines: Representing Technology in the Edison Era.* Stanford, CA: Stanford University Press, 1999.

Gold, Barri J. "The Consolation of Physics: Tennyson's Thermodynamic Solution," *PMLA* 117/3 (2002): 449–64.

Gold, Barri J. *Thermopoetics: Energy in Victorian Literature and Science.* Cambridge, MA: MIT Press, 2010.

Goldberg, Halina. "'Remembering That Tale of Grief': The Prophetic Voice in Chopin's Music." *The Age of Chopin: Interdisciplinary Inquiries.* Ed. Halina Goldberg. Bloomington: Indiana University Press, 2004.

Golder, Frank A. "Russian-American Relations during the Crimean War." *American Historical Review* 31 (1925–26): 462–76.

Gomolicki, Leon. *Dziennik pobytu Adama Mickiewicza w Rosji, 1824–1829.* Warsaw: Książka i Wiedza, 1949.

Goodman, Steve. *Sonic Warfare: Sound, Affect, and the Ecology of Fear.* Cambridge, MA: MIT Press, 2009.

Gooley, Dana. "Saving Improvisation: Hummel and the Free Fantasia in the Early Nineteenth Century." *The Oxford Handbook of Critical Improvisation Studies.* Vol. 2. Ed. George E. Lewis and Benjamin Piekut. Oxford: Oxford University Press, 2016.

Gözübüyük, Şeref, and Suna Kili. *Türk Anayasa Metinleri: Tanzimattan Bugüne Kadar.* Ankara, Turkey: Ankara Üniversitesi Siyasal Bilgiler Fakültesi, 1957.

Grey, Elizabeth. *The Noise of Drums and Trumpets: W. H. Russell Reports from the Crimea*. London: Longmans, 1971.

Guha, Ranajit. *Elementary Aspects of Peasant Insurgency in Colonial India*. Durham, NC: Duke University Press, 1999.

Gusejnova, Dina, ed. *Cosmopolitanism in Conflict. Imperial Encounters from the Seven Years' War to the Cold War*. London: Palgrave Macmillan, 2018.

Halberstam, Judith. *The Queer Art of Failure*. Durham, NC: Duke University Press, 2011.

Hallaq, Wael. *Sharīʿa: Theory, Practice, Transformations*. Cambridge: Cambridge University Press, 2009.

Hamilton, John. *Music, Madness, and the Unworking of Language*. New York: Columbia University Press, 2008.

Hanioğlu, M. Şükrü. *A Brief History of the Late Ottoman Empire*. Princeton, NJ: Princeton University Press, 2008.

Harari, Yuval Noah. *The Ultimate Experience: Battlefield Revelations and the Making of Modern War Culture, 1450–2000*. Basingstoke, UK: Palgrave Macmillan, 2008.

Hardt, Alexander, and Nikolaj Plotnikov, eds. *Diskurse der Personalität: Die Begriffsgeschichte der "Person" aus deutscher und russischer Perspektive*. Paderborn, Germany: Wilhelm Fink, 2008.

Haringer, Andrew. "Hunt, Military, and Pastoral Topics." *The Oxford Handbook of Topic Theory*. Ed. Danuta Mirka. Oxford and New York: Oxford University Press, 2014.

Harris, Philip Rowland. *A History of the British Museum Library, 1753–1973*. London: British Library, 1998.

Harrowven, Jean. *Origins of Rhymes, Songs and Sayings*. London: Kaye & Ward, 1977.

Harvey, David. "Cosmopolitanism and the Banality of Geographical Evils." *Public Culture* 12/2 (2000): 529–64.

Helton, Laura, Justin Leroy, Max A. Mishler, Samantha Seeley, and Shauna Sweeney. "The Question of Recovery: An Introduction." *Social Text 125* 33/4 (2015): 1–18.

Henrywood, R. H., and A. Agarwal. "The Aeroacoustics of a Steam Kettle." *Physics of Fluids* 25 (2013): published online. <https://doi.org/10.1063/1.4821782>.

Herbert, Trevor, and Helen Barlow. *Music and the British Military in the Long Nineteenth Century*. Oxford: Oxford University Press, 2013.

Hewitt, Martin, and Rachell Cowgill, eds. *Victorian Soundscapes Revisited*. Horsforth, UK: Trinity and All Saints/Leeds Centre for Victorian Studies, 2007.

Hirschkind, Charles. *The Ethical Soundscape: Cassette Sermons and Islamic Counterpublics*. New York: Columbia University Press, 2006.

Hofmayr, Isabel. *Gandhi's Printing Press: Experiments in Slow Reading*. Cambridge, MA: Harvard University Press, 2014.

Holley, Rose. "How Good Can It Get? Analysing and Improving OCR Accuracy in Large Scale Historical Newspaper Digitisation Programs." *D-Lib Magazine* 15/4 (2009): published online. <http://www.dlib.org/dlib/march09/holley/03holley.html>.

Hont, Istvan. *Politics in Commercial Society: Jean-Jacques Rousseau and Adam Smith*. Ed. Michael Sonenscher and Béla Kapossy. Cambridge, MA: Harvard University Press, 2015.

Hoogenboom, Hilde. "Sentimental Novels and Pushkin: European Literary Markets and Russian Readers." *The Slavic Review* 74/3 (Fall 2015): 553–74.

Hoskins, Andrew, and Ben O'Loughlin. *War and Media: The Emergence of Diffused War*. Cambridge: Polity, 2010.

Howey, Allan W. "The Widow-Makers." *Civil War Times Magazine* (Oct. 1999), 46.

Hurewitz, J. C. *Diplomacy in the Near and Middle East: A Documentary Record: 1535–1914*. Vol. 1. Princeton, NJ: Van Nostrand Company, 1956.

Hussey, Mark, ed. *Virginia Woolf and War: Fiction, Reality, and Myth*. Syracuse, NY: Syracuse University Press, 1991.

Hutchison, Keith. "W. J. M. Rankine and the Rise of Thermodynamics." *British Journal for the History of Science* 14/1 (1981): 1–26.

Istominam, E. G., and M. A. Poliakov, eds. *Istochniki po istorii russkoi usadebnoi kul'tury*. Yasnaya Polyana, Russia: RGGU, 1997.

Jack, Albert. Illus. Lara Carlini. *Pop Goes the Weasel: The Secret Meanings of Nursery Rhymes*. New York: Perigee, 2009.

Jackson, Sherman A. *Islamic Law and the State: The Constitutional Jurisprudence of Shihāb Al-Dīn Al-Qarāfī*. Leiden, The Netherlands, New York, and Cologne: Brill, 1996.

James, Brian. "Allies in Disarray: The Messy End of the Crimean War." *History Today* 58/3 (2008): 24–31.

Janion, Maria. "Mickiewicz's Jewish Legion." *Hero, Conspiracy, and Death: The Jewish Lectures*. Trans. Alex Shannon. Frankfurt am Main: Peter Lang, 2014.

Janion, Maria. *Płacz Generała*. Warsaw: Wydawnictwo Sic!, 1998.

Jemilev, Mustafa. *A History of the Crimean Tatar National Liberation Movement: A Sociopolitical Perspective*. Trans. Sanoma Lee Kellogg and Inna Pidlusk. Ed. Edye Muslymova. Simferopol, Crimea: Odjak, 2005.

Johnson, James H. *Listening in Paris: A Cultural History*. Berkeley: University of California Press, 1995.

Jones, Edgar, and Simon Wessely. *Shell Shock to PTSD: Military Psychiatry from 1900 to the Gulf War*. New York: Psychology Press, 2005.

Kaldor, Mary. *New and Old Wars*. Cambridge: Polity, 2006.

Kalembka, Sławomir. *Wielka emigracja, 1832–1863*. Warsaw: Wiedza Powszechna, 1971.

Kallberg, Jeffrey. "Hearing Poland: Chopin and Nationalism." *Nineteenth-Century Piano Music*. Ed. R. Larry Todd. New York: Schirmer, 1990.

Kasekamp, Andres. *A History of the Baltic States*. New York: Palgrave Macmillan, 2010.

Kapchan, Deborah, ed. *Theorizing Sound Writing*. Middletown, CT: Wesleyan University Press, 2017.

Keller, Ulrich. *The Ultimate Spectacle: A Visual History of the Crimean War*. Australia: Gordon & Breach, 2001.

Kent, Neil. *Crimea: A History*. London: Hurst, 2016.

Kırımlı, Hakan. *National Movements and National Identity among the Crimean Tatars (1905–1916)*. Ottoman Empire and Its Heritage 7. Leiden, The Netherlands: Brill, 1996.

Kittler, Friedrich. *Discourse Networks 1800/1900*. Trans. Michael Metteer and Chris Cullens. Stanford, CA: Stanford University Press, 1990.

Kittler, Friedrich. *Gramophone, Film, Typewriter*. Trans. Geoffrey Winthrop-Young and Michael Wutz. Stanford, CA: Stanford University Press, 1999.

Knight, William G. *A Major London "Minor": The Surrey Theatre, 1805–1865*. London: Blot, 1997.

Kochnitzky, Leon. *Adolphe Sax and His Saxophone*. New York: Belgian Government Information Center, 1949.

Kopp, Kristin Leigh. *Germany's Wild East: Constructing Poland as Colonial Space*. Ann Arbor: University of Michigan Press, 2012.

Koropeckyj, Roman. "Orientalism in Adam Mickiewicz's *Crimean Sonnets*." *The Slavic and East European Journal* 45/4 (2001): 660–78.

Koropeckyj, Roman. *Adam Mickiewicz: The Life of a Romantic*. Ithaca, NY: Cornell University Press, 2008.

Kozelsky, Mara. "Casualties of Conflict: Crimean Tartars during the Crimean War." *The Slavic Review* 67/4 (2008): 866–91.

Kramer, Lawrence. "Chopin at the Funeral: Episodes in the History of Modern Death." *Journal of the American Musicological Society* 54/1 (2001): 97–125.

Kriegel, Lara. "Who Blew the Balaclava Bugle? The Charge of the Light Brigade and the Afterlife of the Crimean War." *Interdisciplinary Studies in the Long Nineteenth Century* 20 (2015): 1–17.

Krimmer, Elisabeth. *The Representation of War in German Literature: From 1800 to the Present*. Cambridge: Cambridge University Press, 2010.

Krimmer, Elisabeth. "Transcendental Soldiers: Warfare in Schiller's *Wallenstein* and *Die Jungfrau von Orleans*." *Eighteenth-Century Fiction* 19/1 (2006–7): 99–121.

Krimmer, Elisabeth, and Patricia Anne Simpson, eds. *Enlightened War: German Theories and Cultures of Warfare from Frederick the Great to Clausewitz*. London: Camden, 2011.
Lambert, Andrew. *The Crimean War: British Grand Strategy against Russia, 1853–6*. London: Ashgate, 2013.
Landau, Jacob. *The Hejaz Railway and the Muslim Pilgrimage*. Detroit, MI: Wayne State University Press, 1971.
Lastra, James. *Sound and the American Cinema: Perception, Representation, Modernity*. New York: Columbia University Press, 2000.
Lawrence, Vera Brodsky. *Strong on Music: Reverberations, 1850–1856*. Vol. 2. Chicago: University of Chicago, 1995.
Lerner, Daniel. *Hysterical Men: War, Psychiatry, and the Politics of Trauma in Germany, 1890–1930*. Ithaca, NY: Cornell University Press, 2003.
Leskov, Nikolai. "Nesmertel'nyi Golovan." *Istoricheskii vestnik* 12 (1880): 641–78.
Levenback, Karen L. *Virginia Woolf and the Great War*. Syracuse, NY: Syracuse University Press, 1999.
Lévi-Strauss, Claude. "Introduction à l'oeuvre de Marcel Mauss." *Sociologie et Anthropologie*. Ed. Marcel Mauss. Paris, Presses universitaires de France, 1950.
Leys, Ruth. *Trauma: A Genealogy*. Chicago: University of Chicago Press, 2000.
Leys, Ruth. "The Turn to Affect: A Critique." *Critical Inquiry* 37 (2011): 434–72.
Liger-Belaira, Gérard, Marielle Bourget, Clara Cilindrea, Hervé Pron, and Guillaume Polidori. "Champagne Cork Popping Revisited through High-Speed Infrared Imaging: The Role of Temperature." *Journal of Food Engineering* 116/1 (May 2013): 78–85.
Lincoln, Bruce. *The Great Reforms: Autocracy, Bureaucracy, and the Politics of Change in Imperial Russia*. Dekalb: Northern Illinois University Press, 1990.
Locke, Ralph P. *Musical Exoticism: Images and Reflections*. Cambridge: Cambridge University Press, 2009.
Lott, Tommy L. "Descartes on Phantom Limbs." *Mind and Language* 1/3 (Autumn 1986): 243–71.
Lowenthal, David. *The Past Is a Foreign Country*. Cambridge: Cambridge University Press, 1985.
Lucas, John, and Allan Chatburn. *A Brief History of Whistling*. Nottingham, UK: Five Leaves, 2013.
Luhmann, Niklas. *The Reality of the Mass Media*. Trans. Kathleen Cross. Stanford, CA: Stanford University Press, 2000.
Lukács, Georg. "Narrate or Describe?" *Writer and Critic: And Other Essays*. Trans. Arthur Kahn. London: Merlin Press, 1978.
Maag, Georg, Wolfram Pyta, and Martin Windisch, eds. *Der Krimkrieg als erster europäischer Medienkrieg*. Berlin: LIT, 2010.
Macfarlane, Robert. *The Old Ways: A Journey on Foot*. London: Penguin, 2012.
Mahiet, Damien, Mark Ferraguto, and Rebekah Ahrendt, eds. *Music and Diplomacy from the Early Modern Era to the Present*. New York: Palgrave MacMillan, 2014.
Manning, Nathan, ed. *Political (Dis)Engagement: The Changing Nature of the 'Political.'* Bristol, UK: Policy Press, 2017.
Mansel, Philip. *Constantinople: City of the World's Desire 1453–1924*. London: John Murray, 1995.
Markovits, Stefanie. *The Crimean War in the British Imagination*. Cambridge: Cambridge University Press, 2009.
Markovits, Stefanie. "Giving Voice to the Crimean War: Tennyson's 'Charge' and Maud's Battle-Song." *Victorian Poetry* 47/3 (2009): 481–503.
Marsden, Ben. "Rankine, (William John) Macquorn (1820–1872)." *The Oxford Dictionary of National Biography*. Oxford: Oxford University Press, 2004. Online ed., 2011. <http://www.oxforddnb.com/view/article/23133>.
Massumi, Brian. *Parables for the Virtual: Movement, Affect, Sensation*. Durham, NC, and London: Duke University Press, 2002.
McDonald, Lynn. *Mary Seacole: The Making of a Myth*. London: Iguana, 2014.
McLean, Hugh. *In Quest of Tolstoy*. Boston: Academic Studies, 2008.

McLoughlin, Kate, ed. *The Cambridge Companion to War Writing*. Cambridge: Cambridge University Press, 2009.
McPeak, Rick, and Donna Tussing Orwin, eds. *Tolstoy on War: Narrative Art and Historical Truth in "War and Peace."* Ithaca, NY: Cornell University Press, 2012.
Mendieta, Eduardo. "War the School of Space: The Space of War and the War for Space." *Ethics, Place, and Environment* 9 (2006): 207–29.
Messick, Brinkley. *The Calligraphic State: Textual Domination and History in a Muslim Society*. Berkeley: University of California Press, 1993.
Mestyan, Adam. "'A Garden with Mellow Fruits of Refinement': Music Theatres and Cultural Politics in Cairo and Istanbul, 1867–1892." PhD diss. Central European University, Budapest, 2011.
Mickiewicz, Adam. *Dzieła: Wydanie narodowe*. Vol. 10. Warsaw: Czytelnik, 1948.
Mickiewicz, Adam. *Konrad Wallenrod and Grażyna*. Trans. Irene Suboczewski. Lanham, MD: University Press of America, 1989.
Mickiewicz, Adam. *Sonnets from the Crimea*. Trans. Edna Worthley Underwood. San Francisco: Paul Elder, 1917.
Mickiewicz, Władysław. *Żywot Adama Mickiewicza podług zebranych przez siebie materiałów oraz z własnych wspomnień*. 2nd ed. Poznań, Poland: Poznańskie Towarzystwo Przyjaciół Nauk, 1929–1931.
Mieszkowski, Jan. *Watching War*. Stanford, CA: Stanford University Press, 2012.
Milewski, Barbara. "Chopin's Mazurkas and the Myth of the Folk." *Nineteenth-Century Music* 23/2 (1999), 113–35.
Milewski, Barbara. "Mazurkas and National Imaginings." PhD diss. Princeton University, 2002.
Miller, Kathryn B. *Tolstoy and the Genesis of "War and Peace."* Eds. Robyn Feuer Miller and Donna Tussing Orwin. Ithaca, NY, and London: Cornell University Press, 1996.
Monson, Ingrid T. *The African Diaspora: A Musical Perspective*. London: Routledge, 2003.
Moretti, Franco. *Atlas of the European Novel, 1800–1900*. London: Verso, 1998.
Morgan, Elizabeth. "Combat at the Keys: Women and Battle Pieces for the Piano during the American Civil War." *Nineteenth-Century Music* 40/1 (2016): 7–19.
Morpurgo, Michael. "War Horse." *Guardian* (7 Jan. 2012).
Morris, Rosalind, ed. *Can the Subaltern Speak? Reflections on the History of an Idea*. New York: Columbia University Press, 2010.
Morson, Gary Saul. "The Reader as Voyeur: Tolstoi and the Poetics of Didactic Fiction." *Canadian-American Slavic Studies* 12/4 (Winter 1978): 465–80.
Moyn, Samuel. "Empathy in History, Empathizing with Humanity." *History and Theory* 45/3 (Oct. 2006): 397–415.
Myerly, Scott Hughes. *British Military Spectacle: From the Napoleonic Wars through the Crimea*. Cambridge MA: Harvard University Press, 1996.
Nancy, Jean-Luc. *Listening*. Trans. Charlotte Mandell. New York: Fordham University Press, 2007.
Nikoliukin, A. N. *Vzaimosviazi literatur Rossii i SShA*. Moscow: Nauka, 1987.
Novak, David, and Matt Sakakeeny, eds. *Keywords in Sound*. Durham, NC: Duke University Press, 2015.
O'Connell, Robert L. *Of Arms and Men: A History of War, Weapons, and Aggression*: Oxford: Oxford University Press, 1990.
Osterhammel, Jürgen. *The Transformation of the World: A Global History of the Nineteenth Century*. Trans. Patrick Camiller. Princeton, NJ: Princeton University Press, 2014.
Oushakine, Sergei. *The Patriotism of Despair: Nation, War, and Loss in Russia*. Ithaca, NY: Cornell University Press, 2009.
Owen, J. B. *Rules and Regulations, September 12, 1854*. Stratford, UK: Eastern Countries Railway Company, 1857.
Ozenbashli, Enver. "Otrazhenie tragicheskoi sud'by krymskikh tatar v pereselencheskikh pesniakh XIX veka" *Golos Kryma*. Simferopol, 2004.

Ozenbashli, E. *Krymtsi. Zbirnyk prats' z istoriji, etnohrafiji, fol'kloru ta movy* 2nd ed. Simferopol, Crimea: DOLIA, 2006.
Paret, Peter. *Clausewitz and the State: The Man, His Theories and His Times.* Princeton, NJ: Princeton University Press, (1976) 2007.
Parker, James. *Acoustic Jurisprudence: Listening to the Trial of Simon Bikindi.* Oxford: Oxford University Press, 2015.
Partridge, Eric. *A Dictionary of Slang and Unconventional English.* 8th ed. London: Routledge, (1984), 2002.
Peters, Rudolph. "Religious Attitudes toward Modernization in the Ottoman Empire: A Nineteenth Century Pious Text on Steamships, Factories and the Telegraph." *Die Welt des Islams* 26 (1986): 76–105.
Phillips, Rod. *A Short History of Wine.* New York: HarperCollins, 2000.
Picker, John M. *Victorian Soundscapes.* Oxford: Oxford University Press, 2003.
Pieri, Piero. *Storia militare del Risorgimento.* Turin, Italy: Einaudi, 1962.
Pieslak, Jonathan. *Sound Targets: American Soldiers and the War.* Bloomington: Indiana University Press, 2009.
Pilzer, Joshua D. "The Study of Survivors' Music." *The Oxford Handbook of Applied Ethnomusicology.* Ed. Svanibor Pettan and Jeff Todd Titon. Oxford: Oxford University Press, 2015.
Pinch, Trevor, and Karin Bijsterveld, eds. *The Oxford Handbook of Sound Studies.* Oxford: Oxford University Press, 2010.
Plakans, Andrejs. *The Latvians: A Short History.* Stanford, CA: Hoover Institution Press of Stanford University, 1995.
Plakans, Andrejs. *A Concise History of the Baltic States.* Cambridge: Cambridge University Press, 2011.
Pomilio, Tommaso. *Asimmetrie del due: Di alcuni motivi scapigliati.* Lecce, Italy: Manni, 2002.
Ponting, Clive. *The Crimean War: The Truth behind the Myth.* New York: Random House, 2011.
Porter-Szűcs, Brian. *Faith and Fatherland: Catholicism, Modernity, and Poland.* Oxford: Oxford University Press, 2011.
Povinelli, Elizabeth A. *The Cunning of Recognition: Indigenous Alterities and the Making of Australian Multiculturalism.* Durham, NC: Duke University Press, 2002.
Rabb, Intisar. *Doubt in Islamic Law.* Cambridge: Cambridge University Press, 2014.
Radford E., and M. A. Radford. *Encyclopedia of Superstitions.* Ed. and rev. Christina Hole. New York: Barnes and Noble, (1948) 1996.
Raun, Toivo U. *Estonia and the Estonians.* 2nd ed. Stanford, CA: Hoover Institution Press of Stanford University, 2001.
Read, Donald. *The Power of News: The History of Reuters.* Oxford: Oxford University Press, 1999.
Rebagliati, Franco, Furio Ciciliot, and Liliana Betruzzi, eds. *La spedizione d'Oriente: Volontari italiani ed esercito sardo alla Guerra di Crimea, 1855–1856.* Savona, Italy: L. Editrice, 2011.
Reid, James J. *Crisis of the Ottoman Empire: Prelude to Collapse 1839–1878.* Stuttgart: Franz Steiner, 2000.
Reyes, Adelaida. *Songs of the Caged, Songs of the Free: Music and the Vietnamese Refugee Experience.* Philadelphia: Temple University Press, 1999.
Richards, Thomas. *The Imperial Archive: Knowledge and the Fantasy of Empire.* London: Verso, 1993.
Richland, Justin. "Jurisdiction: Grounding Law in Language." *Annual Review of Anthropology* 42 (2013): 209–26.
Riedler, Florian. *Opposition and Legitimacy in the Ottoman Empire: Conspiracies and Political Cultures.* New York: Routledge, 2011.
Robbins, Kevin. "Migrating Music and Good-Enough Cosmopolitanism: Encounter with Robin Denselow and Charlie Gillett." *Migrating Music.* Ed. Jason Toynbee and Byron Dueck. London and New York: Routledge, 2011.
Roche, John. "What Is Potential Energy?" *European Journal of Physics* 24/2 (2003): 185–96.
Rosenfeld, Sophia. "On Being Heard: A Case for Paying Attention to the Historical Ear." *The American Historical Review* 116/2 (2011): 316–34.

Rousseau, Jean-Jacques. *The Government of Poland*. Trans. Willmore Kendall. Indianapolis, IN: Hackett, 1985.

Rowland, David, ed. *The Cambridge Companion to the Piano*. Cambridge: Cambridge University Press, 1998.

Russell, Dave. *Popular Music in England, 1840–1914*. Manchester, UK: Manchester University Press, 1998.

Safran, William. "Diasporas in Modern Societies: Myths of Homeland and Return." *Diaspora: A Journal of Transnational Studies* 1/1 (1991): 83–99.

Said, Edward. *Orientalism*. London: Penguin, (1978) 2003.

Samoylovich, A. N. *Izbrannie trudy o Kryme*. Simferopol, Crimea: DOLIA, 2000.

Saul, David. *The Homicidal Earl: The Life of Lord Cardigan*. Boston: Little, Brown, 1997.

Schafer, R. Murray. *The Soundscape: Our Sonic Environment and the Tuning of the World*. New York: Knopf, 1977.

Schmiesing, Ann. "The Christiania Theater and Norwegian Nationalism: Bjornson's Defense of the 1856 Whistle Concerts in 'Pibernes Program.'" *Scandinavian Studies* 76/3 (Fall 2004): 317–40.

Schuze, Kristen E., Martin Stokes, and Colm Campbell, eds. *Nationalism, Minorities and Diasporas: Identities and Rights in the Middle East*. London: I. B. Tauris, 1996.

Schwartz, Hillel. *Making Noise: From Babel to the Big Bang and Beyond*. Brooklyn, NY: Zone, 2011.

Schwartz, Jessica A. "A 'Voice to Sing': Rongelapese Musical Activism and the Production of Nuclear Knowledge." *Music & Politics* 6/1 (2012): published online. <http://dx.doi.org/10.3998/mp.9460447.0006.101>.

Scott, Derek B. *The Singing Bourgeois: Songs of the Victorian Drawing Room and Parlour*. Milton Keynes, UK, and Philadelphia: Open University Press, 1989.

Scott, Derek B. *Sounds of the Metropolis: The 19th-Century Popular Music Revolution in London, New York, Paris and Vienna*. Oxford: Oxford University Press, 2008.

Scott, James C. *Weapons of the Weak: Everyday Forms of Peasant Resistance*. New Haven, CT: Yale University Press, 1985.

Segell, Michael. *The Devil's Horn: The Story of the Saxophone; From Noisy Novelty to the King of Cool*. New York: Picador, 2005.

Shabana, Ayman. *Custom in Islamic Law and Legal Theory: The Development of the Concepts of 'Urf and 'Adah in the Islamic Legal Tradition*. New York: Palgrave MacMillan, 2010.

Shelemay, Kay Kaufman. *Let Jasmine Rain Down: Song and Remembrance among Syrian Jews*. Chicago: University of Chicago Press, 1998.

Sherfedinov, Ia. *Zvuchit Khaitarma*. Ed. L. N. Lebedinskii. Tashkent, Uzbekistan: Izdatel'stvo literatury i iskusstva imeni Gafura Guliama, 1979.

Siegert, Bernhard. "Doors: On the Materiality of the Symbolic." *Grey Room* 47 (2012): 6–23.

Sikorski, Janusz, ed. *Zarys dziejów wojskowości polskiej do roku 1864*. Vol. 2. Warsaw: Wydawnictwo Ministerstwa Obrony Narodowej, 1966.

Simpson, James. *Creating Wine: The Emergence of a World Industry, 1840–1914*. Princeton, NJ: Princeton University Press, 2011.

Skritskii, N. V. *Krymskaia voina: 1853–1856*. Moscow: Veche, 2006.

Slade, Adolphus. *Turkey and the Crimean War: A Narrative of Historical Events*. Cambridge: Cambridge University Press, 2012.

Slobin, Mark. "Music in Diaspora: The View from Euro-America." *Diaspora: A Journal of Transnational Studies* 3/3 (1994): 243–52.

Smart, Mary Ann. "Liberty On (and Off) the Barricades: Verdi's Risorgimento Fantasies." *Making and Remaking Italy: The Formation of Cultural Identity around the Risorgimento*. Ed. Albert Ascoli and Krystyna von Henneberg. Oxford: Berg, 2001.

Smart, Mary Ann. *Mimomania: Music and Gesture in Nineteenth-Century Opera*. Berkeley: University of California Press, 2004.

Smart, Mary Ann. "Callas Listening." *Women and Music: A Journal of Gender and Culture* 9 (2005): 106–110.

Smith, Anthony. "War and Ethnicity: The Role of Warfare in the Formation, Self-Images and Cohesion of Ethnic Communities." *Ethnic and Racial Studies* 4 (1981): 375–97.

Smith, Bruce. *The Acoustic World of Early Modern England: Attending to the O-Factor*. Chicago: University of Chicago Press, 1999.

Smith, Cecil Woodham. *The Reason Why: The Story of the Fatal Charge of the Light Brigade*. London: Constable, 1953.

Smith, Crosbie. *The Science of Energy—A Cultural History of Energy Physics in Victorian Britain*. Chicago: University of Chicago Press, 1998.

Smith, Mark M. *Listening to Nineteenth-Century America*. Chapel Hill and London: University of North Carolina Press, 2001.

Smith, Mark M. "Listening to the Heard Worlds of Antebellum America." *The Auditory Culture Reader*. Ed. Michael Bull and Les Back. Oxford and New York: Berg, 2003.

Smith, Mark. M *Hearing History: A Reader.* Athens, GA, and London: University of Georgia Press, 2004.

Solie, Ruth A. *Music in Other Words: Victorian Conversations*. Berkeley: University of California Press, 2004.

Spinetti, Federico. *Giuseppe Donizetti pascià: Traiettorie musicali e storiche tra Italia e Turchia*. Bergamo, Italy: Fondazione Donizetti, 2010.

Stauter-Halsted, Keely. *The Nation in the Village: The Genesis of Peasant National Identity in Austrian Poland, 1848–1914*. Ithaca, NY: Cornell University Press, 2001.

Snyder, Timothy. *The Reconstruction of Nations: Poland, Ukraine, Lithuania, Belarus, 1569–1999*. New Haven, CT: Yale University Press, 2003.

Sterne, Jonathan. *The Audible Past: Cultural Origins of Sound Reproduction*. Durham, NC: Duke University Press, 2003.

Sterne, Jonathan. *MP3: The Meaning of a Format*. Durham, NC: Duke University Press, 2012.

Sterne, Jonathan. "Sonic Imaginations." *The Sound Studies Reader.* Ed. Jonathan Sterne. Abingdon, UK: Routledge, 2012.

Stoddart, Helen. *Rings of Desire: Circus History and Representation*. Manchester, UK: Manchester University Press, 2000.

Stoller, Paul. *Sensuous Scholarship*. Philadelphia: University of Pennsylvania Press, 1997.

Suslov, Mikhail D. "'Crimea Is Ours!': Russian Popular Geopolitics in the New Media Age." *Eurasian Geography and Economics* 55/6 (2014): 588–609.

Švābe, Arvēds. *Latvijas vēsture 1800–1914*. Stockholm: Daugava, 1958.

Sweetman, John. *The Crimean War, 1854–1856*. Oxford: Osprey, 2001.

Sytin, I. D. *Zhizn' dlia knigi*. Moscow: Gosudarstvennoe izdatel'stvo politicheskoi literatury, 1960.

Tanrıkut, Asaf. *Türkiye Posta ve Telgraf ve Telefon Tarihi ve Teşkılät ve Mevzuatı*. Ankara, Turkey: Efem Matbaacılık, 1984.

Tarchetti, Iginio Ugo. *Una nobile follia: Drammi della vita militare.* Preface by Roberto Carnero. Milan: Mondadori, 2004.

Tarsaïdzé, Alexandre. *Czars and Presidents: The Story of a Forgotten Friendship*. New York: McDowell, Obolensky, 1958.

Taussig, Michael. *Mimesis and Alterity: A Particular History of the Senses*. London: Routledge, 1993.

Taylor, Diana. *The Archive and the Repertoire: Performing Cultural Memory in the Americas*. Durham, NC, and London: Duke University Press, 2003.

Tentelis, Augusts. *Dokumenti par tautas atmodas laikmetu 1856–1867. g* Riga: Latvijas Vēstures Instituta apgādiens, 1939.

Thompson, Emily. *The Soundscape of Modernity: Architectural Acoustics and the Culture of Listening in America, 1900–1933.* Cambridge, MA: MIT Press, 2002.

Tomaszewski, Wojciech. *Bibliografia warszawskich druków muzycznych, 1801–1850.* Warsaw: Biblioteka Narodowa, 1992.

Tomlinson, Gary. "Italian Romanticism and Italian Opera: An Essay in Their Affinities." *19th-Century Music* 10/1 (Summer 1986): 43–60.

Tomlinson, Gary. *The Singing of the New World: Indigenous Voice in the Era of European Contact.* Cambridge: Cambridge University Press, 2007.

Trochimczyk, Maja. "Sacred/Secular Constructs of National Identity: A Convoluted History of Polish Anthems." *After Chopin: Essays in Polish Music.* Ed. Maja Trochimczyk. Los Angeles: Polish Music Center, 2000.

Troubetzkoy, Alexis. *A Brief History of the Crimean War: History's Most Unnecessary Struggle.* New York: Carroll and Graf, 2006.

Trouillot, Michel-Rolph. *Silencing the Past: Power and the Production of History.* Boston: Beacon, 1995.

Trower, Shelley. *Senses of Vibration: A History of the Pleasure and Pain of Sound.* New York: Continuum, 2012.

Tuğlacı, Pers. *Turkish Bands of the Past and Present.* Istanbul: n.p., 1986.

Uehling, Greta Lynn. *Beyond Memory: The Crimean Tatars' Deportation and Return.* New York: Palgrave Macmillan, 2004.

Uminski, Sigmund H. "Poles and the Confederacy." *Polish American Studies* 22/2 (1965): 99–106.

Valsecchi, Franco. *L'alleanza di Crimea: Il Risorgimento e l'Europa.* Milan: Mondadori, 1948.

Valsecchi, Franco. *L'unificazione italiana e la politica europea dalla guerra di Crimea alla guerra di Lombardia, 1854–1859.* Rome: Istituto per gli studi di politica internazionale, 1939.

Verdery, Katherine. *The Political Lives of Dead Bodies: Reburial and Postsocialist Change.* New York: Columbia University Press, 1999.

Viale, Ettore Bertolè. *Lettere dalla Crimea, 1855–56.* Ed. Umberto Levra. Turin: Carrocci, 2006.

Virilio, Paul. *War and Cinema: The Logistics of Perception.* Trans. Patrick Camiller. London: Verso, (1984) 1989.

Vismann, Cornelia. *Files: Law and Media Technology.* Stanford, CA: Stanford University Press, 2008.

Wachtel, Andrew. "History and Autobiography." *The Cambridge Companion to Tolstoy.* Ed. Donna Tussing Orwin. Cambridge: Cambridge University Press, 2002.

Walicki, Andrzej. *Philosophy and Romantic Nationalism: The Case of Poland.* Oxford: Clarendon, 1982.

Walker, Shaun, and Ian Traynor. "Putin Confirms Crimea Annexation as Ukraine Soldier Becomes First Casualty." *The Guardian,* 19 Mar. 2014. <https://www.theguardian.com/world/2014/mar/18/putin-confirms-annexation-crimea-ukrainian-soldier-casualty>.

Wallace, Alfred Russel. *On the Organic Law of Change: A Facsimile Edition and Annotated Transcription of Alfred Russel Wallace's Species Notebook of 1855–1859.* Ed. James T. Costa. Cambridge, MA: Harvard University Press, 2013.

Walton, Benjamin. "Italian Operatic Fantasies in Latin America." *Modern Italian Studies* 17/4 (2012): 460–71.

Walton, Benjamin. "Quirk Shame." *Representations* 132/1 (2015): 121–29.

Wawrzykowska-Wierciochowa, Dioniza. *Mazurek Dąbrowskiego: Dzieje polskiego hymnu narodowego.* Warsaw: Wydawnictwo Ministerstwa Obrony Narodowej, 1974.

Weber, William. *The Great Transformation of Musical Taste: Concert Programming from Haydn to Brahms.* Cambridge: Cambridge University Press, 2008.

Weintraub, Wiktor. "The Problem of Improvisation in Romantic Literature." *Comparative Literature* 16/2 (1964): 119–37.

White, Hayden. *The Historical Imagination in Nineteenth-Century Europe.* Baltimore: Johns Hopkins University Press, 1973.

Will, Richard. *The Characteristic Symphony in the Age of Haydn and Beethoven.* Cambridge: Cambridge University Press, 2002.

Williams, Brian Glyn. "Hijra and Forced Migration from Nineteenth-Century Russia to the Ottoman Empire." *Cahiers du Monde russe* 41/1 (2000): 79–108.

Williams, Brian Glyn. *The Crimean Tatars: The Diaspora Experience and the Forging of a Nation.* Brill's Inner Asian Library. Leiden, The Netherlands: Brill Academic, 2001.

Williams, Edward V. *The Bells of Russia: History and Technology.* Princeton, NJ: Princeton University Press, 1985.

Winter, Jay. *Sites of Memory, Sites of Mourning: The Great War in European Cultural History*. Cambridge: Cambridge University Press, 2007.

Woldman, Albert A. *Lincoln and the Russians*. Cleveland, OH: The World Publishing Company, 1952.

Wolff, Larry. *Inventing Eastern Europe: The Map of Civilization on the Mind of the Enlightenment*. Stanford, CA: Stanford University Press, 1994.

Wolff, Larry. "Revising Eastern Europe: Memory and the Nation." *Journal of Modern History* 78/1 (Mar. 2006): 93–118.

Woolf, Virginia. "Modern Fiction," reproduced in Virginia Woolf, *The Common Reader: Volume 1*, ed. Andrew McNellie (London: Vintage, [1919] 2003), 146–54.

Yazıcı, Nesimi. "Osmanlı Telgrafında Dil Konusu." *Ankara Üniversitesi İlahiyat Fakültesi Dergisi* 26 (1983): 751–64.

Yazıcı, Nesimi. "Osmanlı Telgraf Fabrikası." *Türk Dünyası Araştırmaları* (1983): 69–81.

Yücel, Hasan-Âli. *Tanzimat, I: Yüzüncü yıldönümü münasebetile*. Istanbul: Maarif Matbaası, 1940.

Yurchak, Alexei. "Little Green Men: Russia, Ukraine and Post-Soviet Sovereignty." *Anthropoliteia*, published online on 31 Mar. 2014. <http://anthropoliteia.net/>.

Zammito, John H. *Kant, Herder, and the Birth of Anthropology*. Chicago: University of Chicago Press, 2002.

Zelče, Vita. "Grāmatniecība un prese", *Latvija 19. gadsimtā*, ed. Jānis Bērziņš. Riga: Latvijas Universitātes Latvijas vēstures institūts, 2000.

Zelkina, Anna. *In Quest for God and Freedom: The Sufi Response to the Russian Advance in the North Caucasus*. London: Hurst, 2000.

Zweig, Stefan. *Casanova, Stendhal, Tolstoy: Adepts in Self-Portraiture*. New Brunswick, NJ: Transaction, 2012.

INDEX

www.ingramcontent.com/pod-product-compliance
Ingram Content Group UK Ltd.
Pitfield, Milton Keynes, MK11 3LW, UK
UKHW020849150926
475595UK00003B/83